The Observational Method in Civil Engineering

The Observational Method in Civil Engineering

Minimising Risk, Maximising Value

Alan Powderham and
Tony O'Brien

CRC Press is an imprint of the
Taylor & Francis Group, an **informa** business

First edition published 2020
by CRC Press
2 Park Square, Milton Park, Abingdon, Oxon, OX14 4RN

and by CRC Press
6000 Broken Sound Parkway NW, Suite 300, Boca Raton, FL
33487-2742

CRC Press is an imprint of Informa UK Limited

British Library Cataloguing-in-Publication Data
A catalogue record for this book is available from the British Library

Library of Congress Cataloging-in-Publication Data
Names: Powderham, Alan J., author. | O'Brien, Anthony, author.
Title: The observational method in civil engineering : maximising value, minimising risk / Alan Powderham and Anthony O'Brien.
Description: First edition. | Boca Raton : CRC Press, 2021. |
Includes bibliographical references and index.
Identifiers: LCCN 2020012577 (print) | LCCN 2020012578 (ebook) | ISBN 9780367361648 (paperback) |
ISBN 9780367361655 (hardback) | ISBN 9780429344244 (ebook)
Subjects: LCSH: Civil engineering–Case studies. |
Construction industry–Management–Case studies. |
Observation (Scientific method)–Case studies.
Classification: LCC TA145 .P68 2021 (print) |
LCC TA145 (ebook) | DDC 624–dc23
LC record available at https://lccn.loc.gov/2020012577
LC ebook record available at https://lccn.loc.gov/2020012578

ISBN: 978-0-367-36165-5 (hbk)
ISBN: 978-0-367-36164-8 (pbk)
ISBN: 978-0-429-34424-4 (ebk)

Typeset in Sabon
by Integra Software Services Pvt. Ltd.

Contents

Foreword

The observational method (OM) is an extremely powerful technique with enormous practical advantages for major infrastructure projects. Used effectively, it can lead to major cost and time savings, increases in safety and more effective collaboration between designers and constructors. It can also serve as a major driver for innovation in the construction industry. Regrettably OM is not used as much as it should be, often because engineers are uncomfortable about properly addressing and managing risk.

This excellent book illustrates the advantages of OM by way of some comprehensive case histories, concerning tunnels, deep excavations and foundations associated with some major projects. A proper understanding of OM can only be achieved by seeing both its advantages and limitations, which the authors clearly present when discussing these case histories.

Early in the book, there is a key chapter on 'The art of achieving agreement' – an all-important aspect of a well-managed project adopting the OM. In this, 'The power of progressive modification' is clearly presented and discussed.

Each of the twelve case histories presents the important aspects of the design and construction, properly emphasising the importance of the geology, the nature of the particular structure and its structural behaviour, soil–structure interaction, the construction sequence, safety aspects and contractual conditions. How the OM was implemented is discussed for each case history. Key aspects presented include appropriate trigger levels, prediction of performance (and the limitations of predictive methods), instrumentation and monitoring and contingency measures. The outcomes of each case history are nicely summarised, with the observed behaviour and overall performance presented and discussed.

A chapter focusses on the advantages and limitations of the OM in the light of the case histories, emphasising the importance of maintaining simplicity and ensuring robustness. A final chapter reflects on the future of OM, commenting on the contractual aspects and peer review processes. A comprehensive list of references is also included.

This informative and excellent book by two very experienced authors will be of interest to civil, structural and geotechnical engineers, both in design and contracting organisations, and also to owners. The wide variety of case histories will serve as comprehensive and authoritative reference sources for those seeking to understand and implement the observational method.

Professor Lord Robert Mair CBE FREng FICE FRS NAE

Authors

Alan Powderham is a former Director of Transportation of Mott MacDonald and a Fellow of the Royal Academy of Engineering, the Institution of Civil Engineers and the Institution of Structural Engineers. Over the past four decades, he has taken a leading role in the development and application of the Observational Method and pioneered many award-winning innovations in the design and construction of major infrastructure projects including the Channel Tunnel and the Boston Central Artery. He has served on many technical committees and, as a Visiting Professor at Imperial College, has supported research in foundation engineering.

Tony O'Brien is Mott MacDonald's Global Practice Leader for Geotechnics and has over thirty years of international experience for transportation, water and energy projects. He has led geotechnical design teams through major projects in the UK, North America, central and south-east Asia. As a Visiting Professor at Southampton University, he has led applied research in earthworks behaviour, soil dynamics and soil–structure interaction. He actively contributes to the work of professional committees and provides industry guidance, notably in the fields of advanced numerical modelling, retaining walls, foundation engineering and earthworks asset management.

Glossary

'*ab initio*' A term introduced by Peck (1969) to denote an application of the observational method (OM) where its use has been envisioned from the inception of the project.

'best way out' An application of the OM implemented during construction when some unexpected and critical development has occurred, placing the project into crisis (Peck, 1969).

Bottom-up A bottom-up construction sequence is defined by the construction of the permanent works from the formation level upwards, casting the foundation slab before the internal walls and slabs above. Temporary props will usually be required to support the external walls until the permanent internal structure provides sufficient support.

Contingency plan A course of action or modification of design, defined in advance, to address an adverse trend which significantly departs from the predicted performance – typically when the monitoring indicates that trigger levels are likely to be exceeded and safety compromised.

Hazard Something with the potential to cause harm.

Moderately conservative A cautious assessment of design conditions – for example, using parameter values that are worse than the probabilistic mean but not as onerous as those assessed as worst credible.

Most probable An assessment of the most likely design conditions using parameters that represent the probabilistic mean of all possible sets of conditions. It represents, in general terms, the design condition predicted to occur in practice.

Progressive collapse This develops when a local over-stress occurs in the soil or structure which, lacking robustness and redundancy, offers no alternate load paths, causing progressive failure leading to overall collapse. The deterioration in conditions tends to occur so rapidly that there is insufficient time to effectively implement contingency plans.

Progressive modification An inherently flexible and comprehensive approach to implementing the OM. It can apply to both 'ab-initio' and 'best way out' scenarios and offers many benefits including enhanced safety combined with maximising benefits. Design changes are introduced sequentially on the basis of observational feedback. (See Chapter 1, Section 1.2)

Risk Risk = hazard severity × likelihood of occurrence (HSE, 1991).

Risk-based design Design based on risk assessment and consideration of risk hierarchy.

Serviceability limit states States that correspond to conditions beyond which specified service requirements for a structure or structural element are no longer met, e.g. its durability is impaired, its maintenance requirements are substantially increased, or damage is caused to non-structural elements. Alternatively, such conditions of the structure that may affect adjacent Structures or services in a like manner (BSI, 1994a).

Top-down A top-down construction sequence exploits the benefits of using the permanent works for temporary support – typically commencing with the roof slab, cast on the ground, providing the top level of propping to the external walls. The top-down sequence is continued downwards beneath the roof slab using intermediate slabs or temporary props as further support, completing the construction with the base slab.

Ultimate limit states States associated with collapse, or with other similar forms of structural failure. They generally relate to the maximum load-carrying capacity of a structure or a structural element.

Worst credible Worst credible conditions relate to conditions which are the worst that the designer considers might realistically occur. With respect to loads or geometric parameters, they are typically values that are very unlikely to be exceeded. In the case of ground strength parameters, they are values that are very unlikely to be any lower.

Abbreviations

ASCE	American Society of Civil Engineers
ATD	above tunnel datum
BS	British Standard
BSCE	Boston Society of Civil Engineers (now the Boston Society of Civil Engineers section of the ASCE)
CDM	Construction (Design and Management) Regulations
CIRIA	Construction Industry Research and Information Association
CPT	cone penetration test
CTA	central terminal area (at Heathrow airport)
DETR	Department of the Environment, Transport and the Regions
DLR	Docklands Light Railway
D/Wall	diaphragm wall
EC1	Eurocode 1
EC7	Eurocode 7
ECC	Engineering and Construction Contract (New Engineering Contract)
EPS	Expanded polystyrene, also known as styrofoam (ultra-lightweight fill)
FOS	factor of safety
GI	ground investigation
HDPE	high-density polyethylene
HEX	Heathrow Express
HSE	Health and Safety Executive
ICE	Institution of Civil Engineers
IStructE	Institution of Structural Engineers
LUL	London Underground Ltd
MC	moderately conservative
MDPE	medium density polyethylene
MP	most probable
NATM	New Austrian Tunnelling Method
NEC	New Engineering Contract 1st Edition
OCR	over-consolidation ratio

Qc	CPT cone resistance
SBP	self-boring pressuremeter
SCL	sprayed concrete linings
SCPT	seismic cone penetration test
SI	site investigation
SLS	serviceability limit state
SPT	standard penetration test
SPT '*N*'	blow count ('*N*') measured during SPT
TBM	tunnel boring machine
TD	tunnel datum
ULS	ultimate limit state
VE	value engineering
VP	verification process
WC	worst credible

Notation

c'	effective cohesion
C_α	coefficient of secondary consolidation (also known as creep coefficient)
C_c	compression index
C_h	coefficient of consolidation, horizontal
C_r	recompression index
C_v	coefficient of consolidation, vertical
E	Young's modulus
E'	drained Young's modulus
E_u	undrained Young's modulus
G	shear modulus
G_o	shear modulus at very small strain ($<10^{-5}$)
G_{hh}	shear modulus, horizontal
G_{hv}/G_{vh}	shear modulus, vertical
i	horizontal distance from tunnel centre-line to the point of inflexion on a settlement trough
k	hydraulic permeability
K	design earth pressure coefficient
K_a	coefficient of active earth pressure
k_h	hydraulic permeability, horizontal
kN	kilonewton
K_o	coefficient of earth pressure at rest
K_p	coefficient of passive earth pressure
K_T	trough width parameter (typically written as K)
k_v	hydraulic permeability, vertical
m	metre
m_v	coefficient of volume compressibility
MN	meganewton (= 102 tonnes force approx.)
N	newton
R_o	radius of tunnel
S	settlement
s_u	undrained shear strength of soil

t	tonne
V_L	volume loss (usually expressed as a percentage of the excavated area of the tunnel)
V_s	area of surface settlement trough
y	transverse distance to tunnel centre-line
z	vertical coordinate
z_o	depth from surface to tunnel axis
ε	axial strain
v	Poisson's ratio
ρ	density
σ	compressive stress
σ_h	horizontal stress
σ_n	normal stress
σ_v	vertical stress
Y_b	bulk unit weight of soil
Y_s	shear strain
Y_w	unit weight of water
φ'	effective friction angle for soil

Introduction

The observational method (OM) is a natural and powerful technique that maximises economy while assuring safety. The prime objective of this book is to highlight the key features and wide ranging potential of the OM through detailed accounts of actual applications. It is not intended to be a formal guide, but some advice and guidelines are offered in Chapters 1, 13 and 14 which supplement the case histories. These feature twelve examples from major infrastructure projects and have been selected to provide a diverse spectrum of applications. They include protection of adjacent structures including buildings and railway systems, bored tunnelling, jacked tunnels, shafts and cofferdams, retaining walls, embankments, deep foundations, ground improvement and groundwater control. They illustrate how the OM is able to:

- create major cost and time savings;
- increase safety;
- achieve more effective collaboration between the client and the design and construction teams;
- enhance the construction industry's ability to learn from experience and improve future practice and stimulate innovation.

Unfortunately, despite this compelling list of advantages, the OM continues to be significantly under-used (Peck and Powderham, 1999). Two key constraints are those imposed by contractual conditions which inhibit design changes during construction and the lack of equitable sharing of benefits and risks. Two other main reasons are the desire for certainty and an inappropriate association of the OM with high levels of risk. Through the case histories, this book demonstrates how, despite multiple and varying constraints, including the concerns of powerful stakeholders, such limitations can be successfully overcome. The practical limits of the effective application of the OM itself are also explored along with the identification of technical constraints and their elimination.

Chapter 1 addresses the 'Art of achieving agreement'. The case histories also include substantial sections on this challenge explaining, frequently in the face of serious doubts, how agreement to implement the OM was secured. A key factor in this has been the promotion and adoption of progressive modification and its use is unreservedly advocated by the authors for the implementation of the OM wherever feasible. Progressive modification was used in all of the case histories featured in this book and has been found to be applicable to all categories of application (see Chapter 1, Figure 1.1). It is also shown how the OM, with its focus on communication and close collaboration between stakeholders, can act as a driver for innovation, whilst minimising project risks and ensuring an acceptable level of safety. The successful implementation of the OM depends on a sound understanding of its advantages and limitations. The concluding chapters address these in detail along with a range of observations and advice on the way forward.

The formal history of the OM effectively commenced with Peck's Rankine Lecture in 1969 in which he introduced the term 'the observational method'. Of course, observational methods in the general sense have been widely adopted throughout the centuries by builders and engineers as a natural and instinctive approach to address uncertainty – though often in ways which would not be considered as acceptably safe nowadays. Peck viewed the term 'the observational method' as having a specific and restricted meaning. This view was shared by Powderham and Nicholson (1996) and further underlined in CIRIA Report 185 (Nicholson *et al.*, 1999). This report produced a detailed definition of the OM:

> *"The Observational Method in ground engineering is a continuous, managed, integrated, process of design, construction control, monitoring and review that enables previously defined modifications to be incorporated during or after construction as appropriate. All these aspects have to be demonstrably robust. The objective is to achieve greater overall economy without compromising safety".*

The report also provided a historical review of the OM following the rationalisation by Terzaghi and Peck (1967). The OM was incorporated into Eurocode 7 (EC7, BSI, 1995) and the Hong Kong Geotechnical Office Guide to retaining wall design (1993). The ninth Géotechnique Symposium in Print (Institution of Civil Engineers (ICE), 1994) featured case histories of the OM from around the world. These were subsequently published in the book, 'The observational method in geotechnical engineering', Thomas Telford, (ICE, 1996). This included a report on the symposium, extensive records of the discussions, a new paper on 'The way forward' and a reprint of Peck's Rankine Lecture and his later commentary (Dunnicliff and Deere, 1984). A review of the OM was presented by Szavits-Nossan (2006) which compared a range of major contributions to the method with regard to the nature of uncertainty and how it has been addressed.

Peck (1969) identified two main categories for an application of the OM which he termed '*ab initio*' and 'best way out'. The former relates to where the OM has been envisioned from the inception of a project or at least well before the start of construction. 'Best way out' applications involve situations where an unexpected critical development has occurred during construction placing the project into crisis. Such scenarios typically involve heightened concerns for safety but the main drivers, while still ensuring acceptable levels of safety, may also be cost or time. The case histories of Limehouse Link and the Heathrow Express cofferdam, in Chapters 4 and 5, respectively, provide examples of the latter situation.

A case could be made for creating additional categories for applications of the OM. For example, in Chapter 2, the case history from the Channel Tunnel does not fit neatly into either category of '*ab initio*' or 'best way out'. Here, while the OM was initiated well into the construction phase, there had been no unexpected development or crisis. So, although something of a hybrid, it was effectively an '*ab initio*' application in nature.

In practice, Peck's two categories share a common technical base since the key principles essentially apply to all applications of the OM. Thus, while it is useful for the reader to know which of the two categories may apply in a given case history, the authors do not propose the creation of any more categories or sub-categories. In fact, we strongly commend the simplicity of just one type wherever feasible – namely that of progressive modification.

Peck, in his Rankine Lecture, set out a sequence of steps which he considered were required for a complete application of the OM. These were:

(a) Exploration sufficient to establish at least the general nature, pattern and properties of the deposits but not necessarily in detail.
(b) Assessment of the most probable conditions and the most unfavourable conceivable deviations from these conditions. In this assessment, geology often plays a major role.
(c) Establishment of the design based on a working hypothesis of behaviour anticipated under the most probable conditions.
(d) Selection of quantities to be observed as construction proceeds and calculation of their anticipated values on the basis of the working hypothesis.
(e) Calculation of values of the same quantities under the most unfavourable conditions compatible with the available data concerning the subsurface conditions.
(f) Selection in advance of a course of action or modification of design for every foreseeable significant deviation of the observational findings from those predicted on the basis of the working hypothesis.

(g) Measurement of quantities to be observed and evaluation of actual conditions.
(h) Modification of design to suit actual conditions.

With the growing interest in the OM, various concerns have been raised about these steps. These reflect the dichotomy faced by any endeavour to rigidly define requirements for an inherently natural and flexible approach – particularly given that the method addresses uncertainty and demands experience and the exercise of engineering judgement. Foremost among the concerns expressed is the risk seen to be implied by step (c) relating to the most probable conditions. These concerns and how to resolve them are further discussed in Section 1.2 'The Power of Progressive Modification' and in Section 13.3 'Solving the issue of the "most probable"'.

As a concluding postscript to Peck's Rankine Lecture, while he noted in 1969 that 'best way out' applications were much the more familiar, we may draw some encouragement from the fact that in the 1994 Symposium by far the most of the examples were '*ab initio*'. The majority of the applications reported in this book are also of this category.

The views expressed in this book are the authors' own but we acknowledge our debt to our colleagues both in Mott MacDonald and to all those around the world who have contributed to the wide range of projects presented in this book.

References

Dunnicliff, J. and Deere, D. V. (editors) (1984). *Judgment in Geotechnical Engineering – The Professional Legacy of Ralph B. Peck*, p. 122, Wiley, New York.

Geotechnical Engineering Office (1993). Guide to retaining wall design, Hong Kong Government.

Institution, B. S. (1995). Geotechnical Design, Part 1: General Rules, Eurocode 7 (EC7), London.

Institution of Civil Engineers (1994). Symposium in Print: the observational method in geotechnical engineering. *Géotechnique*, **XLIV**, No. 4, Dec 1994, London, 611–769.

Institution of Civil Engineers (1996). *The Observational Method in Geotechnical Engineering*, Thomas Telford, London.

Nicholson, D., Tse, C.-M. and Penny, C. (1999). The observational method in ground engineering: principles and applications. *CIRIA Report*, **185**, London, 214.

Peck, R. B. (1969). Advantages and limitations of the observational method in applied soil mechanics. *Géotechnique*, **19**, No. 2, 171–187.

Peck, R. B. and Powderham, A. J. (1999). *Talking Point, Ground Engineering*, February edition, British Geotechnical Society, Institution of Civil Engineers, London, UK.

Powderham, A. J. and Nicholson, D. P. (1996). The Way Forward: The Observational Method in Geotechnical Engineering, Institution of Civil Engineers, pp. 195–204, Thomas Telford, London.
Szavits-Nossan, A. (2006). Observations on the Observational Method, XIII Danube-European Conference on Geotechnical Engineering Ljubljana: Slovenian Geotechnical Society, 2006. str. pp. 171–178.
Terzaghi, K. and Peck, R. B. (1967). *Soil Mechanics in Engineering Practice*, 2nd Edition, pp. pp 294–295, John Wiley & Sons, New York.

Chapter 1

The Art of Achieving Agreement

1.1 Key Factors

One of the most frequently asked questions in Q&A sessions following presentations on the observational method (OM) that the authors have given around the world is: 'How was agreement to use the method achieved?' In practice, the process is quite variable, requiring a flexible approach to address the circumstances particular to a given project and the specific concerns of each stakeholder. How agreement is reached will also be influenced by whether it is an '*ab initio*' or 'best way out' scenario. As illustrated in the featured case histories, the starting position on site for the OM design can have quite widely varying levels of conservatism between projects in order to satisfy specific stakeholder concerns.

A reference set of abstracts is included in Section 1.3, which summarise the key aspects of each case history. These are described in much greater detail in the respective chapter and include a separate section devoted to this issue of achieving agreement to implement the OM.

Achieving agreement can present a substantial challenge. The OM demands commitment by all stakeholders and is fundamental to ensuring success in any application. Consequently, when there is no overriding driver for change, for example, in an '*ab initio*' application, there is typically little incentive to depart from convention or 'business as usual'. The parties are inclined to remain within their respective comfort zones. By contrast, with projects that have reached a crisis, there will be a marked absence of comfort generally. This will be evident in polarised and often strongly conflicting perspectives between the parties. Here the OM would be proposed as the 'best way out' solution. Peck (1969) noted that such applications were much more familiar, indicating, in turn, that resistance to change in the absence of a crisis is quite common. Even with a crisis as a catalyst, it often takes sustained and convincing advocacy by the proposer to secure the 'buy-in' by the other parties. They will probably be in varying degrees of disagreement and will therefore not readily reach a consensus that the OM does indeed offer the best way forward. It will test the commitment, confidence and ability of the proposer. This crucial aspect is discussed further in Chapter 13 on the advantages and limitations of the OM. Resilience is another important quality as it can take detailed development, strong advocacy and, as noted in many of the case histories featured in this book, substantial time and effort to reach agreement. The key factors involved are as follows:

- **Convincing business case**
  The benefits of adopting the OM must be clearly established and communicated to the stakeholders – particularly to the main stakeholder on whose approval the agreement to implement depends. The advantages of the OM need to be sufficiently convincing when evaluated against the established base design case or other potential alternatives.
- **Sound technical basis**
  Applications of the OM inherently involve some form of soil/structure interaction. Both the geotechnical and the structural aspects must be appropriately assessed and understood. This understanding must be manifest. In short, OM practitioners need to have commitment, competence and clarity. This must be evident to the stakeholders.
- **Risk management**
  Maintaining and demonstrating an acceptable level of safety is essential. Moreover, OM practitioners require an active and broad appreciation of the stakeholders' perspectives on risks and constraints including those relating to commercial, programme and contractual issues.
- **Trust**
  Central to any application of the OM are interpersonal relationships in which trust between the parties plays a key role. This trust is not a given but has to be earned.

1.2 The Power of Progressive Modification

One theme common to all of the case histories in this volume was the implementation of the OM through progressive modification. With its additional focus on enhancing and demonstrating safety, whether it will be for '*ab initio*' or 'best way out' situations, progressive modification provides greater comfort to the stakeholders in their concerns about risk. As illustrated in Figure 1.1, progressive modification can address all types of category of the OM. In practice, there is basically no difference technically between the categories in implementing the OM. But there is a big difference contractually and politically since 'best way out' situations, as opposed to '*ab initio*' ones, involve crisis and consequently demand radical solutions. Once agreement has been achieved to use the OM, then each application can effectively proceed on a common conceptual basis. However, it is essential to appreciate that each application of the OM is bespoke and, beyond the broad concepts, each case

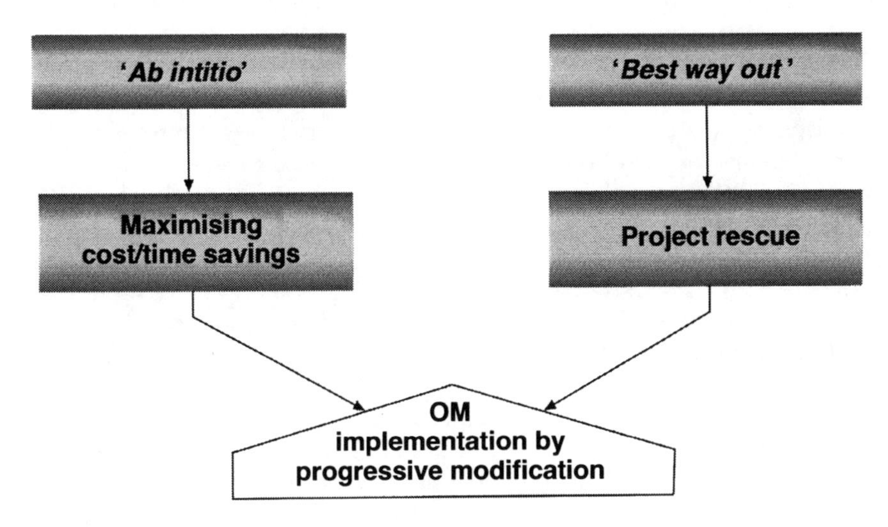

Figure 1.1 OM categories – all can be addressed by progressive modification.

needs to be addressed in detail for the particular conditions. In this, progressive modification brings a very effective flexibility. Consequently, there is no need to commence construction with an unduly optimistic design or one based, with little margins, on estimations of the most probable conditions. Instead, the conservatism of the starting condition can be chosen to accommodate the comfort levels of the stakeholders for each specific case. It is fundamental that every application of the OM should commence from a safe position. Progressive modification, as shown in the featured case histories, does not need to be constrained by any prescriptive requirement to implement and complete an application of the OM with a fixed level of conservatism in the design. It is essential, however, to have a realistic appreciation of the potential range of conditions from the most probable to the most unfavourable. Such understanding is central to establishing the viability of any application of the OM. This assessment requires, as well as site investigation and analysis, experience and engineering judgement. Accordingly, applications of the OM are not routine and it can be very risky to treat them as such – even for apparently very similar situations. OM designs must be carefully developed to suit each case and are consequently not well served by a prescriptive approach. For example, CIRIA Report 185 (1999) recommends that the OM uses:

- *'most probable' and 'moderately conservative' conditions for deformation and load predictions – serviceability limit state (SLS) designs;*

- *'most unfavourable' conditions for ultimate limit state (ULS) designs and for robustness check during risk assessment.*

Such recommendations, while made with the best of intentions to provide consistency for future applications, risk being too prescriptive and are also a recipe for over-complicating the OM (see also the discussion in Chapter 13). Peck was keenly aware of this and, while he included caveats in his Rankine Lecture, expressed his dissatisfaction with his attempt to formalise the method which he felt was too prescriptive (Dunnicliff and Deere, 1984). There is always the initial issue of how to establish in advance what the likely conditions for most probable and most unfavourable are likely to be. There are inevitably unknowns and our predictions have often been inaccurate. A recent comprehensive review of databases of retaining wall case studies concluded that 'ground movements cannot be predicted exactly' (Gaba *et al.*, 2017). Considerable scatter was observed in maximum retaining wall deflections for similar wall types in similar ground conditions. The scatter was attributed to a range of factors, including the impact of variations in construction sequence, with the timing of establishing wall supports and local construction details. These construction issues are not amenable to prediction by a designer. The uncertainties associated with underground construction mean that, even with unlimited analytical power at their disposal, engineers should not expect to be able make accurate predictions of soil/structure interaction behaviour. While such surveys question our abilities to predict the most probable conditions, such a disparity also implies a warning about those for the most unfavourable. Moreover, while it is generally understood how the terms 'most probable', 'moderately conservative' and 'most unfavourable' may be addressed in terms of design parameters for soils, it is less clear how they relate to soil/structure interaction and the range of structural responses that may be involved. For example, as described in Chapter 3, the key issue for the Mansion House was the degree and nature of the response of the building to imposed settlements. While it was expected to be within SLS conditions, it was quite unacceptable to the building owners to commence the OM with a design based on estimates of the most probable or even those for moderately conservative conditions. In this case, only a level of slight risk of damage was acceptable. Agreement to implement the OM was therefore achieved by commencing on site at the much lower level of very slight risk. This recognised that the level of risk would progressively increase as tunnelling construction progressed and ground movements increased within the zone of influence. The success of this case history presents a compelling endorsement for implementation through progressive modification where, in practice, the starting position for the OM was demonstrated to be one of virtually negligible risk.

It is important to note that design modifications do not always have to be contingent measures of a corrective kind, although this is often assumed. It is implied, for example, in the relevant clauses relating to the OM in Eurocode 7. This assumption is more explicitly expressed by Muir Wood (2000) who also finds the procedures set out by Peck in his Rankine Lecture (1969) to be unnecessarily cumbersome. It is likely that Peck would have had some sympathy with this concern as indicated in his commentary on his Rankine Lecture (Dunnicliff and Deere, 1984). The concern is understandable given the tendency to treat Peck's list of procedures too prescriptively. However, the authors have found throughout their own experience of applying the OM that, if considered flexibly, Peck's procedures provide an excellent guide. A nuanced reading of Peck's paper can reveal his pragmatic approach underlined by the need to exercise engineering judgement to each case individually.

Rather than design modifications being restricted to corrective contingencies, progressive modification inherently offers the potential for a fully flexible process. Through its step-by-step approach, it creates the opportunity to maximise value by facilitating the introduction of beneficial design changes in a safe and controlled manner. And, if the OM is commenced from a demonstrably safe conservative base, most of the design changes are likely to be beneficial. Moreover, this managed incremental approach also facilitates a very effective way to deal with unexpectedly unfavourable conditions. Thus, with this ability to satisfy safety concerns while also maximising value, progressive modification creates a fertile basis on which to achieve agreement to implement the OM. The advantages of adopting progressive modification are highlighted in the case histories and are further discussed in Chapter 13.

1.3 Summaries of Featured Case Histories

1.3.1 The Channel Tunnel (1988–1991)

A hybrid '*ab initio*' and first overt application of what has become known as progressive modification: It was applied to cut and cover works in Gault Clay. Comprising bottom-up and top-down construction with contiguous piled walls, it involved tunnel portals and crossover works within an active and very complex landslip. It was also the first example of the use of blinding struts in the OM.

The OM is an inherently natural way to address uncertainty and one that engineers would instinctively adopt (Peck, 1969). In that context, this first case history could be considered as an exemplar. It developed naturally to address the basic question: 'How could we do this better?' Only in retrospect did the realisation dawn that it was essentially an example of the OM. It did not fall neatly into either an '*ab initio*' or 'best way out' application, although in spirit it was closer to the former. There was no crisis to act as a catalyst for corrective change. On the contrary, the application of the OM simply evolved from the early rapport established between the contractor and designer. It fostered creative teamwork facilitated by the harmonious interaction of aligned objectives and was introduced and managed by the original designer.

1.3.2 The Mansion House (1989–1991)

'Best way out' application: Protection of a Grade I listed masonry building of national importance from potential damage caused by tunnelling. This involved complex soil/structure interaction from a range of both shallow and deep bored tunnels constructed in London Clay within the zone of influence of the building. Its success marked a significant development of the implementation of the OM through progressive modification. The prime objective was to safely protect the building from any unacceptable damage while, at the same time, avoiding the contingencies of expensive and time consuming preventative works. This was achieved by commencing construction with negligible risk and demonstrating, at each stage in a sequential process, that the accumulative risk would be maintained within acceptable levels. It was the first application of real-time electronic data capture of the critical observations combined with the traffic light system for the OM. It was introduced and managed by an independent specialist team.

1.3.3 Limehouse Link (1991–1993)

'Best way out' application to cut and cover highway tunnels: This case history, like the Mansion House, highlights key obstacles to achieving agreement to implement the OM. Here the immediate problem was the contractual constraints to introducing design changes during construction. Such limitations make the OM a complete non-starter. It was unlocked by adding a value engineering clause to the contract and was the first example in the United Kingdom of the OM being introduced through such a clause. This was added as a variation to the design and build contract well after construction had started. The OM was applied on multiple fronts involving top-down construction with diaphragm walls in London Clay, River Terrace Deposits and the Woolwich and Reading beds of the Lambeth Group. It extended the technique of using blinding struts which was further advanced and optimised in the range of applications of the OM at Heathrow Airport. The comprehensive success of the OM for the top-down construction also led its application to the sheet-piled walls of a cofferdam in Limehouse basin. This involved bottom-up construction through soft deposits of hydraulic fill above London Clay and provided an instructive example of reaching a practical limit in the OM. It was introduced by the specialist consultant to the main contractor.

1.3.4 Heathrow Express Cofferdam (1994–1995)

'Best way out' application: Following the collapse of the station platform tunnels in the Central Terminal Area in October 1994 during construction for the Heathrow Express, Heathrow Airport became something of a hot-spot for applications of the OM. This is reflected in the selection of case histories reported in this volume – each chosen to highlight different aspects of the OM. There is an encouraging irony here given that the understandable initial reaction to major failures is one of conservatism. That the OM, with its focus on innovation and eliminating over-conservatism, was so extensively adopted speaks volumes for the positive and creative approach of the client, BAA. The recovery design solution, featuring a 60 m diameter, 30 m deep cofferdam, was specifically developed to facilitate the application of the OM through progressive modification. Constructed through made ground and highly disturbed London Clay following the collapse, it utilised an innovative form of stepped secant and contiguous piles in a bottom-up sequence of construction. The specialist OM team from the main design consultant was appointed by the client.

1.3.5 Heathrow Airport Multi-Storey Car Park (1995–1996)

'*Ab initio*' application for building protection against damage from tunnelling: This involved a multi-storey car park which, unusually, had one half founded on pad footings and the other on piles. However, the unexpected sensitivity in the building's structural response to the induced movement was even more unusual – if not unique. The measured ground settlements were well within the predicted range and the risk of damage prior to tunnel construction was assessed as very slight to slight. This was one of a number of structures in the airport that were being monitored in a series of applications of the OM at Heathrow. It was assessed to have the lowest level of risk in a group of three very similar car parks exposed to tunnelling. Yet it was the only one to suffer any recorded damage. Here the OM failed through not identifying the critical observations in advance. With the benefit of hindsight, there were two factors which combined to cause a sudden and brittle failure in the lower columns of the car park. It would have required a structural modification to eliminate such a mode of failure in advance of tunnelling. A specialist OM team from the main design consultant was appointed by the client.

1.3.6 Boston Central Artery (1991–2001)

An '*ab initio*' application involving the jacking of major interstate highway tunnels, combined with global ground freezing, beneath an operating railway network of seven interconnected tracks: The inherent complexity and nature of the project demanded an observational approach. The challenge was exacerbated by very difficult ground conditions comprising contaminated fill containing extensive and major obstructions overlying soft Boston Blue Clay. It was a key example demonstrating the power of using progressive modification where seven distinct sources of ground movement were managed through just one set of OM traffic lights. Value engineering and close teamwork between all parties through partnering enabled the allowable control limits set initially to be progressively and substantially increased. The OM was introduced and managed by a specialist sub-consultant to the main designer.

1.3.7 Irlam Railway Bridge Embankment (1997–1999)

'*Ab initio*' application featuring an innovative solution for the replacement of a 150-year-old railway bridge: It involved the United Kingdom's first expanded polystyrene (EPS) embankment to support a railway which had to be kept operational except for a 100 hour possession. The bridge was supported on timber piles which were vulnerable to disturbance from construction activities and vibrations from heavy rolling stock. The bridge was underlain by highly compressible industrial waste and soft alluvium. It was natural candidate for an OM application through progressive modification. Construction of a preload embankment was necessary to minimise long-term settlement beneath the permanent embankment. The OM was introduced to control the rate of embankment construction and reduce the risk of disturbance to the timber piles. The application was highly successful in managing the risks inherent in working adjacent to an old structure. The railway remained fully operational throughout the works and the EPS embankment was successfully completed during the possession. The OM was introduced and managed by a specialist team from the main designer.

1.3.8 Heathrow Airport Airside Road Tunnel (ART) (2001–2003)

'*Ab initio*' application: In theory achieving acceptance to implement the OM should have been relatively straightforward since the client BAA encouraged cost and programme saving innovations, especially through the OM. However, the cut and cover construction for the ART was in a sensitive and spatially constrained location close to aircraft taxi ways. A linear sequence with trial sections, as used so successfully for the

Channel Tunnel and Limehouse Link, could not be accommodated. The first element was the TBM launch chamber which was confined within a footprint of just 20 by 30 m. The solution lay in changing the direction of travel by 90°, enabling the OM to operate progressive modification in a downwards, rather than horizontal, step-by-step sequence. This application also demonstrated the potential for the method on small projects. It featured the innovative use of laser-controlled excavation with blinding struts as a potential contingency measure. This innovation was then adapted to create a stepped cut and cover construction sequence to address a sensitive interface with an operating London Underground tunnel. The OM was introduced and managed by a specialist team from the main designer.

1.3.9 Wembley Stadium Arch (2004–2005)

'Best way out' application involving unique use of the OM to control deformation of pile groups to enable the safe raising of the iconic arch at the new Wembley Stadium: The arch was fabricated at ground level and rotated into its final position by jacking off 15 pile groups constructed in stiff clay. The arch is one of the largest and most slender structures of its type in the world and raising it into its permanent position generated a complex set of loads on the pile groups. Applying the OM was seen by all parties to offer the best potential solution to minimising risks, while avoiding additional project delays. Prior to agreement being reached, the practicability of applying the OM was fully evaluated through further design analyses and a series of workshops. Although the process of raising the arch created a complex set of scenarios, it was considered essential to maintain simplicity in applying the OM. Simple limits for critical movements and contingency measures were established and enabled the OM team to demonstrate that risks were maintained at a tolerable level throughout the arch raising. The OM was introduced and managed by a specialist team from the main designer.

1.3.10 Crossrail Blomfield Box (2012–2014)

'*Ab initio*' application for groundwater control: This is a vital activity for many underground works, yet is fraught with uncertainty, since it may be critically influenced by minor but unknown geological details, as noted by Terzaghi (1955). Dewatering was required to control the risk of base instability (due to high groundwater pressures in sand layers at depth) and to reduce the forces acting on the retaining walls. Variations in ground conditions, including faulting across part of the box, necessitated the use of different groundwater control systems during the excavation. Agreement to implement the OM was achieved through the recognition that its application through progressive modification offered the most effective way forward. It enabled the dewatering to be successful despite significant geological variability being encountered towards the base of one of the deepest excavations on Crossrail. The OM provided assurance of site safety while also delivering time and cost savings. It was introduced and managed by a specialist team from the main designer.

1.3.11 Crossrail Moorgate Shaft (2014–2015)

'Best way out' application to address delay caused by unexpectedly difficult obstructions encountered during construction: The Moorgate shaft is an irregular polygon in plan, roughly 35 by 35 m, formed with 1.2 m thick diaphragm walls, 55 m deep. It is located within a highly confined area surrounded by several sensitive structures (above and below ground), including a 100-year-old tunnel located only 4 m away. Timely completion of the works was critical for the central London section. The OM was implemented through progressive modification supplemented by an iterative design technique termed 'the verification process' (VP). The VP was considered necessary to provide additional confidence to the stakeholders, including a Cat 3 checker, that construction could proceed fully assured from the base case design. The OM enabled the construction sequence to be considerably modified and for temporary props to be eliminated over the final 12 m of the shaft excavation. The OM monitoring also identified a complex permeation grouting/temperature related interaction which increased deflections locally in one wall of the shaft. It provided early indication of this issue which was thus able to be effectively addressed and the shaft was completed without further delay. A specialist OM team from the main design consultant was appointed by the client.

1.3.12 Heathrow Airport Terminal 5 Tunnels (2005–2006)

'*Ab initio*' application for tunnel protection: This case history illustrates the reverse of the situation of those pertaining to the Mansion House and the Heathrow multi-storey car park. Here it was the need to protect tunnels from ground movements imposed from the construction of buildings above. This case history is not presented within the main sequence featured in this book but is included in Chapter 13 on the advantages and limitations of the OM. It involved the displacement and deformation control of deep bored tunnels constructed in London Clay with expanded reinforced concrete linings. It provides an instructive example of the vital need to recognise the limitations of the OM prior to its implementation. This constraint related to the inability to reliably obtain critical observations. Agreement to implement the OM was achieved by finding a way to remove this limitation. Its resolution involved the introduction of structural and construction sequence modifications in advance to enable the safe and assured application of the OM through progressive modification. The specialist OM team from the main design consultant was appointed by the client.

References

Dunnicliff, J. and Deere, D. V. (editors) (1984). *Judgment in Geotechnical Engineering – The Professional Legacy of Ralph B. Peck*, p. 122, Wiley,, New York.

Gaba, A., Hardy, S., Doughty, L., Powrie, W. and Selemetas, D. (2017). Guidance on embedded retaining wall design, CIRIA Report C760, London, p. 455.

Muir Wood, A. M. (2000). *Tunnelling – Management by Design*, E and FN Spon, London and New York.

Nicholson, D., Tse, C.-M. and Penny, C. (1999). The observational method in ground engineering: principles and applications, CIRIA Report 185, London, p. 214.

Peck, R. B. (1969). Advantages and limitations of the observational method in applied soil mechanics. *Géotechnique*, **19**, No. 2, 171–187.

Terzaghi, K. (1955). Influence of geological factors on the engineering properties of sediments, Economic geology, 50th anniversary volume, pp. 557–618.

Chapter 2

Channel Tunnel Cut and Cover Works (1988–1990)

2.1 Introduction

The application of the observational method (OM) to the Channel Tunnel cut and cover works was both metaphorically and literally ground breaking. Viewed from a perspective of some 30 years later this may seem a convenient hindsight. In fact the achievements described here developed almost spontaneously in a design and construction environment that was already full of change (Powderham, 1994). The OM delivered an unexpected bonus in what was already an iconic and world famous project. As shown in Figure 2.1, the cut and cover works included three portals. One each on either side of Castle Hill with the so-called New Austrian tunnelling method (NATM) bored tunnels and the third at Sugarloaf Hill (SH) with the tunnel boring machine (TBM) tunnels driven underland from the coast at Dover. It was at these three portals where the OM was applied – being initiated at Castle Hill East (CHE) and then opportunistically implemented next at SH and then finally at Castle Hill West (CHW).

A good rapport had developed early between the teams of consultant designers and the contractor assigned to this section of the project. As it moved from design to construction, potential improvements were continually being considered. This applied to the temporary and permanent works and, being a design and build contract, it inherently provided the contractual flexibility to introduce and implement such changes. This facilitated a natural opportunity to introduce the OM purely by seeking a more economic and safer way of construction. Certainly, the wider and far-reaching implications of this initiative were not anticipated during these early stages. Simply stated, a basic collaborative effort to reduce the amount of temporary structural steelwork at one location produced the acorn from which the OM grew and blossomed. It was introduced to address a specific programming issue at one of the tunnel portals. The unexpected nature of the initial success saw it develop progressively with significantly increasing benefits at the two other portals in these cut and cover works. The scale of the combined success of these

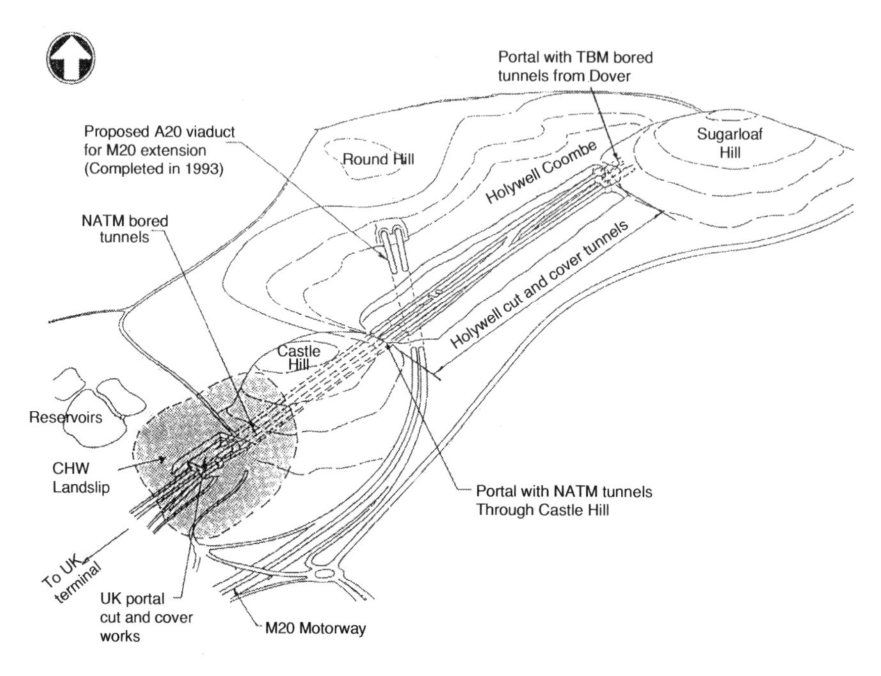

Figure 2.1 Castle Hill West to Sugarloaf Hill: overall layout of cut and cover and bored tunnels.

endeavours also set the stage for what has been described as a renaissance for the OM in a series of major projects that followed the Channel Tunnel, many of which are described in the other case histories addressed in this book.

2.2 Key Aspects of Design and Construction

2.2.1 Overview

The cut and cover works constituted a major part of the construction in and adjacent to the UK terminal for the Channel Tunnel. It is pertinent to note that the Channel Tunnel was one of the first and very prominent examples of such a major project being commissioned through a design and build contract in the United Kingdom. The main design for the permanent works was undertaken by the consultant engaged as designer for the design and build contractor consortium, Transmanche Link. The contractual environment facilitated a range of changes to be incorporated during the detailed design and, more pertinently in the context of the OM, during construction. Some of these changes had a substantial impact on the cut and cover works. The most significant derived from environmental concerns to protect sites of special scientific interest (SSSI) in Holywell. The preliminary design, prepared for the Parliamentary Bill to gain Royal Assent for the project, had an alignment in Holywell that provided enough space between the tunnel portals at CHE and SH to include two crossovers between rail tracks. These two crossovers are essential for the operation of the railway. However, this alignment took the construction right through one of the main SSSI in Holywell Coombe. It was important environmentally to safeguard this SSSI, and so a more southerly alignment that completely avoided it was developed. However, this had a profound effect on the cut cover works since there was now inadequate distance between the bored tunnel portals in the escarpments to accommodate the double crossover in Holywell as provided in the preliminary design at tender stage. The solution was to split the crossover into two separate arms – one in Holywell and the other within a substantially enlarged cut and cover portal construction at CHW. The new alignment is shown in Figures 2.1 and 2.2. However, at CHW, this more than doubled the total excavated volume and accordingly the amount of unloading that the construction would now create in the in the toe of the major landslip at this location (see Figures 2.3–2.5). So, apart from safeguarding SSSIs, these changes generated a fertile environment for creative coordination between the designer and the contractor. This, in turn, stimulated further design changes directed at enhancing the ease, speed and safety of construction and so led to the series of applications of the OM.

2.2.2 *Geology*

The three tunnel portals, at CHW, CHE and SH, were located on gently sloping ground at the base of the escarpment of the North Downs. Three short valleys or coombes cut into the scarp, thus isolating Castle Hill, Round Hill and SH.

The escarpment, which has a slope of about 1 in 3, consists of Lower Chalk capped by the harder basal bed of the Middle Chalk. The Lower Chalk, Glauconitic Marl and Gault Clay outcrop in succession at the base of the escarpment. The regional dip is about one degree to the north east, although this may increase near to the escarpment due to stress relief. The Gault Clay is heavily over-consolidated and of high plasticity. Although this clay is generally very stiff or hard, weathering has substantially reduced its strength at the surface and there is a history of instability at outcrops. These presented the risk of significant ground movements being initiated by the construction at the portals. Groundwater within the coombe areas are maintained close to ground level by infiltration from the base of the chalk escarpment. Springs occur at the base of the scarp along the line of the outcrop of the Glauconitic Marl.

Figure 2.2 Holywell Coombe, looking west from Sugarloaf Hill to Castle Hill. Photo courtesy of Eurotunnel.

Figure 2.2 shows the escarpment of Castel Hill rising some 70 m above the coombe, contrasting against the depth of the excavation in the central section. One of the main SSSI is shown fenced off in the foreground of this photograph. It was through this that the preliminary alignment passed. The cut and cover tunnels are shown under construction well to the south of this environmentally important area.

2.2.3 Design Uncertainties

The ground conditions and complex geology presented a range of design uncertainties. These included escarpment stability and the sensitivity of the Gault Clay with its high plasticity and potential to soften and swell when unloaded during the construction period. These aspects, along with the steeply varying topography at the portals, raised questions regarding the dominant direction of lateral earth pressure loadings on the cut and cover works along with the influence of construction methods and sequences. There was also a lack of useful case history data for such construction in Gault Clay.

Figure 2.3 UK portal construction at Castle Hill West showing the imposing presence of the landslip.

Photo courtesy of Eurotunnel.

In Holywell, the relationship of the vertical alignment of the tunnels and the existing topography created substantial variations of cut and fill. The Gault Clay underlying the cut and cover tunnels was thus subjected to unloading in the areas of cut and bulk loading in the areas of fill. This, in turn, would lead respectively to zones of heave and settlement. Consequently, any abrupt change in support conditions to the cut and cover tunnels presented the risk of adverse effects on the railway tolerances in both the medium and long term. It was thus necessary to impose limitations on the depth of piled wall embedments to minimise differential settlements between the piled wall structures and the open cut sections.

The range of lateral loading considered in the design for the contiguous piled walls is discussed in Section 2.4.6. It should be noted that the high geological value of 2 shown for K_o, relating to the undisturbed condition

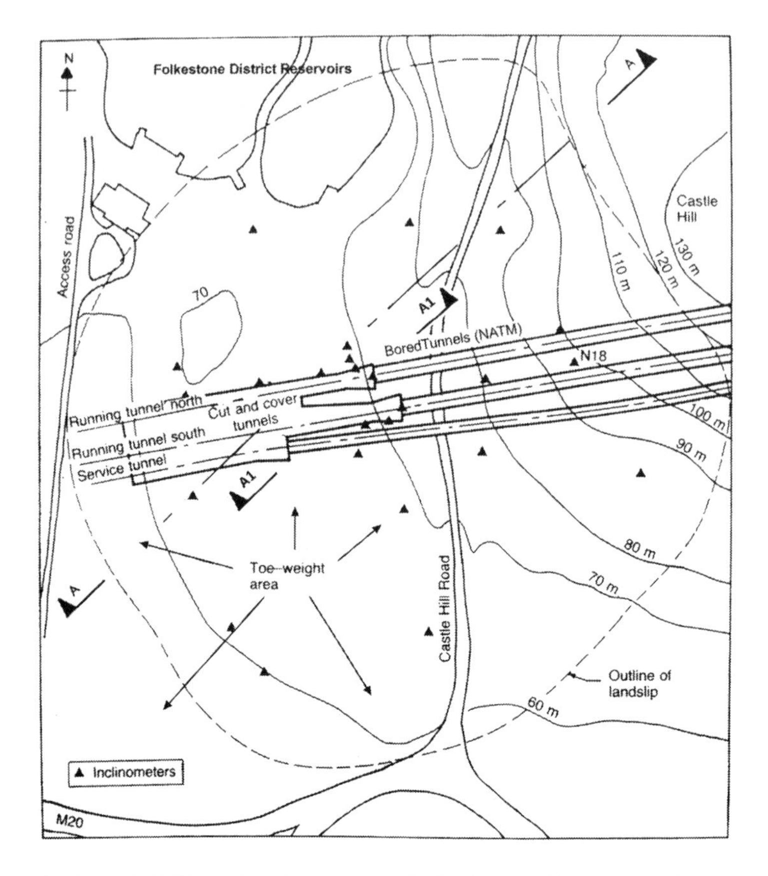

Figure 2.4 Castle Hill West: plan showing tunnels, landslip, and instrumentation.

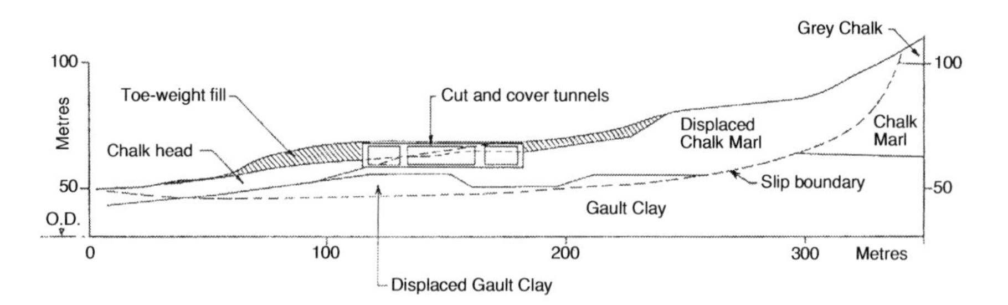

Figure 2.5 Castle Hill West: section A-A through landslip.

of the over-consolidated Gault Clay, was not directly used as a design parameter for lateral loading. Rather, it is shown drawn to scale for comparison with various design loadings. It is a simplified representation, as it does not, for example, indicate the reduction in K_o relating to decreasing over-consolidation ratios with depth. However, it does highlight its relative magnitude even when compared with the maximum loading assessed for the landslip (see Section 2.4.10).

A value of $K = 1$ was adopted as the design coefficient for lateral loading in the long-term for the Gault Clay. At the time, the issue of K values was proving a stimulating subject of discussion between the structural and geotechnical design teams. Contrasting views were forwarded for design values of K, on either side of unity, for over-consolidated clays. The Limehouse Link case history provides another example where Stevenson and De Moor (1994) quote values for K_o of 1.5 for the London Clay and 2.5 for the clay strata of the Woolwich and Reading beds.

There is no generally definitive answer to this question of appropriate design values for K. They are site specific and depend on many interactive effects which are influenced by geological history as well as the design and construction methods and sequences.

All of these aspects of design uncertainty applied with greater significance to the major landslip at CHW with its unique and onerous loading conditions (see Section 2.4.10). This large and complex landslide is the most evident visually among a number of such features along the escarpment in this area. It was postulated that these failures occurred, or were re-activated, during the Late Glacial period but are considered to have remained marginally stable under present conditions. The construction works for the UK portal of the Channel Tunnel necessitated comprehensive risk assessments and management to ensure the maintenance of an acceptable factor of safety for the stability of this landslip. This and further details of the design uncertainties are discussed in Sections 2.4.3 and 2.4.9.

2.3 Achieving Agreement to Use the OM

2.3.1 Creating Opportunity

Throughout this design and build project, the design and construction methods were subject to ongoing changes. The design was thus developed iteratively with construction. This demanded close coordination between design development and construction progress. However, it is important to note in this context that such coordination can in practice be predominately administrative and does not automatically ensure that designs are fully related to construction. To achieve the latter requires an adequate understanding by the designer of the specific construction methods including the plant and sequences employed.

The OM was not originally envisaged in the design of the cut and cover works. It was, however, being applied to the substantial issues of slope stability and drainage at this site. So it could be viewed as a conducive environment for introducing further applications of OM. But such introductions do not just happen. They need advocacy and incentives – such as enhancements in safety, time or cost. The path to the additional implementation of the OM, which could even be characterised as serendipitous, is outlined in the next section.

2.3.2 Finding the Key

Monitoring trials to initiate the OM took place at CHE in July 1988; the last phase of the method was started at CHW in May 1989. The OM, as applied to the cut and cover works, was thus initiated well after construction had commenced. The approach was essentially opportunist since there was no crisis to act as a catalyst for a 'best way out' solution and it could not strictly be described as an '*ab initio*' application either, although essentially it was closer in spirit to the latter. (See discussion in Section 2.4.1.)

So what was the key that unlocked these very significant opportunities and led to the extensive applications of the OM to the cut and cover structures? Given that the OM typically addresses temporary conditions and temporary works, the relating of design to construction is particularly vital in its application. This is considerably enhanced when the designer has an ongoing presence on site – and indeed should be a prerequisite for any application of the OM. Moreover, soil/structure interaction is usually a dominant factor and this brings an intrinsic and substantial component of uncertainty.

Initial site visits enabled a rapport to be developed between the design team and the contractor's agent on site. As noted, the first section of cut and cover construction was at CHE and, from observations of its early progress, it was evident that the design of the temporary support propping of the contiguous piled walls was quite conservative. The intermediate and lower levels of this steelwork obstructed access to commence the NATM tunnelling through Castle Hill from the east. Given the pressures of programme, the contractor expressed an interest as to what scope there might be in significantly reducing the amount of the obstructing temporary steelwork. A salient point here is that this was first raised, almost as an aside, during a routine site visit and may well have passed unremarked and without effect. There is a strong dynamic between consultants and contractors which can be either notably negative or productively positive. The nature of such relationships was engagingly addressed by Terzaghi (1958).

For the cut and cover works, the early rapport that had been established took this synergy in a very positive direction. Somewhat in jest and perhaps merely to test the designer's sense of humour, the contractor asked what the props were there for. The reply, expressed in similar currency, was that they were there to hold the load cells in place. The irony of this response was immediately appreciated as these were some of the few elements of instrumentation that had been somewhat reluctantly approved for installation in the cut and cover works. The contractor accordingly observed that it should be possible, on that basis, to eliminate the propping. The rejoinder to this was that, pending approval for substantially more instrumentation, there should indeed be promising potential to reduce the amount of such temporary steelwork support at this first section of contiguous piled wall construction of CHE.

The OM as such was not specifically mentioned at this stage and the process that developed was essentially an inherently natural engineering approach to address uncertainty. It was only in retrospect much later that this process was seen to essentially be that of the OM. And so, from a seemingly minor interchange of what essentially was site banter grew the whole range of OM applications to the cut and cover works. It should be noted that the immediate driver was the contractor's priority to gain access to the portal at CHE to enable the earliest start of boring the NATM tunnels through to CHW. At this first stage, the focus was simply directed at gaining access to the three tunnel portals at CHE. There was no initial grand plan to globally address the potential for reducing the temporary support at the other portals at SH and CHW. The awareness of this potential developed progressively. While the observations of the construction at CHE soon underlined the potential for the analogous conditions at SH, it was a far greater step to consider how such benefits might be realised at CHW.

2.4 Implementation of the OM

2.4.1 Development of Progressive Modification

The applications of the OM developed for the Channel Tunnel cut and cover works involved both bottom-up and top-down construction. Substantial benefits in both cost and time were achieved along with enhanced safety. This success was augmented progressively at each of the three locations comprising a total of nine tunnel portals – three each at CHE, SH and CHW. It concluded with the greatest savings being achieved at the most challenging of the three locations – namely that within the landslip at CHW. The applications of the OM included the innovative use of blinding struts, modifications to both temporary and permanent works and addressing long-term conditions. The latter related to the ongoing maintenance of acceptable tolerances for the alignment of the railway. These tracks served both the Channel Tunnel Shuttle and the high-speed continental main line and consequently had tight limits on rail tolerances.

A basis for applications of the OM was set out by Peck in his 1969 Rankine lecture and his key ingredients were essentially echoed in the emerging development of the OM for the cut and cover works. (The full list of Peck's eight ingredients is provided in the Introduction.) The programme of construction for the three locations of CHE, SH and CHW fortuitously facilitated the implementation of the OM on an evolving step by step basis – an approach which duly became known as 'progressive modification'. In fact, all of the examples of the OM featured in this volume were implemented through progressive modification. It is also important to note that while many of the applications of the OM in the various case histories may be viewed as adventurous or even dramatic, both in their technical achievements and the savings realised, a key driver to implement the OM was safety – and enhanced safety is a key feature of progressive modification. (See further discussion of this in Chapters 1, 13 and 14.)

2.4.2 Construction Sequence, Form and Geometry

The vertically retained cut at all three locations at CHE, SH and CHW used permanent walls formed with contiguous reinforced concrete-bored piles. The varying spatial relationship between the three tunnels and the associated construction sequences and methods created a wide variety of geometrical arrangements for these walls.

To enhance access for the bored tunnelling works, the retained cut sections at CHE and SH were originally designed for a bottom-up construction sequence. However, at CHW top-down construction was considered essential from the start because of landslip stability requirements. As noted, these became even more critical with the increased unloading resulting from the relocation of one of the railway crossovers from Holywell to CHW. The maintenance of the stability of the CHW landslip, both in the short and long term, was addressed by a separate '*ab initio*' application of the OM (Avgherinos *et al.*, 1993).

The geometry of the cut and cover works was generally rectilinear but included circular shafts to act as reception chambers for the bored tunnelling machines at SH (Figures 2.6 and 2.13). The breakthrough of one of the running tunnel TBMs is shown in Figure 2.7.

The piling subcontract started in May 1988 at CHE, followed by SH and then completing at CHW. Bulk excavation and construction of the cut and cover tunnels followed in the same sequence, starting at CHE in June 1988. The amount and arrangement of temporary propping as designed was not exceptional. It did not impose an unusual construction problem and no unexpected development had occurred to complicate construction unduly. The OM simply offered an opportunity to bring substantial improvements to a basically conventional situation. Such potential is not uncommon. Unfortunately, it is far less often realised. Here, two key factors were present to enable this

Figure 2.6 Cut and cover construction in Holywell viewed from Sugarloaf Hill. Photo courtesy of Eurotunnel.

potential to be realised: the motivation and rapport established between the design and construction teams and the contractual flexibility afforded by the design and build contract to readily introduce design changes.

2.4.3 Design Constraints

In relation to the potential benefits offered by the OM, it is relevant to consider the conservatism appropriate to the original design. The constraints which influenced its development, particularly in relation to temporary works, were as follows:

(a) Uncertainties regarding the ground conditions, particularly in the performance of the Gault Clay. This included the depth and severity of weathering and the susceptibility of the clay to softening during the construction period.

Figure 2.7 Circular reception chamber at Sugarloaf Hill showing breakthrough of a running tunnel TBM.

Photo courtesy of Eurotunnel.

(b) The magnitude of short-term lateral loads and passive support to the piled walls. In addition to the uncertainties noted in (a), there was the influence of construction methods and sequences, including how long a given section of temporary propping might have to remain in place.
(c) The lack of useful case histories and data for similar construction in these ground conditions.
(d) Escarpment stability: A major and very complex landslip had been identified at CHW (Griffiths, 2017). Geomorphological mapping had also indicated potential slip conditions at CHE and SH. It was principally to limit the unloading effects from excavation that vertically retained cut was used at each escarpment.
(e) The embedment of the contiguous piles was limited to 2 m to avoid a hard transition between the abutting open cut construction for the tunnel boxes in the central area of Holywell. This, in turn, limited the passive support during construction placing more reliance on the temporary propping.

Against this background it was considered inappropriate for the OM to proceed directly with a design based on predictions of the most probable conditions, thus diverging somewhat from a literal adoption of the steps set out by Peck (1969). (See discussion on this issue in Sections 1.2 and 13.3.) To initially assess potential, a preliminary monitoring trial was undertaken at CHE. The results were encouraging and a process that was subsequently recognised as the OM was then implemented by introducing beneficial design changes in a controlled step by step basis. Three levels of temporary propping, vertically spaced 2.5 m apart, were provided in the base case design. At CHE and SH this comprised structural steelwork, while, at CHW, there were two levels of temporary steel propping since, in the top-down construction sequence, the roof slab provided the first level. However, the most significant difference at CHW, as noted in Section 2.4.10, was that the dominant wall loadings were governed by the criteria to ensure maintaining the stability of the landslip, rather than those relating to a more conventional retaining wall design as at CHE and SH. With the increasing success achieved progressively at these first two locations, the key question that inevitably arose was could such benefits translate to similar opportunities at CHW with its notably more onerous ground loading environment? Certainly, such a question to remove temporary support on such a scale within an active landslip would have hardly been contemplated prior to the achievements at CHE and SH. That success not only underpinned the technical basis but also, most importantly, demonstrated the effectiveness of the procedural approach through progressive modification enhancing understanding and trust between the design and construction teams.

2.4.4 Addressing Uncertainty

In the assessment of the effects of the most unfavourable conditions, the magnitude and nature of passive support were critical. These would depend on the influence of fissuring and the rate of softening and dissipation of negative porewater pressures. Such uncertainties were addressed by rapid excavation and protection of the formation with blinding concrete. In highly over-consolidated clays, the stress path in vertical unloading leads, in extension, towards a condition of limiting passive failure. Consequently, large lateral movements of the toe of a wall embedded in such soils are not needed to mobilise a high percentage of passive support. However, a brittle response leading to sudden loss of passive support and excessive wall movements had to be avoided. The approach adopted was to limit wall movements and thus keep soil strains low by excavating in short sections to exploit soil arching. Measurements of wall movements and back-analysis of prop loads indicated average soil strains of less than 0·2%. From field observations and research into the performance of over-consolidated clays, a case could also be made for a ductile rather than a brittle deformation response for the passive support, particularly under short-term total stress conditions. A limit of one month was placed on the construction of each bay to establish permanent lateral support at formation level. To ease the time pressure on completing the base slab, this support was achieved by installing an enhanced blinding strut. Two additional constraints were imposed to control of the potential for progressive collapse:

(a) Tension members were placed at the level of the top props to control rotation about the mid-height props.
(b) The length of a bay during excavation to formation level was limited and the blinding strut cast sequentially in 5 m lengths.

These provisions were made to limit the potential for unacceptable deflections of the walls, under the most unfavourable conditions, to a very local zone of wall. The courses of action for significant variation in the observed values from those predicted were to:

(a) install the lower steelwork props if wall movements were greater than expected;
(b) extend bay lengths if wall movements were less than expected.

2.4.5 Instrumentation and Monitoring

The instrumentation to monitor the performance of the structures comprised inclinometers in the piles and adjacent ground, load cells and tape extensometers. This was supplemented by standard surveying techniques to measure local changes in line and level. Emphasis was placed on the generation of meaningful and manageable data which could be easily checked, correlated and repeated. Wall movements and trends were identified as the critical observations. Prop loads provided secondary and less useful information and throughout the project their measured values remained very low and were typically less than 15% of their design capacity. Observed wall movements remained well within the acceptable values.

2.4.6 Castle Hill East – The OM Stage 1: Initiation

At CHE the three portals with the bored tunnels were aligned in close proximity to each other (Figures 2.6, 2.9 and 2.10). All of these cut and cover works had been designed for bottom-up construction. However, this maximised the amount of temporary propping. As noted in Section 2.3.2, during the initial phase of its installation at CHE, the contractor questioned the possibility of reducing this steelwork. This led to the development of the OM bringing beneficial design changes in both the temporary and the permanent works. The basis for the application of the method was as follows:

(a) In addition to the original site investigation, the preliminary assessment of the ground conditions and loadings involved visual inspection of the ground during excavation and the measurement of prop loads and wall deflections.
(b) A design was developed adopting conditions judged to be more likely than those assumed in the base case design. Essentially, this exploited short-term soil strengths using lower active and higher passive pressures. It was based on simple plane strain analysis and thus assumed no benefit from three-dimensional effects. This revised design basically had to establish a reasonable case for implementation of the OM. A more optimistic basis for the design was not necessary at this stage.
(c) The analysis using mixed total and effective stress was developed from the basis set out by Padfield and Mair (1984). Using the parameters shown in Figure 2.8, this indicated the potential for eliminating the bottom prop. The ground water table was set at the top of the clay, although the inevitable seepage through the gaps between the contiguous piles made this a conservative assumption. With only the top prop in place, the factor of safety was estimated to be only 1.3. This is very low factor for mixed analysis, particularly when assuming full passive pressure on short embedment lengths. Padfield and Mair recommended a minimum factor of safety of 2.0 when using moderately conservative soil parameters along with the requirement to include an allowance for

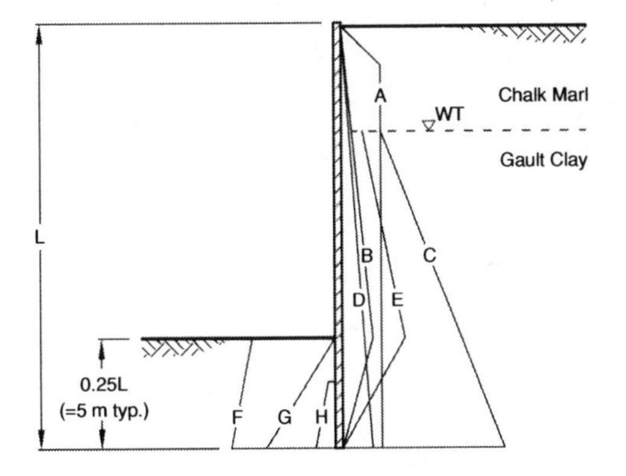

A	Maximum slip force (200 KN/m²)
B	$K = 0.5$ Design earth pressure (temporary)
C	$K_o = 2$ Earth pressure 'at rest' (geological)
D	$K = 0.37$ Active, $ø' = 24°$, $C' = 0$
E	$K = 1$ Design earth pressure (long term)
F	Full passive-total stress
G	Full passive-effective stress, $ø' = 24°$, $C' = 0$, $K = 3.4$
H	1/3 passive-effective stress, $ø' = 24°$, $C' = 0$, $K = 3.4/3$

Figure 2.8 Notional earth pressure diagrams relating to propped wall design at CHE and SH.

softening of the top metre by applying a reduction factor to the undrained shear strength. Thus, although considered adequately conservative as a starting position for the OM, the design was operating outside normal guidelines and practice. This emphasises the importance of critical observations and the rigour and control necessary in the use of the OM. As noted, although the design developed for the OM was significantly less conservative than a conventional one, it still did not rely on criteria judged to be the most probable. The OM design was thus judged to be reasonably robust, which, in turn, made it adaptable to the progressive modification approach. As discussed in Section 2.4.1, this provides a reserve to potentially counter the uncertainties which inevitably arise in soil/structure interaction and structural performance. The potential disadvantages of designing too closely to estimates of most probable conditions were indicated in the construction of the NATM tunnels (see Section 2.5).

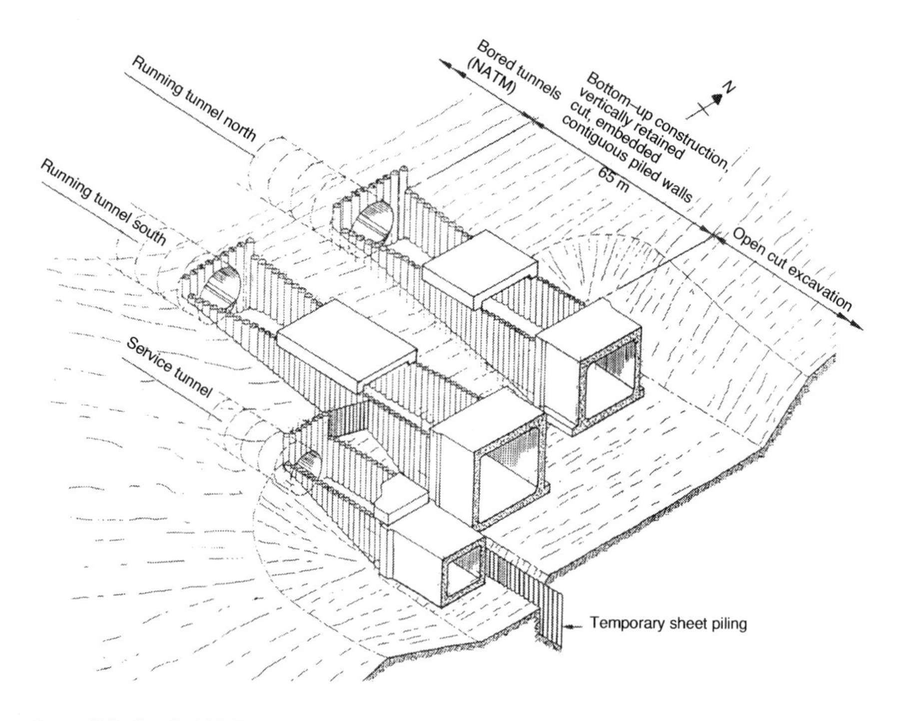

Figure 2.9 Castle Hill East cut and cover portal structures with NATM tunnels.

2.4.7 Measured Performance at CHE

Wall deflections remained consistently well below the assigned maximum of 15 mm, being generally in the range of 3–7 mm. A deflection of 15 mm was around half of the acceptable limit and effectively constituted the red trigger level – that is, one if reached, or deemed likely from adverse trends, would require implementing contingency measures. These would essentially require returning to the original design and installing all of the temporary propping. The implementation of the OM for these Channel Tunnel works was effectively the forerunner of the subsequent examples described in this book. Its application was developmental and consequently less sophisticated than in those case histories that followed where, for example, the traffic light system using green, amber and red control zones was introduced and comprehensive real-time monitoring was pioneered. The original development of the traffic light system for the OM is described in the Mansion House case history where it was first introduced. At CHE observations indicated that, rather than triggering contingencies, it would be possible to advantageously modify the design by relaxing the constraint on bay lengths. However, this benefit was not realised at CHE as construction was too advanced before the OM was introduced. Nevertheless, a reduction of around 20% comprising some 250 tonnes of temporary steelwork was achieved at this first location Figures 2.10 and 2.11).

Figure 2.10 Castle Hill East Service Tunnel and Running Tunnel South under construction.

Figure 2.11 Castle Hill East: cut and cover interface portal with NATM tunnels.

2.4.8 Sugarloaf Hill – The OM Phase 2: Development

Given the analogous construction, the basis of the application of the OM was similar to that established at CHE. The arrangement of the contiguous piled sections of the tunnels at SH is shown in Figure 2.12. To maximise benefit from the OM, the design of the permanent works in the rectilinear sections was changed from bottom-up to top-down construction. This offered two main advantages. First, it was possible to exploit the permanent works for temporary lateral support as the roof slab eliminated the need for one level of props. Second, and more importantly, it helped to address the problem of the worst-case scenario by providing a moment connection with the walls during the critical excavation phase to formation level. This converted the risk progressive collapse with to that of limited structural overstress and thus enhanced the potential to introduce longer bay lengths.

The design change to top-down construction did not require major changes to reinforcement details. However, it did need extensive reanalysis and, as it directly affected the permanent works, involved more

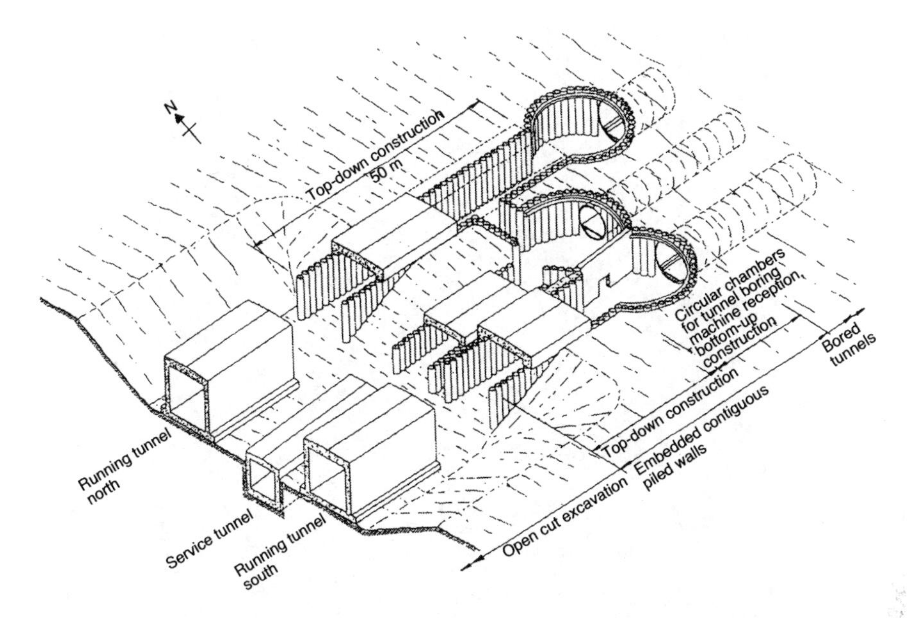

Figure 2.12 Sugarloaf Hill cut and cover portal structures with TBM tunnels.

comprehensive approval procedures. With the roof slab brought into play for temporary support, it proved possible through the OM for the original three levels of steelwork to be reduced to just one level at mid-height. And this was placed only in the most heavily loaded sections of the tunnel, adjacent to the circular shafts. A 60% reduction in the amount of temporary steelwork was thus achieved. The combined savings achieved including CHE were over 700 tonnes. The sense of achievement was readily reflected in the increasing rapport between the design and construction teams. This positive atmosphere was very evident on site as the benefits went significantly beyond the basic material savings in temporary steelwork. With considerably less steelwork to erect and remove, the greater freedom within the working space within the cut and cover tunnels enabled faster construction. Safer working conditions were also created as the need to manoeuvre the heavy sections of steelwork in confined spaces was substantially reduced.

2.4.9 Castle Hill West – The OM Phase 3: Conclusion

The nature of the success of the OM at CHE and SH predictably focussed attention on the potential at CHW. Although a top-down sequence had been developed in the original design for this portal construction, the

Figure 2.13 Cut and cover progress in Holywell showing the top-down construction and two of the TBM reception chambers at Sugarloaf Hill in the foreground.
Photo courtesy of Eurotunnel.

amount of temporary steelwork propping required was still very substantial. Its design and installation was complicated by the variable geometry and much larger spans. The landslip conditions increased the challenges bringing additional uncertainties relating to soil strength and the probability of high lateral loading on the structure (Figure 2.14). The low factor of safety of the slip also imposed an additional need to limit ground movements. To help compensate for the substantial increase in unloading caused by the incorporation of a crossover in the UK portal structure, the innovative use of heave reducing piles were included in the design (Figures

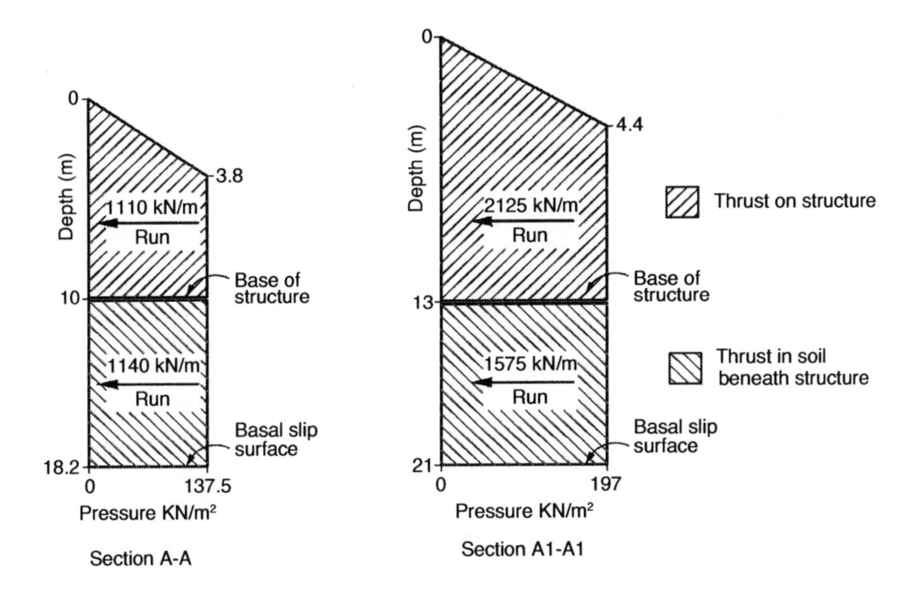

Figure 2.14 Landslip thrust loadings (sections locations shown in Figure 2.4).

2.15 and 2.16). They were installed as low cut-off piles from the same platform as the contiguous piles for the walls (Figure 2.17). This platform was 12 m above the base slab and so the heave reducing effects would be mobilised right from the start of excavation below the platform. The benefits of this approach, though not precisely quantifiable, were twofold. Firstly, the short-term movements in the landslip due to the unloading from excavation would be reduced. This would aid the overall robustness of the design and accordingly improve the potential for the elimination of temporary propping using the OM. Secondly, the heave reducing piles would also limit long-term movements enabling the base slab to provide a very stable support for the trackwork. These piles were 0.6 m in diameter, 15 m long and spaced at 1.8 m centres. A total of 486 were installed. In addition to the qualitative benefits discussed earlier, approximately two-thirds of their cost was offset by economies to the base slab, which was supported by these piles instead of spanning directly between the walls over lengths of up to 18 m (Duggleby *et al.*, 1991).

2.4.10 Landslip Loading

The landslip at CHW consisted of an upper mass, which was in an active state, resting against the passive lower zone. For any substantial excavation made within the landslip, it was necessary to ensure that the thrust

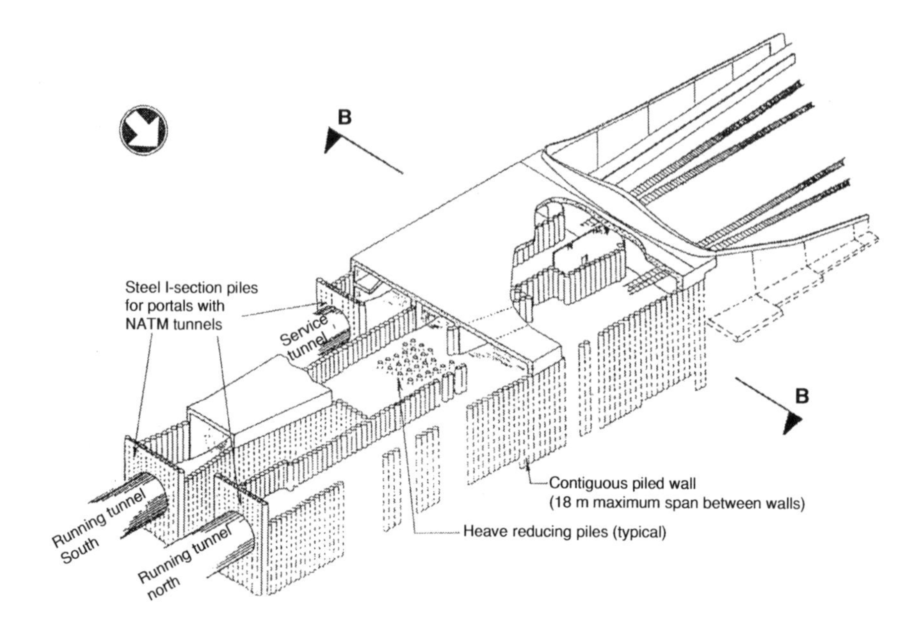

Figure 2.15 UK Portal Castle Hill West.

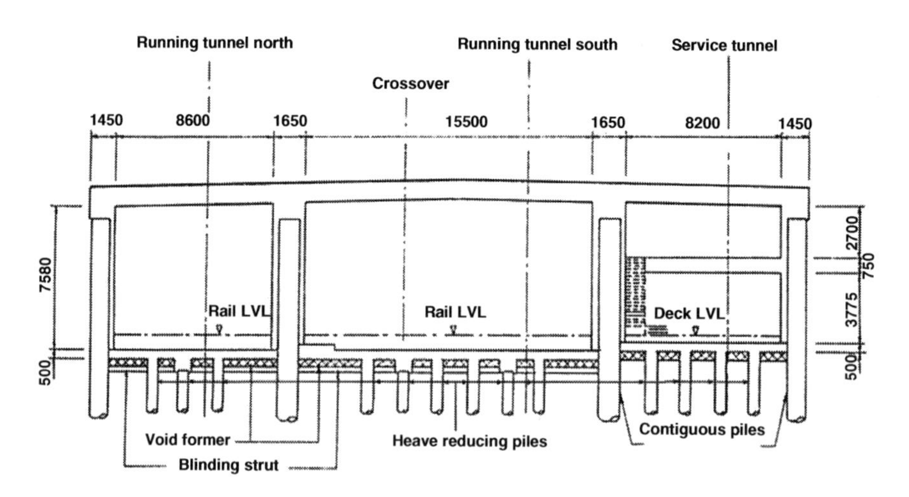

Figure 2.16 UK Portal cross section BB.

Figure 2.17 Castle Hill West piling platform for UK Portal and crossover construction. Photo courtesy of Eurotunnel.

required to fully support the upper mass was maintained. Normal design criteria for lateral loading of excavations do not cover this situation. The magnitude of the thrust magnitude and its line of action were estimated using statically correct slices in a stability analysis based on the method of Sarma (1973). This requires the percentage of available shear strength actually mobilised to support the weight of the upper slope at the section analysed. Assuming too high a percentage of strength mobilisation would underestimate the actual thrust. As a primary objective of the design was to minimise displacement, it was desirable to maintain the pre-construction mobilisation of strength. At the locations shown in Figure 2.4, this was estimated as 83% at section A-A and 87% at section A1-A1. The values are below 100% as the cut and cover tunnels lie a little outside the less stable central third of the slip. The pressure distributions corresponding to these thrusts were then derived as shown in Figure 2.14.

With maximum lateral loadings estimated around 200 kN/m^2, these were considerably more than those for a conventional design in undisturbed deposits. In the original design, the thrust above formation level had to be carried across the opening in the short term by the temporary propping and in the long term by the completed structure. The thrust below formation level would be transmitted through the unexcavated soil.

As noted earlier, the extensive use of heave reducing piles below formation level substantially maintained the soils ability to transmit this thrust. Furthermore, a principal aspect in the success of the OM in Holywell had been the exploitation of short-term soil strength by faster construction. Such advantages were not completely negated by the conditions presented by the landslip. However, three-dimensional effects needed to be considered comprehensively since success would be dependent on soil arching and effective load paths.

Excavation was started in short bay lengths of 5 m in the least loaded areas at the west end of the cut and cover. One level of mid-height temporary steel propping was used initially. This was similar to the procedure developed at SH which had involved changing the design there from bottom-up to a top-down sequence. This established a very useful precedent in addressing the far more complex and much greater scale of the top-down construction at CHW. At the western end, satisfactory observations enabled the bay length to be doubled to 10 m. Work was then progressed to the eastern section adjacent to the portals with the bored tunnels. This presented the greatest challenge since it was the zone of the highest lateral load in the landslip. Nevertheless, the application of the OM reduced the temporary steel propping to just one level at mid-height. With the base slab constructed at both the western and eastern ends, it then proved comfortably possible to complete the large central area without any temporary steel propping at all (Figure 2.18). The original requirement of 2,250 tonnes of temporary

Figure 2.18 Castle Hill West showing clear construction space without mid-height propping.

Photo courtesy of Eurotunnel.

steelwork for CHW at the outline design stage was reduced to a final amount of 280 tonnes.

2.5 Connection with the NATM Tunnels

2.5.1 Interface Coordination

As discussed in Chapter 14 on the advantages and limitations of the OM, while avoiding unnecessary conservatism, robustness and ease of construction are important considerations. An illustrative example of this features the use of structural steel I-sections for the portal walls between the cut and cover and NATM tunnels. Figure 2.11 shows an example for one of the running tunnels at CHE. The I-sections were installed in bores augured with the piling rig and set in a 20:1 sand/cement mix. This friable but adequately supportive matrix allowed easy access to the steel sections during the subsequent excavation. This enabled their sequential removal in a safe and controlled manner during the construction of the permanent connection between the cut and cover and NATM tunnels. The use of steel sections in this way was an effective alternative to the more typical installation with concrete piles reinforced with rebar. The I-sections can be easily flame cut through and safely removed sequentially in manageable elements with no vibration. Reinforced concrete piles by contrast typically require breaking out with substantial percussive plant involving heavy vibration. Such vibration may cause significant issues in unstable ground – such as was present in the outcrops of the Gault Clay (see Section 2.2.2).

2.5.2 The OM and 'Most Probable' Conditions

The NATM tunnels through Castle Hill involved a separate application of the OM. Its implementation for the cut and cover works at the tunnel portals thus involved abutting applications of the method though with different teams. This interface required effective coordination and, as noted earlier, the benefit of using I-sections for the portal walls, was an example of addressing this. Although the process of progressive modification is inherent for NATM construction, the OM design for the tunnels through Castle Hill was based on predictions of the most probable conditions. This contrasted with the approach adopted for the cut and cover works as described in Section 2.4.3 where the OM design used conservative criteria. In the event, the estimates of the most probable conditions for the NATM construction proved too optimistic. The starting design for these tunnels assumed that construction could be progressed with full face excavation. However, in practice, instability of the Gault Clay required this to be changed to a heading and bench sequence and with inclined rather than vertical excavation faces. Steel lattice girders were also installed aligned with these inclined faces to prevent blocks in the Gault Clay from loosening. Thus, for the NATM tunnels, the progressive modifications were contingency measures. By contrast in the OM application for the cut and cover works, starting from a more conservative base, the modifications were not contingencies but additional savings. The NATM tunnelling through Castle Hill is reported in detail by Nicholson *et al.* (1999).

2.6 Conclusions

The total reduction in temporary steelwork for the cut and cover works overall was around 2,700 tonnes. Dramatically, over 80% of this was achieved for the construction within the most complex and challenging ground conditions presented by the CHW landslip. This location featured by far the largest structural spans and greatest loadings. In addition to these material savings, construction safety was enhanced through eliminating the need to install and remove heavy structural steelwork in confined spaces. This also significantly reduced construction times. Thus, from just a modest beginning, the applications of the OM for the cut and cover tunnels demonstrated its huge potential and, in particular, how benefits can be maximised by implementation through progressive modification.

References

Avgherinos, P. J., Jobling, P. W. and Varley, P. (1993). The Channel Tunnel – Use of Geotechnical Monitoring to Assist the Design and Construction of Tunnels through an Ancient landslip at Castle Hill, Options for tunnelling, pp. 689–698. Elsevier, Amsterdam.

Duggleby, J. C., Avgherinos, P. J. and Powderham, A. J. (1991). Channel Tunnel: foundation engineering at the U.K. portal. DFI Conference, Stresa. Rotterdam; Balkema, pp. 373–382.

Griffiths, J. C. (2017) Case Study 9.13: UK Channel Tunnel portal at Castle Hill, Engineering Geology and Geomorphology of Glaciated and Periglaciated Terrains – Engineering Group Working Party Report, p. 916.

Nicholson, D., Tse, C.-M. and Penny, C. (1999) The observational method in ground engineering: principles and applications, CIRIA Report 185, London, 214 pp.

Padfield, C. J. and Mair, R. J. (1984). Design of retaining walls embedded in stiff clays, CIRIA Report 104, London, UK.

Peck, R. B. (1969). Advantages and limitations of the observational method in applied soil mechanics. *Géotechnique*, **19**, No. 2, 171–187.

Powderham, A. J. (1994). An overview of the observational method: development in cut and cover and bored tunnelling projects. *Géotechnique*, **44**, No. 4, 619–636.

Sarma, S. K. (1973). Stability analysis of embankments and slopes. *Géotechnique*, **23**, No. 3, 423.

Terzaghi, K. (1958). Consultants, Clients and Contractors. *BSCE*, **45**, No. 1, 239–279.

Chapter 3

Mansion House (1989–1991)

3.1 Introduction

The Mansion House has been the official residence of the Lord Mayors of London for over 200 years. As one of grandest surviving Georgian town palaces in London, apart from providing living accommodation, it features magnificent spaces for receptions including the Egyptian Hall which hosts, among other major events, the annual speech from the Chancellor of the Exchequer.

This imposing Palladian style masonry structure has a Grade 1 national heritage listing (Figure 3.1). The building in its original form was constructed

Figure 3.1 Mansion House in 1990s: north and west elevations.

over the period 1739–1753 to the design of George Dance the Elder (Figure 3.2). This design included some of the finest mid-18th century interiors anywhere in England, which were themselves inspired by those of leading architects of the day such as Inigo Jones. The glory of these works has been somewhat overshadowed by ongoing criticisms and concerns about the structure of the building throughout its eventful history. Located at the heart of the city, it was built on the site of a livestock market over the buried River Walbrook. Today, it shares a five-way junction with Royal Exchange and the Bank of England. The building is around 60 by 30 m in plan and generally consists of five storeys with a vaulted masonry arch basement under the northern two-thirds. The Bank underground station complex below the building was significantly extended in the early 1990s to include the terminal station for the Docklands Light Railway (DLR).

Figure 3.2 Mansion House in the 18th century.

3.2 Key Aspects of the Design and Construction

3.2.1 Foundations in the Building's History

Throughout its life, the building has been subject to extensive modifications, including major structural alterations. Much of this work related to or directly affected the foundations. With its western flank sitting partially over the soft deposits of banks of the River Walbrook, it faced a challenging start. Even during its construction, it suffered substantial subsidence including a significant differential component. This issue of the variability of the founding conditions was highlighted in a survey undertaken as part of the site investigation for the project reported in this case history. It revealed that the building has had to accommodate a 200 mm drop in level between the south-east and the north-west corner (which extends the furthest over the Walbrook). Part of this would have been caused by the construction of the Central Line tunnels at the beginning of the 20th century as discussed later. The original building had two clerestory roof extensions, constructed over the Egyptian Hall and the Ballroom, colourfully nicknamed 'Noah's Ark' and the 'Mayor's Nest' (Figure 3.2). The vaulted ceiling of the ballroom after the reconstruction is shown in Figure 3.3.

Figure 3.3 Interior view of ballroom at northern end of building looking east, 1989.

Apart from their controversial aesthetics, the additional weight of these substantial structures added to the ongoing problems of subsidence and was respectively demolished in 1795 and 1842.

Historically, there has been extensive tunnelling below the Mansion House. Construction of tunnels for the London Underground Central Line, which started in 1900, led to some substantial underpinning at the northern end of the building. These tunnels, which included the large diameter platform tunnels, were constructed at half the depth of those of this case history of the DLR. Although there is no detailed information on the actual ground movements generated, such major construction at that time, with less stringent controls than demanded today, would have inevitably imposed high ground strains and settlements. A back analysis, undertaken as part of the historical assessment for the current project, indicated that the short-term settlement of the building arising from the construction of the Central Line tunnels would have been around 40–50 mm.

The city was a prime target for air raids during the Second World War and the most dramatic event in the building's colourful history occurred during the night on 11 January 1941. Bank Station suffered a direct hit from a high-intensity bomb. A total of fifty-six people were reported killed in the tremendous blast – some at street level and others below ground in the station booking halls and escalators and right down at depth on the platforms of the Northern Line tunnels. The resulting surface crater was known as the largest in London and the damage so extensive that it required the army to erect a Bailey bridge across it. This temporary structure, which remained in place for 4 months, was officially opened by the Lord Mayor on 3 February 1941. Contemporary photographs, such as that shown in Figure 3.4, were suppressed from publication in order not to lower public morale. A plaque of remembrance placed in Bank Station was unveiled by the Lord Mayor of London on 11 January 2017.

3.2.2 The Challenge Presented by the DLR

During the late 1980s, the extension of the DLR from London's Tower Hill to Bank Station involved substantial bored tunnelling works to be undertaken beneath the Mansion House.

The Parliamentary Bill granting Royal Assent for the construction of the extension to the DLR included a clause that required the approval of the City Engineer before tunnelling beneath the Mansion House could commence. Such a clause was unusual and it was not expected that such a powerful sanction would be exercised in practice. The first stage of the associated works for the DLR had required the construction of a small diameter tunnel – the Waterloo and City Line pedestrian link – directly beneath the building (Figures 3.5– 3.7 and 3.13). The measured settlement directly

Figure 3.4 Bomb crater, of January 1941, at Bank Station looking across from the Mansion House towards the Bank of England. (Photo: Imperial War Museum collection.)

after its construction seemed quite acceptable being just 5 mm and slightly less than half of the 'greenfield' prediction (Rankin, 1988). Thus at this stage, it was evident that the soil/structure interaction with the building had advantageously modified the settlement trough. Preliminary settlement assessments indicated that the longer-term component after three years would be similar, giving a total by that time of around 10 mm. However, by mid-1989, during the year immediately following the construction of this link tunnel, the observed settlement trend below the building appeared to be ongoing with an indication that it was likely to exceed the 3-year-term prediction (Figure 3.5). The complex origin of this trend only became fully evident much later from the detailed analysis of the data generated from the application of the observational method (OM). However, in 1989, this unexpected settlement trend raised serious concerns about the risk of unacceptable damage to the Mansion House that might arise from the construction of the substantially larger tunnels of the DLR. Consequently, in accordance with the parliamentary clause, all remaining tunnelling within the zone of influence beneath the building was placed on hold by the City Engineer. A detailed evaluation of the implications to the building was then initiated.

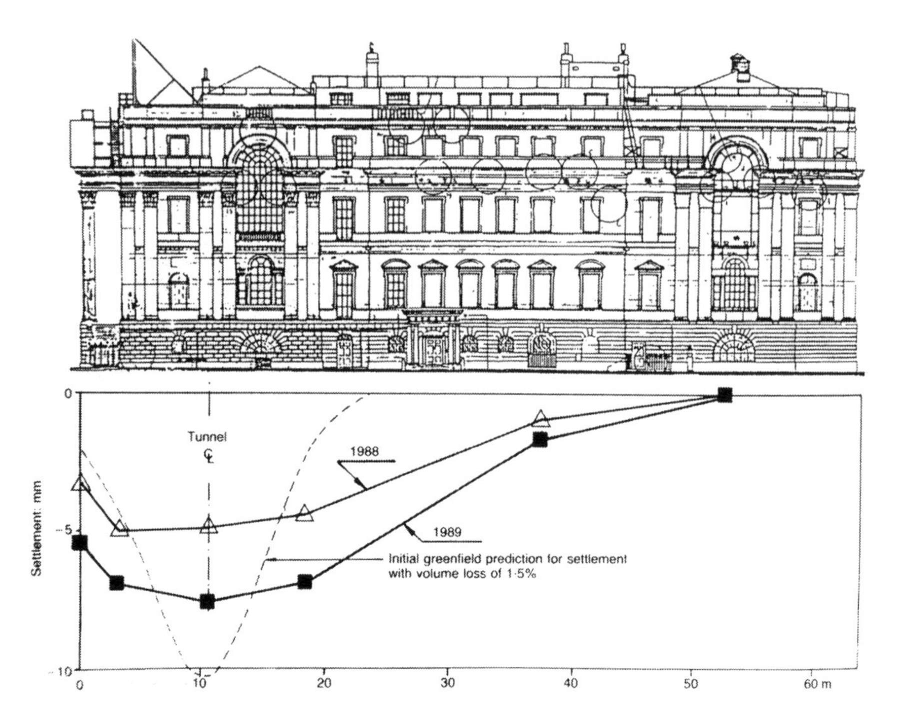

Figure 3.5 Observed settlements between July 1988 and June 1989 measured along west wall.

This assessment considered a wide range of alternative methods to protect the building. These were:

- Shielding of the foundations of the building from the imposed settlement trough from tunnelling by the formation of a structural curtain wall.
- Localised foundation strengthening, such as underpinning and ground treatment.
- Building strengthening such as a system of structural ties.
- Elimination of the settlement effects on the building by compensation grouting.
- Complete underpinning of the building combined with a global jacking system to compensate for the imposed settlements.

All of these preventive methods to mitigate the perceived risks from the tunnelling would have required major structural intervention and/or extensive ground treatment. Moreover, even if any of these options appeared reasonably feasible, each one would have introduced new risks of damage to the building. At the request of the London Underground, a review team was formed to undertake an independent assessment. This team reported to the project director of the DLR and to the City Engineer and his consultants.

3.3 Achieving Agreement to Use the OM

3.3.1 Addressing Complexity

As can be seen in Figures 3.6 and 3.7, the range of tunnelling works beneath the Mansion House, both historically and that then current for the DLR, was extensive. Faced with this complexity and the heightened concerns of risks to the building, the challenge was to develop a solution focussed on minimising these risks and the prevention of unacceptable damage. Any solution proposed had to be acceptable to the building owners. The ideal solution would also be one that secured the earliest restart to tunnel construction. The resolution of these challenges duly led to the adoption of the OM (Powderham, 2002). It was implemented on the basis of progressive modification. Compared to the Channel Tunnel cut and cover works, described in Chapter 2, the Mansion House presented a radically different scenario for an application of the OM. As noted, because of the perceived risk of unacceptable damage to the

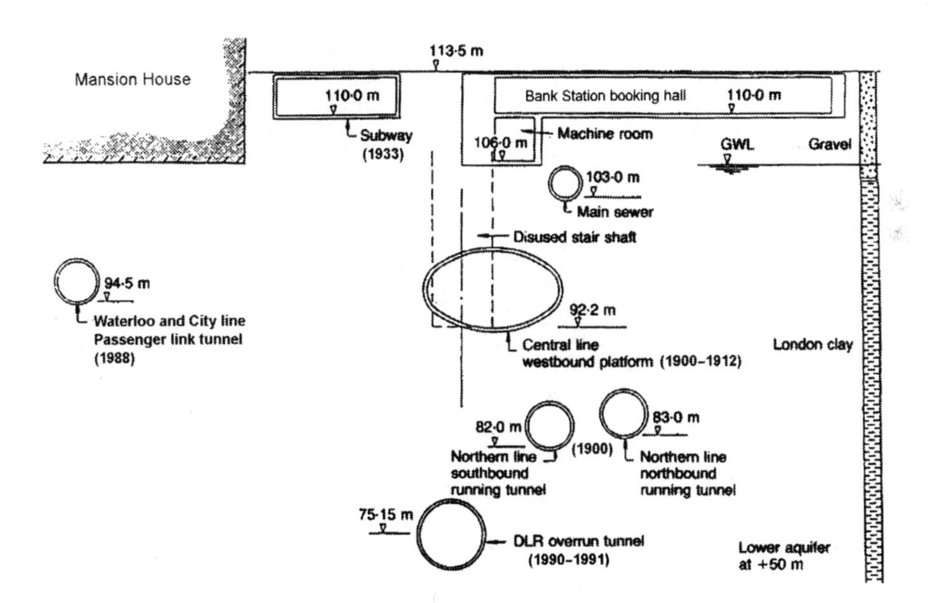

Figure 3.6 Tunnelling works beneath the Mansion House from 1900 to 1990s.

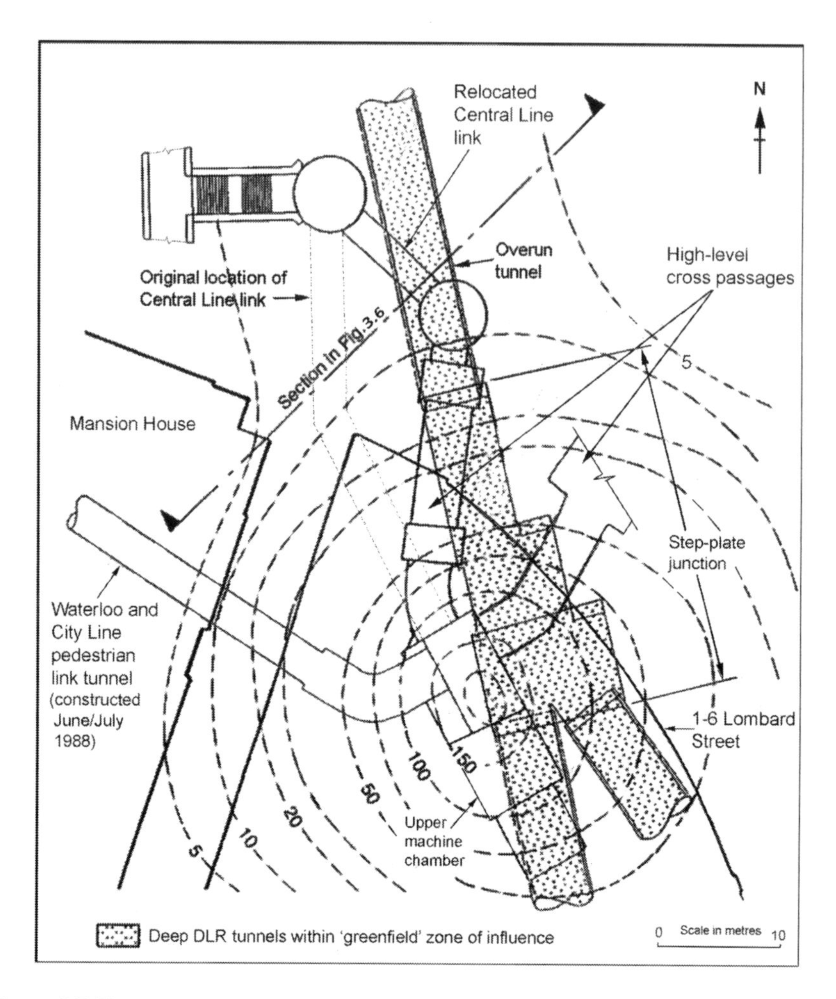

Figure 3.7 Plan arrangement of tunnels for the DLR and associated works beneath the Mansion House. Also shown are the surface settlements (in mm) computed for 'greenfield' conditions for the combination of the over-run tunnel and step-plate junction. These are the immediate post-construction values based on a volume loss of 2%.

Mansion House, the construction of the DLR tunnels beneath the building had been halted, lacking the approval of the City Engineer.

Given the catalogue of historical and present complexities described earlier, the building and its soil/structure interaction were hardly amenable to comprehensive assessment purely through numerical analysis – however sophisticated (Powderham, 2003). Such definitive idealisation remains effectively beyond our capabilities and certainly cannot supplant the role of engineering judgement. Typically, for such situations, there was a range of significant unknowns that related and added to the layers of interactive complexity. These would include, for example, the present state of stress in the structures involved, the full nature and condition of the foundations building including previous strengthening works, and the effects of the geological history on the ground including the influence from earlier adjacent construction. This emphasises the significant limitations of numerical analysis to predict the likely response of the building. Instead, the overtly empirical approach of implementing the OM was adopted. The assessment of the probable behaviour of the building indicated an acceptably low level of risk of damage. Given the mass and stiffness of the building, main component of the response of the building to tunnelling-induced settlement was predicted to be in free-body rotation – that is, one that would intrinsically induce no deformation.

From an OM perspective, this was clearly a 'best way out' situation. As noted in Section 3.2.2, a wide range of substantial preventive measures to protect the building were under consideration. However, each option risked the introduction of new unintended consequences that could compromise the safety of the building – and thus potentially rendering the cure worse than the complaint. Moreover, the full development and implementation of any of these alternatives would extend the already substantial delay to tunnelling. Relationships between the parties were understandably tense. Overall, it was a very complex situation both technically and contractually. The DLR project had reached a crisis.

3.3.2 Creating the Solution through the OM

The question posed to the independent review team was: 'Is there another way in which the risks could be eliminated or controlled within an acceptably low level but also allow the earliest recommencement of tunnelling?' The immediate need was to establish the confidence of the building owners that there was a better alternative to those involving substantial physical preventive measures. Central to this would require that maintaining risks within an acceptably low threshold could be convincingly demonstrated. The OM offered a promising way forward by applying it on a progressive basis in concert with the successive stages of tunnelling. Originally, following the completion of the Waterloo and City Line pedestrian link, there were five other stages of tunnelling to be constructed within the theoretical or contractual zone of influence for the Mansion House. (The derivation of this zone of influence, hereafter termed contractual, is given in Section 3.4.2.) These five stages comprised the Central Line link and the four elements of the deep DLR tunnels – namely the over-run tunnel and the three enlargements of the step-plate junction

(Figures 3.7 and 3.13). The Central Line link was at a similar depth, at around 15 m, to that of the Waterloo and City Line, and since also being a pedestrian link, its alignment was reasonably flexible. It was therefore agreed to locate it beyond the contractual zone of influence as shown in Figure 3.7, thus leaving just the deep DLR tunnels within it. The proposal was to commence these remaining four stages with the over-run tunnel and, pending acceptable performance, to be followed successively with the three enlargements of the step-plate junction. The computed settlement contours for all four of these stages under 'greenfield' conditions are shown in Figure 3.7. Simple but conservative assessments were made for the resulting angular distortion. Thus at this initial stage, no benefit was assumed from any soil/structure interaction. This would mean that the building would distort with its foundations fully conforming to the induced settlement. In other words, the angular distortion would follow the same profile as that computed for the 'greenfield' trough. In practice, any soil/structure interaction would extend and flatten the settlement trough, reducing both the maximum settlement and the ground slopes and hence the angular distortion. Given the massive nature of the foundations of the Mansion House, such interaction was inevitable and probably likely to be substantial. This had already been demonstrated during the construction of the Waterloo and City Line link where the immediate settlement and ground slopes were considerably less than those estimated for 'greenfield' conditions (Figure 3.5). The separate but related issue of the ongoing settlement trend above this tunnel still remained and needed monitoring. However, there was no evidence of new damage directly attributable to this trend. It was therefore not considered to be a critical issue that required resolution prior to the identification of an acceptable way forward for the remaining tunnelling works. As discussed in Section 3.5, the resolution of this issue was an important and very informative by-product from the OM. It is relevant to note here that an agreement had been reached to reduce the diameter of the Waterloo and City Line link tunnel from 4 to 3.05 m due to the building owner's original concerns about potential damage. This was well before the present issues with the DLR which, with the step-plate junction, included substantial tunnelling with diameters over 11 m.

Understandably, at this preliminary stage, to convince the building owners of the efficacy of any solution, especially an innovative one, was a major challenge. Given the combined complexities of the building structure and the sub-surface conditions, a compellingly simple explanation was required. Was there a key criterion that could be demonstrated to embrace the complexity and be practically monitored and clearly communicated in a timely way? It had been established from case history data that the level of risk of building damage from induced settlement related to the combined effects of two key criteria – namely horizontal strain and angular distortion. While horizontal strains generated in the ground from tunnelling are feasible to monitor, it was not considered practical to try to measure their transmission into the building structure. It was, however, agreed that, while it would demand very accurate instrumentation, the angular distortion could be reliably monitored. Nevertheless, it was still essential to account for the risks arising from horizontal strain. This was addressed by adopting the chart developed by Boscardin and Cording (1989). This approach enabled trigger levels to be related to degrees of risk which embraced the combination of horizontal strain and angular distortion (Figure 3.8).

3.3.3 Introduction of the OM Traffic Light System

Using the aforementioned criteria facilitated the establishment of a simple traffic light system which focussed on the measurement of one key parameter – that of angular distortion. This was the first known development of the traffic light system in applications of the OM. Recognising its effectiveness, it was promptly adopted and developed for the concurrent applications of the OM at Limehouse Link (see Chapter 4). Thereafter, this approach with a single set of traffic lights, green, amber and red, based on one carefully identified critical observation, was the guiding principle for the subsequent applications of the OM featured in this volume. It is

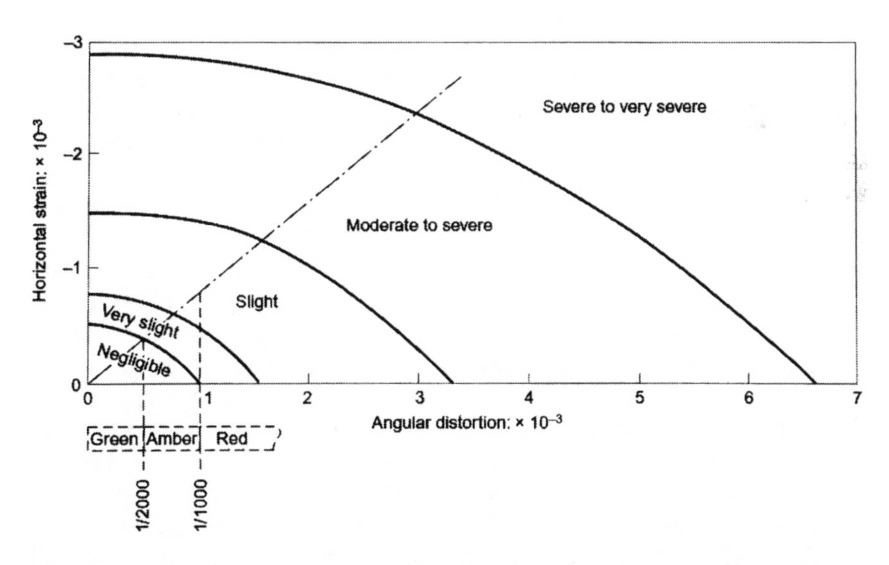

Figure 3.8 Damage risk combining angular distortion and horizontal tensile strain (Boscardin and Cording, 1989).

essential to appreciate that to be effective, the simplicity of adopting just one set of traffic lights must be concurrently related to the broad range of supplementary observations pertinent to the response of the structure in question. For the Mansion House this integrated process is set out in the flow chart shown in Figure 3.9. The objective was to demonstrate that the risk of damage would be maintained at or below a level acceptable to the building owners. In conjunction with the monitoring of angular distortion, comprehensive and ongoing condition surveys would be undertaken. The building owners would have 'real-time' access to the monitoring data and the authority to place a hold on the next stage of tunnelling in the event of any development of adverse trends becoming evident. This progressive approach would permit the maintenance of an acceptable level of safety to be clearly demonstrated.

Achieving this agreement took over three months and involved extensive design development with several high-level presentations to the building owners and their consultants. Beyond safeguarding the building, the additional objective was to avoid the need for any contingency measures. In 'best way out' applications, such as for the Mansion House, apart from rescuing the project, success is also measured by the avoidance of unnecessary contingencies.

3.4 Implementation of the OM

3.4.1 Managing Risk

The application of the OM guided the overall approach to manage and minimise the risk of damage to the Mansion House (Powderham and Tamaro, 1995). Developing the traffic light system enabled an iterative and progressive process to limit and maintain this risk within acceptably low levels and which, crucially, could be clearly demonstrated. The implementation of the OM required approval by all parties. The assigned levels of risk in terms of the Boscardin and Cording chart ranged from negligible and extended just beyond very slight as shown in Figure 3.8. As noted, these values were derived from the predictions for 'greenfield' conditions. This conservative starting point was duly accepted by the building owners and agreement to implement the OM approved. It thus provides a further example of implementing the OM with a significantly conservative design rather than one based on estimates of the most probable conditions.

Contingency measures would be initiated if necessary should any significantly adverse trends were detected (Figure 3.9). The triggering of such measures would need to be coordinated with the successive stages of tunnel construction. As it would not be practicable or advisable for a trigger to halt tunnel construction during a given stage, it was essential that an acceptably low level of risk for proceeding to the following stage

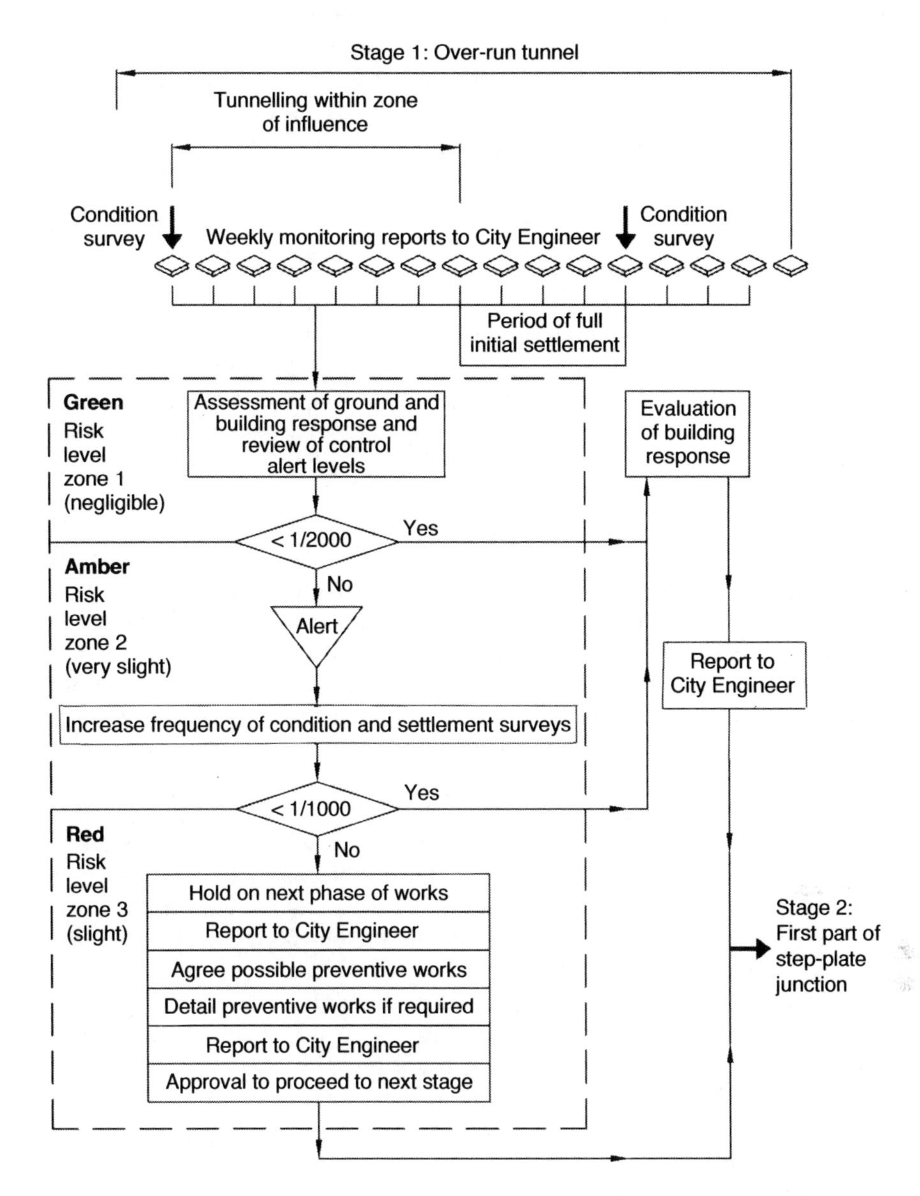

Figure 3.9 Flow chart for risk levels and respective responses within the traffic light system.

could be demonstrated prior to the commencement of each stage. Consequently, noting that the induced effects would be accumulative, each stage of tunnelling was selected with appropriately low levels of risk. The initial value set for the angular distortion for the over-run tunnel, which was the first stage of tunnelling for the OM, was 1/2,000. This value was based on 'greenfield' criteria as described in Section 3.4.2. Applying these criteria also predicted a maximum settlement of 3.5 mm in the north-eastern corner of the building. In terms of the risk of damage, an angular distortion of 1/2,000 would constitute a negligible level (Figure 3.8). With the same criteria, the combined effects of the over-run tunnel with the three subsequent stages of the step-plate junction, the predicted values for angular distortion and maximum settlement were 1/1,000 and 4.5 mm, respectively. It can be seen from Figure 3.8 that by completion of all four stages of tunnelling for the deep DLR tunnels, the level of risk was predicted, on the basis of angular distortion, to reach the lower end of the region for slight damage. Thus, given acceptable performance, it would be possible to demonstrate that each of the four stages of tunnelling would commence from an acceptable level of risk and by completion each stage would add an acceptable incremental increase to that risk.

3.4.2 The Profound Influence of Soil/Structure Interaction

As noted, the initial estimates of the induced effects on the building were based on 'greenfield' conditions. This is conventionally considered conservative since the structural stiffness of buildings intrinsically acts to reduce the actual distortion. However, given the urban setting, the conditions could hardly be characterised as 'greenfield'. Even the most cursory perusal of Figures 3.6 and 3.7 will readily convey that the environment was far from a green field. It was obvious that a significant amount of soil/structure interaction would be generated in practice. Not only would the stiffness of the Mansion House be brought into play but there was also the very substantial presence of the building of 1–6 Lombard Street (Figure 3.10). As shown in Figure 3.7, the epicentre of the remaining four stages of tunnelling for the deep DLR tunnels lay directly beneath this building. Moreover, all of the elements of the shallow tunnelling works still to be constructed, comprising the relocated Central Line link and the high-level cross-passages, while outside the contractual zone of influence for the Mansion House, were well within the influence zone for 1–6 Lombard Street. This also being a very substantial and stiff building would inevitably modify the trough of any induced settlement from these tunnelling works. It was also evident that part of this modified settlement would be likely to spread underneath the Mansion House – thus transferring settlement from outside its own zone of influence. Based on engineering judgement, an essentially simple strategy was developed to deal with the complexities presented by soil/structure interaction. Resolution of this, and a full appreciation of the extensive influence of the soil/structure interaction, duly required careful and detailed assessment of the data generated by the application of the OM (see Section 3.5).

Figure 3.10 Spatial relationship of Mansion House (on right) and 1–6 Lombard Street.

The complexities created by the soil/structure interaction arising from both the Mansion House and 1–6 Lombard Street were addressed as follows:

- For the Mansion House the assigned values for angular distortion would be set from predictions for 'greenfield' settlement operating within the contractual zone of influence. This zone was defined by planes set at 45° to the horizontal fanning outwards from the foundations of the building – or by an area bounded 20 m laterally from the building perimeter – whichever was the greater.
- The starting position for the OM design would be conservative and not include any beneficial effects of soil/structure interaction. Thus the building foundations would be assumed to conform completely to the shape of the 'greenfield' settlement trough. The values of angular distortion to inform the risk assessment would be derived from these criteria.
- The four stages of tunnelling would be defined to ensure that each successive stage commenced with an acceptably low value of angular distortion.
- It was appreciated that the Mansion House in reality would be subjected to settlements generated from the lateral extension of the settlement trough originating beneath 1–6 Lombard Street. The potential

for such effects were already evident in the ongoing settlement trend following the completion of the Waterloo and City Line pedestrian link tunnel beneath the Mansion House. It can be seen in Figure 3.7 that a significant proportion of this tunnel was constructed directly beneath 1–6 Lombard Street.

- In practice, the soil/structure interaction would thus create two components of settlement for the Mansion House. One would be from the intrinsic action of the Mansion House which would flatten the settlement trough beneath it. The other would be the additional settlement generated from 1–6 Lombard Street. This latter component would, in practice, be subject to the trough flattening effects of both this building and the Mansion House.
- It was accordingly judged that the conservatism, within the starting assumption for angular distortions based on the 'greenfield' predictions, should adequately compensate for the additional component of settlement being transferred from 1–6 Lombard Street.

Given the amount of data already generated from the earlier periods of tunnelling, it was relatively straightforward to estimate most probable geological conditions for tunnelling and the likely 'greenfield' ground movements that this would generate. The key issue was, of course, the actual response of the building to the range of imposed ground movements. The assessment of this in advance of tunnelling was based on field data from case history studies and related to the current condition of the building as established from structural surveys. For the worst-case scenarios, the most unfavourable conceivable deviations from these conditions involved higher volume losses from tunnelling than the assumed value of 2% and/or an undue sensitivity of the building to the imposed settlement. Critically, this would arise from existing planes of weakness or new ones developing in the masonry structure. This could then lead to a response to settlement being induced in a bending mode – for example, with vertically orientated planes opening up in a hogging deformation. However, it was considered more likely that the building would respond in a shear mode of deformation (Boscardin and Cording, 1989). As there was no damage reported for the Mansion House during the application of the OM, some graphic examples of shearing mode are provided by the Gardiner Building. This is located on Long Wharf, in Boston, Massachusetts. Interestingly, it dates from the same period as the Mansion House and has also experienced an eventful history – particularly with regard to settlement. It is a brick masonry Colonial style warehouse built in 1763 and rebuilt in 1812. Long Wharf was once crowded with such buildings, but the Gardiner remains as the wharf's oldest surviving structure. Founded on timber piles over water, it has been subjected to periods of significant differential settlement. At different times, these have created waves of settlement troughs in both directions along the length of the building as is evident from the shear cracks orientated at 45° in opposite directions. Typically, openings, such as windows and doors, in a masonry wall have a much greater weakening effect on shearing as opposed to bending stiffness. This again is evident in the Gardiner building noting the predominance of the diagonal cracks between the windows and above the doors (Figures 3.11 and 3.12).

In practice, the tendency towards responding in a shear mode is also enhanced since settlement troughs are not instantaneously generated (as would typically apply to the 'wished in place' scenario of numerical analysis). Rather, they are imposed in a travelling wave that increases in concert with the progress of the tunnelling. Consequently, as the wave first meets with the building, it engages low length to height (L/H) ratios which present high bending stiffness. Burland and Wroth (1974) showed that the diagonal strain (i.e. shear strain) is critical for L/H ratios less than unity. Thus, unless there are inherent zones of weakness affecting bending stiffness, a building will tend to respond in a shear mode to settlement and the first evidence of visual damage will be that of diagonal cracking. Such a response mode is more desirable than one in bending since (a) it still presents a stiff building response which enhances the beneficial effects from soil/structure interaction and (b) any cracking generated is generally much easier to repair. It should be emphasised, in this context, that the damage classifications were developed for brickwork or

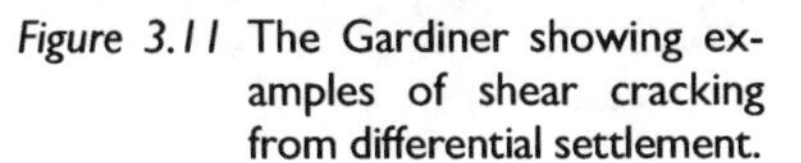

Figure 3.11 The Gardiner showing examples of shear cracking from differential settlement.

Figure 3.12 Detail of front wall of the Gardiner.

blockwork and stone masonry buildings and are based on the ease of repair of visible damage. The example of the Gardiner building also illustrates the ease of repair. With the shear cracks fully filled the building now functions perfectly as the Chart House restaurant (Figure 3.11)

In special cases, such as for a building with valuable or sensitive finishes, the ranking of severity of damage may not be appropriate (Burland *et al.*, 2001). This latter caveat would clearly apply to the Mansion House. For this reason, the approach to managing the risk of damage adopted conservative targets with the highest set at the lower end of slight damage (Figure 3.8). In passing, it should also be noted the damage classification does not apply to reinforced concrete structures. Moreover, it is essential to appreciate that there are profound differences between the structural performance of masonry compared to a reinforced concrete or steel frame structure (Heyman, 1995). They operate in accordance with very different criteria. This dichotomy was dramatically highlighted in the case history of the Heathrow Airport multi-storey car park in Chapter 6.

Contours of the initial 'greenfield' settlements, for the combined effects of the four stages of tunnelling, are shown in Figure 3.7. As noted, these were computed for a volume loss of 2%. This was considered a reasonably conservative value, given the variation in tunnelling construction involved. All tunnelling in the influence zone of the Mansion House was undertaken by hand methods using simple shields. Tunnelling measures to minimise volume loss included continuous shift working and close control of ring installation, initial grouting and re-grouting behind segments.

3.4.3 Instrumentation and Monitoring

The primary requirement was to monitor and record the detailed response of the building to induced settlement from the tunnelling works. The principal system of instrumentation used to obtain these observations was carefully selected arrays of horizontally and vertically aligned electro-levels supplemented by precise levelling (Figures 3.13 and 3.14).

The electro-levels, developed at the UK Building Research Establishment, had a range of ± 3° with a specified resolution of 1 arc second (Price *et al.*, 1994). In practice, a repeatability of 2.5 arc seconds was reported, enabling the system to record slope changes to an accuracy in the order of 1/80,000. This was comfortably more than adequate to fulfil the task required. It should be noted that, although the capabilities of electro-levels had been in development during the 20 years preceding the project, the scale of their use for such a high-profile application of building instrumentation was innovative. Their performance is particularly sensitive to the installation procedures. These must be very precise. Even a seemingly minor inadequacy in how an electro-level is attached to its respective beam could, for example, make the readings unduly susceptible to temperature variations. Correct installation is one key aspect in

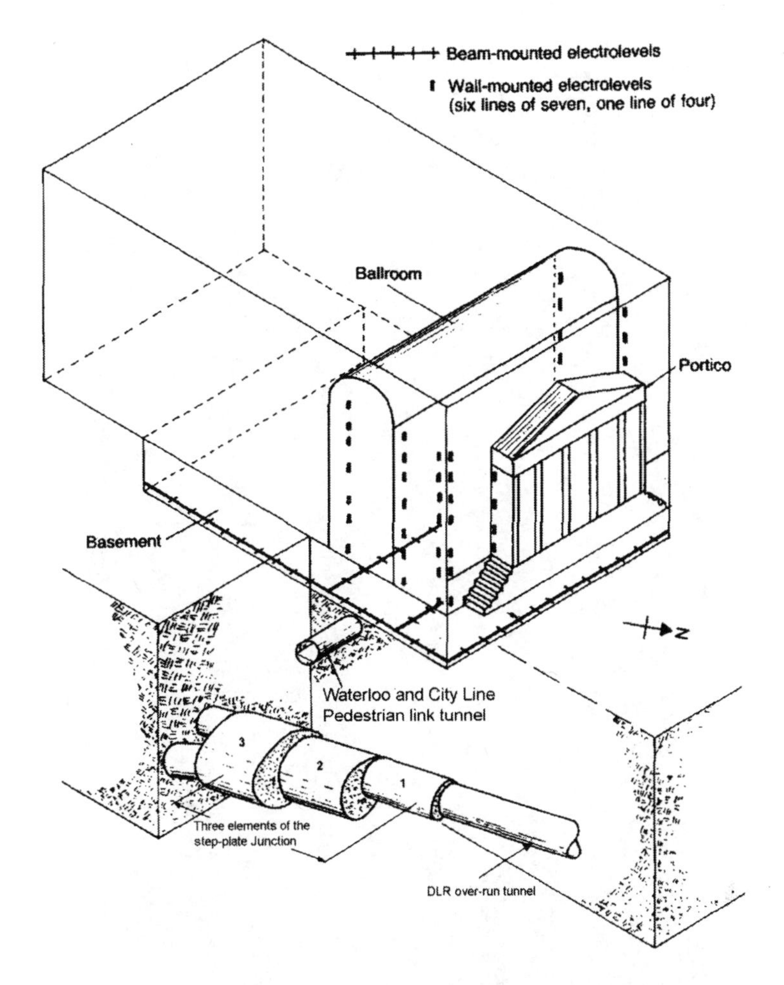

Figure 3.13 Primary instrumentation showing location of electro-level strings.

the procurement of instrumentation. Comprehensive advice on effective procurement is given by Dunnicliff and Powderham (2003). The readings were also most accurate within the middle third of their range. The OM team was keenly aware of such issues and the critical need for the monitoring system to provide reliable and consistently accurate readings. The success of the electro-levels at the Mansion House promptly heightened awareness of their potential and they were subsequently more frequently selected to monitor sensitive structures.

For the Mansion House a total 101 electro-levels were attached to the building, with fifty-five in the basement in four horizontal strings. Each of

Figure 3.14 West elevation showing height of ballroom and external electro-levels.

these beam-mounted electro-levels recorded changes in slope between adjacent reference pins set about 3 m apart. The other forty-six electro-levels were fixed individually in vertical lines to the external faces of seven of the principal masonry columns at the north end of the building (Figure 3.14).

3.4.4 Temperature-Induced Cyclic Movements

The performance of the system of electro-levels in real time proved so consistently accurate that it was possible to demonstrate that the temperature-induced lateral movements caused in the building as the sun moved around it on a hot day were dimensionally more than the building settlement induced by the construction of the overrun tunnel. A plot of the variation in lateral deflection over a daily cycle caused by such temperature changes is shown in Figure 3.15.

These temperature-induced cyclic movements also highlighted the limitations of undertaking a conventional spatial survey with a manual theodolite because of the time required to complete a set of readings – especially since this involved setting up in the busy thoroughfares around the Mansion House. (In this context, to achieve the required accuracy, it shortly became commonplace after Mansion House to use remote reading total station theodolites. These overcome such daily cyclic variations by providing frequent and simultaneous multiple readings.)

Subsurface instrumentation also included horizontal and vertical strings of electro-levels installed in inclinometer tubes. These were to monitor ground movements and provide a useful correlation with any tunnelling-induced effects detected in the building. Secondary systems included a water-levelling system and spatial surveys. The level survey points external to the building are shown in Figure 3.16. These had to be correlated with the internal reference points on the electro-level strings within the basement of the building.

Further details of the instrumentation, its performance and interpretation of the readings are reported by Forbes *et al.* (1994) and Price *et al.* (1994).

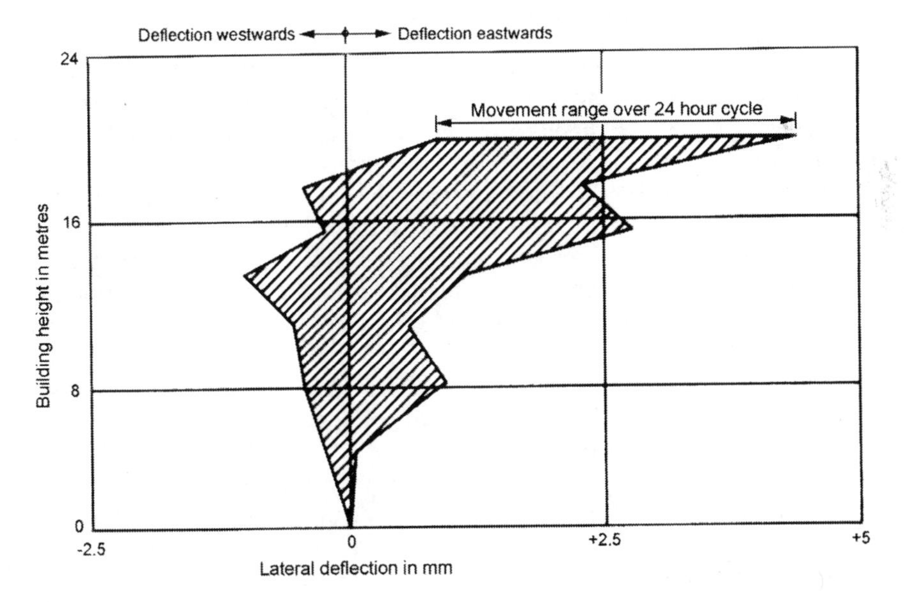

Figure 3.15 Diurnal cyclic movement measured by vertical electro-level string on west elevation.

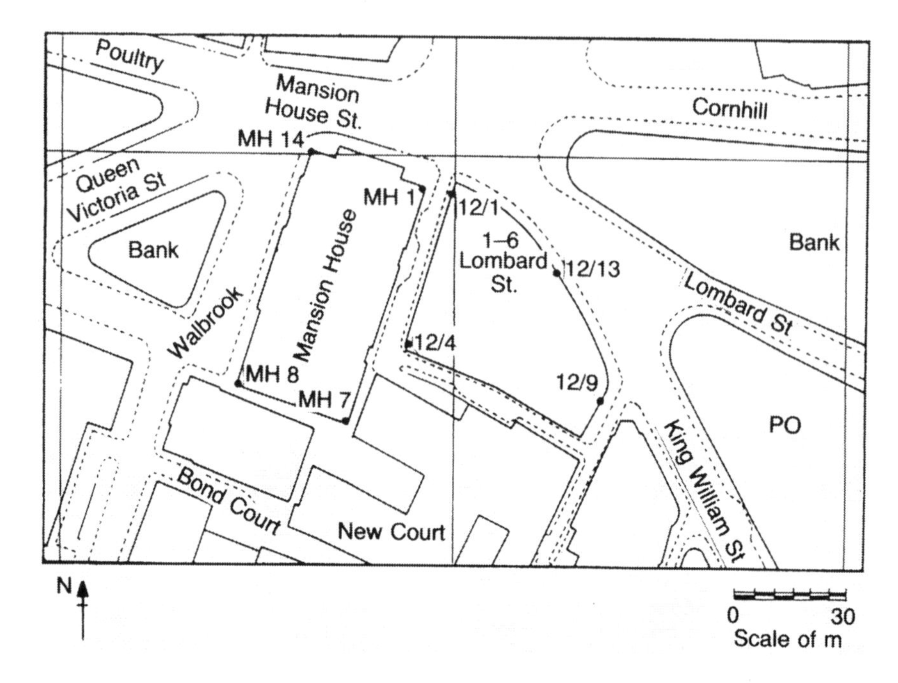

Figure 3.16 Location plan showing external level survey points (e.g. MH1 and 12/9).

3.4.5 Development of Progressive Modification

The implementation of the OM was based on the first overt application through progressive modification. With its basis set in defined construction stages, progressive modification enables progress from an established position of safety to a new one based on demonstrably robust predictions. As explained in Chapter 2, such a process had developed naturally in the application of the OM to cut and cover tunnels. For the Mansion House it was the cumulative effect of tunnelling beneath the building that was being progressively modified and evaluated through staged construction as described below. The procedure was as follows:

(1) An assessment of the building and its foundation conditions was undertaken. This included a thorough condition survey and a comprehensive review of historical records.
(2) Detailed consultations were made with the main contractor to review tunnelling methods and performance. Particular attention was given to the sequence of tunnelling and the level of risk relating to each of the remaining four stages.

(3) The potential for tunnel realignment was also reviewed. There was no potential for this for the deep DLR tunnels and so was limited to the Central Line passenger link. This was at a relatively shallow depth and would thus not require moving far to be outside the contractual zone of influence (Figure 3.7). This left only the deep DLR tunnels for the over-run and the step-plate junction within the contractual zone of influence.
(4) The courses of action planned in advance for significant deviations from anticipated behaviour were carefully assessed and developed in close coordination with the contractor and approved by the building owner.
(5) The overall process was set out in a flow chart as shown in Figure 3.9. This chart set out the iterative process in which the effects of each successive stage on the building would be carefully monitored and assessed.
(6) Approval to proceed to the next stage was dependent on the results of the preceding stage.
(7) The overrun tunnel was undertaken first.

3.4.6 Critical Observations, Trigger Levels and Contingencies

The three zones of green, amber and red, as shown in Figure 3.8, set the basis for the 'traffic-light' approach. As noted in Section 3.3.3, this enabled the level of risk to be focussed on one critical factor – that of angular distortion. This had to be carefully assessed in conjunction with the ongoing condition surveys:

(1) Trigger levels were set to initiate specific contingency measures. These related to the boundaries marking negligible, very slight and slight risk of damage to the building.
(2) These boundaries were obtained by consideration of both angular distortion and horizontal tensile strain. The latter was relevant because the building would be subjected to a hogging deformation on the limb of the settlement trough.
(3) With the risk level taking account of both effects, as shown in Figure 3.8, it was considered necessary only, and indeed practicable, to directly monitor the angular distortion.
(4) Thus defined zones of levels of risk were established with trigger levels for angular distortions of 1/2,000 and 1/1,000 setting the boundaries as described below.
(5) Measurements within first zone up to the first trigger level of 1/2,000 meant that tunnelling could proceed according to programme – providing no untoward adverse effects were evident from the condition surveys of the building.

(6) The second zone between the trigger levels of 1/2,000 and 1/1,000 initiated the contingent action of higher frequency of reporting and additional condition surveys. Again, with satisfactory observations, approval would allow the next stage of tunnelling to proceed in sequence.
(7) The third zone beyond a deformation of 1/1,000 meant suspension of the next stage of tunnelling, pending a comprehensive assessment of the effects on the building. If this proved significantly adverse, the options of installing one or more from the range of preventive works, listed in Section 3.2.2, would be evaluated.

The avoidance of such contingency measures involving major structural intervention to the building was of course one of the key objectives of the implementation of the OM through progressive modification. One of the more preferred options was structural strengthening in the form of steel ties. The locations and the ability to install them within a reasonably short period were carefully assessed in advance on site. Steel ties had been installed at the southern end of the building during Victorian times and were reported to have introduced some undesirable side effects by creating new zones of stiffness leading to differential movements. Such effects are analogous to the risks associated with partial underpinning. Local strengthening may act in opposition to the inherent rhythms of a building such as those arising from temperature-induced effects. If some form of physical intervention was deemed necessary, during the implementation of the OM, the viability of any such measure would be assessed in the light of the observed response of the building up that stage. Thus, it would have been on a far more informed basis than that which originally pertained before any tunnelling works. The existing instrumentation would have then been used to assist in monitoring the effects of its installation and subsequent performance.

3.5 Results

3.5.1 Overview

Tunnelling was completed without further delay in the sequence proposed. The measured settlements remained well within predicted values. The deformation of the building was very low, with angular distortion not exceeding 1/7,000. This result clearly confirms the success of the OM strategy and demonstrates that each of the four stages of tunnelling were commenced and completed maintaining acceptably low levels of risk to the Mansion House. In fact, the angular distortion remained well within the negligible zone throughout. The maximum recorded settlement at the north-west corner of the building up to the completion of the step-plate junction in February 1991 was around 20 mm as shown in Figure 3.17. However, considerably less than half of this was assessed to be directly attributable to short-term tunnelling effects of each of the four stages of tunnelling as discussed in Section 3.5.2.

3.5.2 Detailed Assessment

The quality and quantity of data obtained from the instrumentation installed for the OM enabled a very detailed assessment of the magnitudes and trends of settlement over time. The field measurements from the monitoring programme enabled the settlements to be differentiated into distinct 'events' related to the four stages of tunnelling (Figure 3.17). From the data, these 'events' were inferred to have a typical 'immediate' construction element and a longer time-dependent one. This latter element was more unusual since it developed over the tunnelling construction period of around thirty months. This ongoing settlement trend was attributed to soil/structure interaction – principally that from the lateral spreading of the settlement troughs generated by 1–6 Lombard Street. Thus even when there was no tunnelling within the contractual zone of influence of the Mansion House, the building was still exposed to induced settlement from the ongoing tunnelling works beyond this zone.

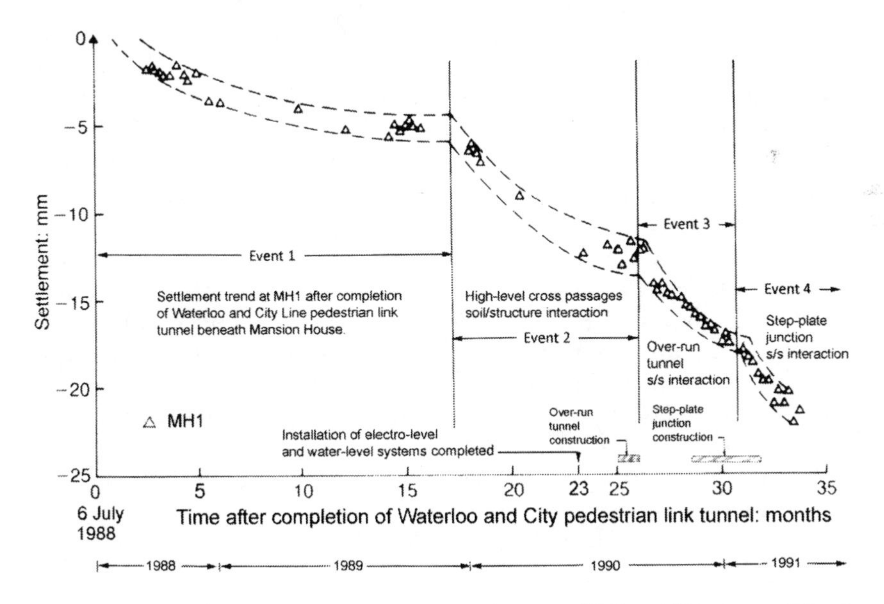

Figure 3.17 Inferred settlement 'events' at survey station MH1 in NE corner of Mansion House.

In terms of the risk to the Mansion House and its response to induced settlement, it is instructive to compare the initial estimates with the results. The maximum settlement of the building predictably occurred in the north-eastern corner of the building as recorded by the external levelling reference point MH1. It was possible to correlate the readings of MH1 at this location with those from the electro-level and the water-level systems. These may be summarised as follows:

- The value computed under 'greenfield' conditions, at MH1, was around 7 mm as shown in Figure 3.7. This set the conservative starting position for the OM, for the four stages of the deep DLR tunnelling. The over-run tunnel contributed around half of this.
- Close inspection of the data revealed a series of ongoing settlement trends which continued even during the intervals when there was no tunnelling within the zone of influence. These trends had rates from 0.5 to 0.7 mm/month and were evident as a series of four 'events' (Figure 3.17).
- The total settlement recorded at MH1 for the construction of the over-run tunnel between months 25 and 26 was 3.5 mm. That this is the same as that computed under 'greenfield' conditions is, in the given context, essentially coincidental. Removing the contribution from the background trend, the immediate settlement directly attributable to the construction of the over-run tunnel was just 1.5 mm.
- Similarly, the immediate settlement directly attributable to the construction of the three stages for the step-plate junction was estimated to be only 1.6 mm, while the total settlement recorded by the electro-level system during this period was around 10 mm.
- Thus, the directly comparable values between the 'greenfield' and the actual immediate settlements from tunnel construction, inferred at MH1, during the entire period for the OM were 7 and 3.1 mm, respectively.
- It is interesting to note that the total construction settlement of 3.1 mm is less than the maximum range recorded for temperature-induced movements (Figure 3.15). Moreover, the temperature movements relate to a daily cycle, while the construction settlements took place over 6 months. Short-term movements tend to present greater risk of potential damage than longer-term ones as there is much less time for creep and other mitigating effects to act.
- By April 1991, the overall settlement at MH1 had reached 22 mm.
- It was estimated at this stage that the total settlement at MH1 would reach 27 mm by early 1993 and around 30 mm by 2000.

The analysis of the data is described in more detail by Forbes *et al.* (1994).

3.5.3 The Benefit of Hindsight

From the detailed analysis of the data of settlement trends, it became evident that the overall settlement trend observed beneath the ballroom during 1988–1989, which generated the original concerns, was not wholly attributable to the Waterloo and City Line pedestrian link tunnel. In fact, the settlements intrinsic to this tunnel may have contributed quite the minor part. As previously noted in the initial risk assessment, a significant mitigating factor is that the time-dependent lateral extensions of settlement troughs arising from soil/structure interaction and for long-term settlements in general tend to generate much flatter slopes than those caused in the short-term during construction. Consequently, structural distortions are substantially reduced presenting considerably less risk to buildings. The analysis of the detailed data from the OM provided good evidence for this. In general, there has not been a history of reported of building damage attributable to long-term settlement trends from tunnels in over-consolidated cohesive deposits. Ongoing monitoring of the building subsequent to the DLR tunnelling for the Mansion House has not revealed any associated damage. A study was undertaken some seven years later (Anketell-Jones, 1998) in which the original survey pins used for monitoring the DLR tunnelling were re-surveyed. This enabled the differential settlements along the outer walls of the Mansion House to be reviewed. This study reached similar conclusions regarding the 'events' of settlement to those reported by Forbes *et al.* (1994). The only new aspect reported to emerge over this timescale was a small uplift of just a few millimetres of the south-western corner of the building (Anketell-Jones and Burland, 2003). No associated damage was detected. Such results reinforce the views advanced in this case history that massive structures on raft foundations tend to respond primarily to induced settlement through free-body movement. This, when the building is situated over the limb of a trough, as in the case of the Mansion House, is characterised by rotation.

3.6 Conclusions

The results demonstrated that for the Mansion House, the dominant component of the short-term settlement was accommodated in free-body rotation as predicted. No damage to the building associated with these works was evident. Checks to ascertain this involved detailed condition surveys including analytical photogrammetry. The latter was used to detect any effects to the barrel-vaulted roof which forms the ceiling of the ballroom. The ornate plasterwork of the ceiling of this barrel-vaulted room was likely to be very sensitive to differential movement (Figure 3.3). The surveys prior to the DLR revealed the extensive presence of fine diagonal cracks within this plasterwork (Figure 3.18). They were not reported as recent in origin but appeared compatible with possible torsional distortion of the barrel vault that may have arisen from historical settlements around the front façade of the building. Analytical photogrammetry would have revealed any new cracks or even minor extensions to existing ones. None were reported. This data clearly demonstrated that by the end of tunnelling construction in early 1991, the response of the building structure had effectively shielded the roof of the ballroom from any visible damage. The Mansion House underwent a major refurbishment between 1991 and 1993. This included a complete restoration of the ballroom ceiling. As noted, the OM team had estimated that by February 1993 the settlement in the NE corner of the building at MH1 would have reached 27 mm and 30 mm by 2000. Anketell-Jones and Burland (2003) reported that by 2001 the re-survey confirmed that the settlement had not exceeded 27 mm – a full decade after completion of the DLR step-plate junction. It was thus evident that contrary to the original concerns in 1989, only quite minor long-term settlement had followed after tunnel construction. Figure 3.19 shows a section of the restored ceiling of the ballroom. The photo was taken in 2011 almost 20 years after the restoration. Close inspection revealed the plasterwork to be in perfect condition visually with no cracking evident in its surface.

Figure 3.18 Detail of plasterwork in ballroom ceiling showing examples of the extensive cracking visible in 1989.

Figure 3.19 Restored ceiling of ballroom ceiling. Photo taken in 2011.

The electro-level system performed particularly well. It was very sensitive and yet robust and reliable. The high degree of accuracy achieved proved most effective in identifying the short-term tunnelling-induced response separately from ongoing background effects (Forbes *et al.*, 1994). This was most important in the assessment of associated risk to the building.

It had thus been possible by implementing the OM through progressive modification to show that the risk of damage was maintained well below the assigned acceptable limits and that the safety of the building was thus assured. The substantial costs and delays in implementing major protective works to the foundations were also avoided. The estimate for the curtain wall was £3 million, and that for the full underpinning scheme was £13 million (Frischmann *et al.*, 1994).

References

Anketell-Jones, J. E. (1998). A study of the effect and possible causes of recent and historical damage to the Mansion House, City of London, MSc dissertation, Imperial College, London.

Anketell-Jones, J. E. and Burland, J. B. (2003). The mansion house revisited, Proc. CIRIA International Conference, 2001: 'Response of buildings to excavation-induced ground movements', Special Publication 201.

Boscardin, M. D. and Cording, E. J. (1989). Buildings response to excavation-induced settlement. *ASCE Journal of Geotechnical Engineering*, **115**, No. I, January, 1–21.

Burland, J. B., Standing, J. R. and Jardine, F. M. (2001). Assessing the risk of building damage due to tunnelling – lessons learnt for the Jubilee Line Extension, London, Proc Geotechnical Engineering. Ho and Li (editors) Swets and Zeitlinger.

Burland, J. B. and Wroth, C. P. (1974). Settlement of buildings and associate damage, Proceedings of a Conference on Settlement of Structures, Cambridge, pp. 611–654.

Dunnicliff, J. and Powderham, A. J. (2003). Better procurement practice for geotechnical instrumentation and monitoring, Proc. CIRIA International Conference, 2001: 'Response of buildings to excavation-induced ground movements', Special Publication 201.

Forbes, J., Bassett, R. H. and Latham, M. S. (1994). Monitoring and interpretation of movement of the Mansion House due to tunneling. *Geotechnical Engineering*, **107**, April, 89–98.

Frischmann, W. W., Hellings, J. E. and Snowden, C. (1994). Protection of the Mansion House against damage caused by ground movements due to the Docklands Light Railway extension. *ICE Geotechnical Engineering*, **107**, April, 65–76.

Heyman, J. (1995). *The Stone Skeleton*, Cambridge University Press, United Kingdom.

Powderham, A. J. (2002). The observational method – learning from projects. *Proceedings of the Institution of Civil Engineers, Geotechnical Engineering*, **155**, No. 1 59–69.

Powderham, A. J. (2003). Protecting historic infrastructure using the observational method, Proc. Int. Geotechnical Conf. Reconstruction of Historical Cities and Geotechnical Engineering, St Petersburg, September 2003.

Powderham, A. J. and Tamaro, G. (1995). *The Mansion House London – Risk Assessment and Protection*, ASCE Construction Engineering Management, USA, pp. 266–272.

Price, G., Longworth, T. I. and Sullivan, P. J. E. (1994). Installation and performance of monitoring systems at the Mansion House. *ICE Geotechnical Engineering*, **107**, April, 77–87.

Rankin, W. J. (1988). *Ground Movements Results from Urban Tunnelling: Predictions and Effects*, Engineering Geology Special Publication No.5, Geological Society., London, Engineering geology of underground movement (editors. Bells *et al.*), pp. 79–92.

Chapter 4

Limehouse Link (1991–1993)

4.1 Introduction

Limehouse Link is a major highway tunnel in London's Docklands. Providing 1.8 km of underground dual carriageway, it constituted a key component in a range of infrastructure initiatives to improve transport between the City of London and the then new enterprise zone in Docklands (Figure 4.1). The Link was designed as a cut and cover tunnel constructed by both topdown and bottom-up techniques. It passes beneath densely populated residential and (at the time) declined industrial areas. The client, the London Docklands Development Corporation, selected the route alignment to minimise demolition and disturbance and opted to place most of it underground to reduce environmental impact and maximise development potential above the tunnel. The planning and execution of this complex project presented many environmental, technical and managerial challenges. It was constructed through some of the most difficult and variable ground conditions in London. Construction of this £250 million project commenced in November 1989 but by early 1991 was seriously over budget.

4.2 Key Aspects of Design and Construction

4.2.1 Overview

The project involved complex subsurface construction in a congested urban site with substantial physical, environmental and planning constraints. The significance of these challenges soon became evident with continually increasing delay and rising costs. To address this deteriorating situation, a variation agreement between the client and the main contractor was subsequently negotiated. In particular, this included the addition of a value engineering (VE) clause to the contract in March 1991. The inclusion of the VE clause was pivotal in enabling the observational method (OM) to be introduced and, crucially, accepted by all parties. This, in turn, created the major opportunity to increase the rate of construction while achieving substantial cost

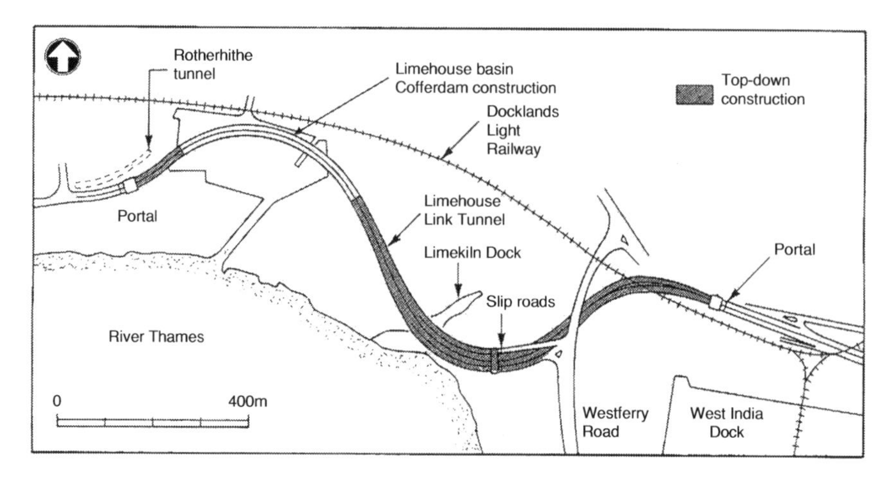

Figure 4.1 Plan and location of Limehouse Link.

savings. The principal need was to recover the severe delay to the project. It could thus be characterised as a 'best way out' application of the OM, and was, as in all of the successful case histories described in this volume, implemented through the process of progressive modification.

The prime objective of using the OM at Limehouse was the complete elimination of the extensive heavy steelwork for the mid-height propping system (Glass and Powderham, 1994). This was designed to support the diaphragm walls during the temporary condition (Figures 4.2 and 4.3). This had to be erected and dismantled within the confined spaces of the tunnel beneath the roof slab. So, any reduction enhanced operational safety. However, even a substantial reduction in the amount, although still very desirable, would have brought disproportionately less benefit. This was because it was the global creation of unhindered working space beneath the roof slab without any temporary props that enabled the dramatic increase in construction progress. Figure 4.4 shows the clear space established beyond the propping trial following its successful conclusion. This clear zone is below one of the access openings in the roof slab and, somewhat poetically, is consequently bathed in sunlight.

Most of the tunnel was designed for top-down construction between reinforced concrete diaphragm walls, but it included a 600 m section of bottom-up construction in a cofferdam through Limehouse Basin. The OM was applied to both forms of construction, although with notably greater success in the top-down sections. The conditions in the Basin were so radically different that the application there effectively constitutes

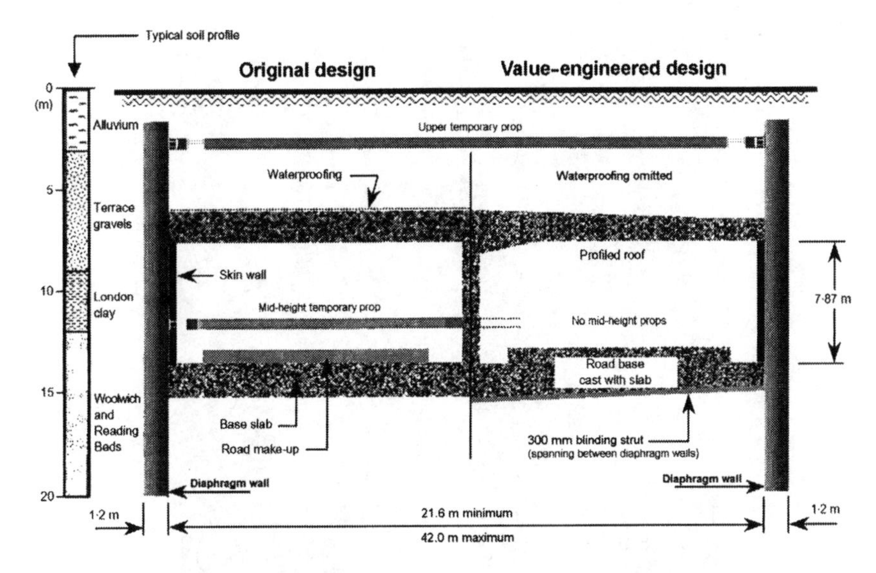

Figure 4.2 Typical cross section and range of widths for top-down tunnel construction.

a separate case history for the OM. It provided a cogent example of how the practical limits for application of the method can be reached.

Key factors in the application of the method were the speed and control of construction, the accuracy and reliability of measurements, and the ability to reliably implement predetermined contingency measures to control any adverse trends and thus avoid unacceptable wall movements. In this, implementation through progressive modification played a key role (Powderham, 1998). This enabled construction to progress from a position of established safety towards less well-understood conditions, with safe, controlled procedures.

4.2.2 Geology

The ground conditions were variable and challenging with a high water table and further complicated by the presence of major obstructions. The surficial soils consisted of man-made fill and alluvium above River Terrace Gravels. These overlie the London Clay, Woolwich and Reading Beds and Thanet Sands. The materials near the surface are highly variable but essentially relatively weak soils. The Terrace Gravels are typically medium dense or dense, well-graded and sandy. The London Clay is overconsolidated and of high plasticity. It is firm, becoming stiff to very stiff

Figure 4.3 First propping trial.

with depth. The Woolwich and Reading Beds are variable. They consist primarily of over-consolidated silty clays and clayey silts but with some discrete more permeable sandy layers. The clays in these beds vary from low to high plasticity. They also include harder strata and, in places, limestone boulders. Groundwater is generally encountered about 5 m below ground level in the superficial soils. At depth piezometric pressures have been affected by under-drainage to the Chalk beneath the Thanet Sands. The effect is first detectable in the more permeable strata of the Woolwich and Reading Beds. The relative location of the tunnel with respect to a typical soil profile is shown in Figure 4.2. The numerous major

Figure 4.4 Clear working space after trial.

obstructions included mass and reinforced concrete and heavy timber piling. The soil strata are summarised in Table 4.1.

The Woolwich and Reading Beds, because of their variability, presented the greatest uncertainty in assessment and sampling proved difficult in these soils. Self-boring pressure meter tests were used to assess in situ horizontal stresses. In the original design, a value for K_0 of 1.5 was adopted for the London Clay and of 2.5 for the clay strata of the Woolwich and Reading Beds. Reduced values of K_0 in the granular strata were applied to allow for the disturbance arising from the construction of the diaphragm walls.

4.2.3 Original Design

The basis for the original design of the diaphragm walls is described in the paper by Stevenson and de Moor (1994). This design mainly utilised

Table 4.1 Description of soil strata (from Stevenson and de Moor, 1994).

Stratum	*Thickness*	*Description*
Made ground	2–8 m	Generally poorly compacted, mostly granular building debris sometimes mixed in with the underlying alluvial clays
Alluvium	0–5 m	Predominantly soft silty sandy clay, with occasional peat bands and distinct sand layers
Thames Gravel	1–8 m	Medium dense to dense sandy gravel
London Clay	1–5 m	Firm, becoming stiff to very stiff, fissured silty clay of high plasticity, underlain in some areas by a dense clayey sand and gravel layer, identified as the Basal beds
Woolwich and Reading Beds unit A	2–5 m	Very stiff to hard fissured multi-coloured silty clay of medium to high plasticity
Woolwich and Reading Beds unit B	1–3 m	Dense to very dense silty fine and medium sand with occasional thin bands of silty clay
Woolwich and Reading Beds unit C	2–5 m	Very stiff, to hard fissured, silty clay to clayey silt of medium to high plasticity containing varying amounts of shells, and silty sand laminations
Woolwich and Reading Beds unit D	1–3 m	Bands of weak to moderately strong limestone and stiff to hard sandy silty clay of low to medium plasticity containing fragments of limestone
Woolwich and Reading Beds unit E	2–6 m	Very dense silty fine sand with thinly bedded clays and varying amounts of gravel with occasional bands of very stiff silty sandy clay
Thanet Sand	Approx. 15 m	Very dense grey silty fine sand

diaphragm walling for the outer walls in top-down construction. There were also short sections of secant piled walls to facilitate the use of low headroom techniques beneath existing bridges. During top-down excavation, the original design required temporary lateral support to be provided above the roof slab and at mid-height between the roof and base slabs (Figure 4.2). The developer had accepted an alternative design at tender for the section through Limehouse Basin. This employed a bottom-up method within a cofferdam to replace that of the original design which had used secant pile walls (see Section 4.5). In all areas temporary steel props were designed by the contractor to specified loads (Table 4.2). All prop loads were to be monitored and records issued to the supervising officer, who acted as the engineer for the developer. The tunnel geometry changes throughout the length of the tunnel with transverse spans between the outer walls varying from around 22 to 42 m. The widest spans are in an area where slip roads leave and join the tunnel to form

Table 4.2 Comparison of original design and actual prop loads.

Location	Chainage	Top prop loads: kN			Intermediate prop loads (kN)		
		Design kN/m	Actual kN/m	Actual as % of design load (pre-load)	Design kN/m	Actual kN/m	Actual as % of design load (pre-load)
Branch Road	138–250	–	–	–	2,000	117	6 (5.0)
Limehouse Basin	250–713	300–750	250	47 (11)	2,900	250	9 (3–4)
Ropemakers 1	713–850	480	250	52 (11)	2,900	429	15 (6.5)
Ropemakers 2	850–972	870	267	31 (10)	2,900	–	–
Dundee	972–1300	570-640	250	41 (15)	2,900	313	11 (4.2)
North Slip Road	972–1150	–	–	–	1,900 2,900	–	–
DLR	1,300–1,600	390	167	42	2,400	–	–
North Quay	1,600–1,691	–	–	–	2,400	238	10 (3.0)

a grade separated junction. The arrangement of the temporary mid-height props showing the variation, including the specified design loads and spans, between different areas – namely at Ropemaker Fields and Dundee Wharf – are shown in Figures 4.6 and 4.7. Tubular props up to 1,320 mm in diameter at 4.5 m spacing were to be provided in addition to twin 914 mm universal beam spreaders and walings along each face of the diaphragm wall. The temporary works relating to the original design would have required a total of 5,400 tonne of steelwork for the mid-height propping system for the entire length of the top-down section.

4.2.4 Value Engineering

VE can be defined as 'a creative organised approach whose objective is to optimise cost and/or performance of a facility or system' (Dell'Isola, 1982; Institution of Civil Engineers, 1996). The procedure focusses on the identification and elimination of unnecessary costs. Consequently there is a natural synergy between VE and the OM (Powderham and Rutty, 1994). Early design and procurement of the extensive propping system were essential to meet key dates in the construction programme. Consequently, substantial quantities of materials were under fabrication or already delivered to site well before the OM proposal was accepted by the developer and his designers. The logistics and planning of procurement, delivery, handling and storage of the temporary steelwork were major issues and made them an obvious target for consideration. For the VE there were four key factors (Farley and Glass, 1994):

(a) Major delays to construction had been caused by late access to areas of the site and the installation of secondary glazing to properties likely to be affected by construction noise.
(b) A supplementary variation agreement to the contract was agreed between the developer and the contractor to recover as much of the delay as possible. Various measures were introduced including a cost incentive clause using VE. This provided a mechanism whereby the contractor could propose design changes to reduce delay and costs, with the benefits shared between the contractor and the developer.
(c) The OM was proposed to and accepted by the supervising officer on behalf of the developer as one of a series of design changes facilitated through the VE clause.
(d) The application of the OM was also the main factor in recovering delay while simultaneously providing the major contribution to the substantial cost savings achieved.

The comparison between the original and value-engineered designs is shown in Figure 4.2.

4.3 Achieving Agreement to Use the OM

4.3.1 Overcoming Traditional Barriers

This case history is particularly instructive in highlighting some of the typical road blocks to achieving agreement to accept and implement the OM. The prevailing contractual environment at the time was not inherently favourable. Consequently, while the success of the OM on the cut and cover works for the Channel Tunnel had been comprehensively demonstrated, it was effectively an outlier. Nevertheless, it seemed reasonable to expect that awareness for such opportunities with similar construction would have accordingly been heightened. This potential appeared especially apt for the Limehouse Link since the same contractor was involved and it was the same designer proposing the adoption of the OM for both of these projects. However, such easy assumptions do not readily translate to consequent success. Experience has continually shown that it takes time for such awareness to permeate through and be recognised within a large organisation. There is also further inertia constraining such innovation becoming established throughout company culture. If the key decision makers for new opportunities were not involved and suitably cognisant of previous successes, they may resist the most enthusiastic advocacy. There was also another constraint in play here. Even with the recognition of the recent track record, the contractual conditions strongly mitigated against the contractor offering such innovation through the OM in his tender. Without a VE clause, all cost savings deriving from design changes introduced by the contractor would accrue to the developer. At the same time, the contractor would be potentially exposed to any risks associated with such design changes. Clearly, that is quite a disincentive. Another potential flaw in such an assumption is through neglecting to appreciate that each application of the OM will tend to present different and potentially unique conditions and challenges. This emphasises the limitations of trying to adopt an overly prescriptive approach. In this context, it was very encouraging to see the fostering of a much more harmonious approach within the next few years. This was heralded by the introduction of the New Engineering Contract (ICE, 1993) and the report 'Constructing the Team' (Latham, 1994). These initiatives were further reinforced with the report 'Rethinking Construction' (Egan, 1998). The next case history in this volume, the Heathrow Express Cofferdam, highlights how this refreshing approach was so effective in recovering that project from a crisis.

On this project, even with its own crisis as a catalyst, it needed the introduction of the VE clause to create the opportunity for a complete review of the existing design. Crucially, this now included the proactive involvement of the contractor. As any cost savings deriving from VE would be shared between the developer and the contractor, the contractual landscape was dramatically transformed from one of confrontation to a climate of mutual cooperation and teamwork. It was thus now possible to relate design in a direct and detailed way to specific construction methods and sequences. The strengthening of this relationship was essential to the introduction of the OM, a fundamental requirement of which is the facility to alter the design during construction (Peck, 1969). Without the provision of a VE clause, it is almost certain that the typical constraints arising from the existing traditional form of contract would have prevented implementation of the method and consequently all of the benefits that it enabled.

Limehouse Link was confronted with the stresses of running increasingly over time and budget. By comparison, for the application of OM to the Channel Tunnel cut and cover works, there was no crisis and the OM was essentially introduced into the design and construction process on an opportunistic and relatively intuitive basis. This was significantly aided by the contractual conditions of design and construct. These naturally foster a closer connection between designers and contractors than the traditional form of contract employed for the Limehouse Link. Before the introduction of VE at Limehouse, there were no comparative levels of rapport established to explore and facilitate the acceptance of design changes –

especially any as radical as implementing the OM. Even after the VE clause was added, any design changes had to be agreed by all parties and, to be frank, for reaching agreement to implement the OM, the key decision makers for the project were initially far from optimistic. Certainly there was a huge advantage to be gained if the mid-height props could be eliminated. In this, the immediate key challenge was the resolution of the technical complexity pertaining to the actual loading in these props. This loading would depend on a combination of many interacting factors. These included construction sequences, rate and depth of excavation, preloads, load paths (both in the soil and the structure), widely varying soil strata and spatial arrangements, temperature variations and other time-dependent effects. There was a wealth of data from the existing instrumentation, particularly from the strain gauges on the props, but its interpretation was proving complex and controversial. So the overriding concern was that, despite best intentions, although there would be extensive evaluation and debate, no definitive agreement would be reached – especially given the very limited time frame to introduce and implement the OM. Consequently, advocacy for the OM had first to start by winning over a wide audience with quite low expectations. Getting distracted with all the inherent complications can prove quite toxic to reaching agreement. So, while embracing the complexity, the challenge is to find a simple yet compelling way of conveying and resolving the key issues. Progress may then be made towards agreement – even with an understandably sceptical audience. It is pertinent to note that the title of Ralph Peck's last paper on the subject was 'The OM can be simple' (Peck, 2001). The Mansion House case history in Chapter 2 provides just one of the compelling examples in this volume where a wide range of complex and interacting issues was resolved through a focus on a key critical factor.

4.3.2 The Key to Consensus

The breakthrough for Limehouse Link was the proposal to unequivocally demonstrate that the mid-height props were not needed. This was achieved not by endeavouring to produce a new creative and hopefully convincing interpretation of the data, but rather to navigate around the entire issue of loading data entirely. The performance specification for the diaphragm walls included an allowable mid-height deflection of 50 mm, giving a maximum convergence of 100 mm. So, for the initial justification to implement the OM, the proposal was to erect a trial section of the mid-height propping in place beneath the roof slab. However, rather than try to measure loads and hopefully show somehow that they were not significant, the strategy was to simply leave a gap of 25 mm between the wall and one end of each prop in the trial section. Thus, if the gap did not close before full excavation to formation level was secured with a blinding strut, then the mid-height props were then demonstrably unnecessary. The whole process for the application of the OM is described in detail in Section 4.4. The key to unlock the constraints to agreement for implementing the OM lay in the resolution of the complexity of the soil/structure interaction with the simple introduction of a physical gap. It might be said that this was a gap through which the whole set of the OM cart and horses was able to be driven.

4.4 Implementation of the OM

4.4.1 Extending the Process of Progressive Modification

The implementation of the method for the Limehouse Link broadly followed the process summarised by Peck (1969). Design uncertainties, combined with a lack of case history data, needed careful assessment. This particularly concerned the perceived level of risk – both to the safety of the tunnel construction and to adjacent structures. It was therefore considered inappropriate to start on site with a design directly based on an assessment of the most probable conditions – that is, a design which would have marginal factors of safety at best in the event of significantly adverse variations from the assumed most probable conditions. Instead, as described later, starting from the original design, changes were introduced sequentially based on observations of actual performance. The OM was thus implemented through a process of progressive modification. The background to the development of applications of the OM through this process is discussed in Chapter 1.

The potential modes of soil–structure interaction, relating to each method and sequence of construction and the varying ground conditions, were carefully assessed. Conceptual evaluation was supplemented by simple plane strain analysis to assess likely factors of safety under actual conditions.

Parametric studies were undertaken to provide an indicative range of wall deflections. However, no undue reliance was placed on this pre-construction analysis. The validity and safety of the assumptions were tested during construction in a controlled step-by-step sequence. This was essential for the ongoing assessment of not only the structural implications for the tunnel and its support works but also any associated risks deriving from induced ground movements on adjacent structures. Monitoring and back-analysis were coordinated with construction progress.

4.4.2 Sequence for Each Implementation of the OM

(a) Examination and logging of the actual ground conditions and properties during top-down excavation, including that for the diaphragm walls and central piles, were undertaken.

(b) Based on this information, and using the original site investigation data, the most probable conditions and the most unfavourable deviations from them were then assessed. The focus was on the likely short-term performance from the soil/structure interaction. This centred on the probable generation of much lower earth pressures on the active side combined with fully mobilised passive support resulting from the unloading of the cohesive soil (Figure 4.5). The concept was analogous to that developed for the Channel Tunnel cut and cover tunnels (Chapter 2). It was supported by careful field observations and a judgement on what level of conservatism was required to justify the implementation of the OM. However, as noted, the guiding approach was to achieve this within a comfortable margin of safety. In other words, to avoid dependence on potentially optimistic criteria based on predictions of the most probable conditions. Such predictions tend to vary quite widely and are likely to need ongoing revision from the observations of actual performance.

(c) Proceeding on this basis, using site-specific data at given chainages, mixed effective and total stress analyses were undertaken. A reduction factor of 0.7 on the passive pressures was applied but no allowance was made for softening in the upper layers or for over-excavation. The resulting factors of safety on bending moments in the diaphragm walls were in the range of 1.3–1.5. These are rather lower than would normally be adopted. It was noted that Padfield and Mair (1984) recommended a factor of safety of 2 for total stress analysis and expressed caution on

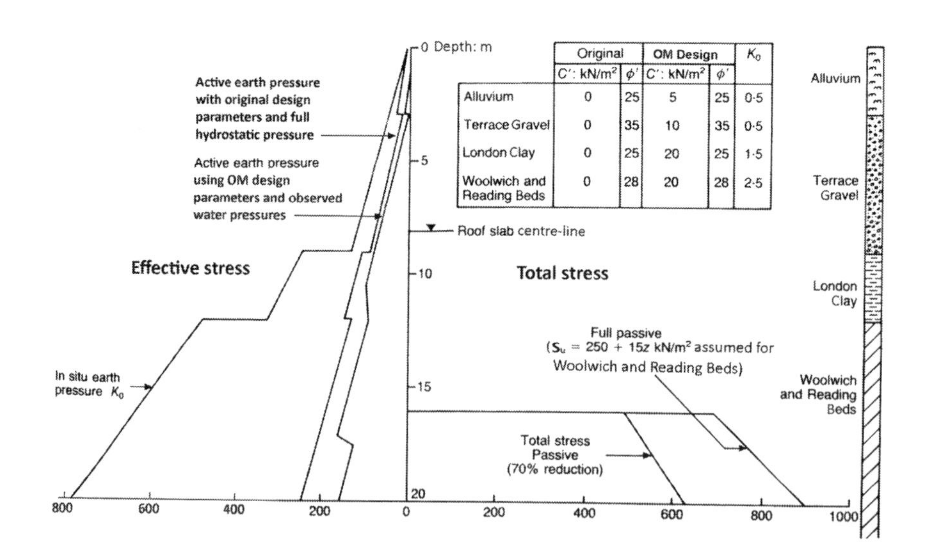

	Original		OM Design		K_0
	C′: kN/m²	φ′	C′: kN/m²	φ′	
Alluvium	0	25	5	25	0·5
Terrace Gravel	0	35	10	35	0·5
London Clay	0	25	20	25	1·5
Woolwich and Reading Beds	0	28	20	28	2·5

Figure 4.5 Top-down construction: basis for earth pressures assumed for OM design.

the implications of mixed analysis when used outside direct experience with specific clays. Thus, the designs developed for the OM were significantly less conservative than would result using the criteria of Padfield and Mair for moderately conservative parameters. However, it was considered that the assumptions for the OM design would still be significantly more conservative than those based on the then current estimate for the most probable conditions. As subsequently demonstrated by measurements of actual wall movements, this consistently proved to be the case despite the wide range of soil conditions, spatial arrangements and excavation depths. It is pertinent to note here that restricting the term 'most probable' to apply only to soil properties risks being over-prescriptive. The actual conditions and nature of the soil/structure interaction will inevitably depend on more than just the soil properties. A further benefit of implementing the OM through progressive modification is that it is not essential to rely on developing designs for commencing the OM closely based on estimates of the most probable conditions. It also helps to avoid controversy since there is always likely to be a range of predictions across the parties involved as to what constitutes the most probable conditions. These issues are discussed further in Chapters 1 and 14. For Limehouse it was necessary in the back-analysis to increase c' from 0 to 20 kN/m ^{2} and to reduce K_o to unity to achieve a reasonable correlation with the measured wall movements. This was based on a simple plane strain analysis which, conservatively, took no account of three-dimensional effects. While the precise values of propping forces developed within the roof slab were not known, this basic analysis indicated actual factors of safety in excess of 3.

(d) However, the OM designs were not implemented immediately. That would have effectively sanctioned proceeding directly with no mid-height props. At this stage, given the contractual environment, that would have been pre-emptively too far to satisfy all stakeholders. Instead, the OM was initiated at each working front with the props of the original design in place but with the gap creating a 'soft prop' as noted in Section 4.3.2. In every case, the trial was successfully concluded with the gap between the props and the diaphragm walls all remaining open.

(e) The next step was the selection of the quantities to be measured, at which locations and at what frequencies. These observations included ground movements and changes in porewater pressure, but the key focus for risk control was the rate and magnitude of wall movement. Predicted values for wall movements were based on the estimates for conditions that were significantly less onerous than those adopted for the original design.

(f) The implications for the most unfavourable conditions were also assessed. This included the worst-case scenario which would have been a critical inadequacy of passive support to the diaphragm walls without the props in place, thus leading to unacceptable wall and ground movements. To assess and contain such a risk, each new working front was commenced by limiting the exposure to the walls within a 5 m bay length. Adverse deflection trends could be tracked and actions taken to arrest them before any unacceptable conditions were reached.

4.4.3 Observations, Trigger Levels and Contingencies

The critical observations were the measured wall movements. The construction method, sequence and cycle time were selected so that there was sufficient flexibility to allow for the taking and assessment of these measurements.

A traffic light system, following its original development at Mansion House, was established setting green, amber and red ranges for wall

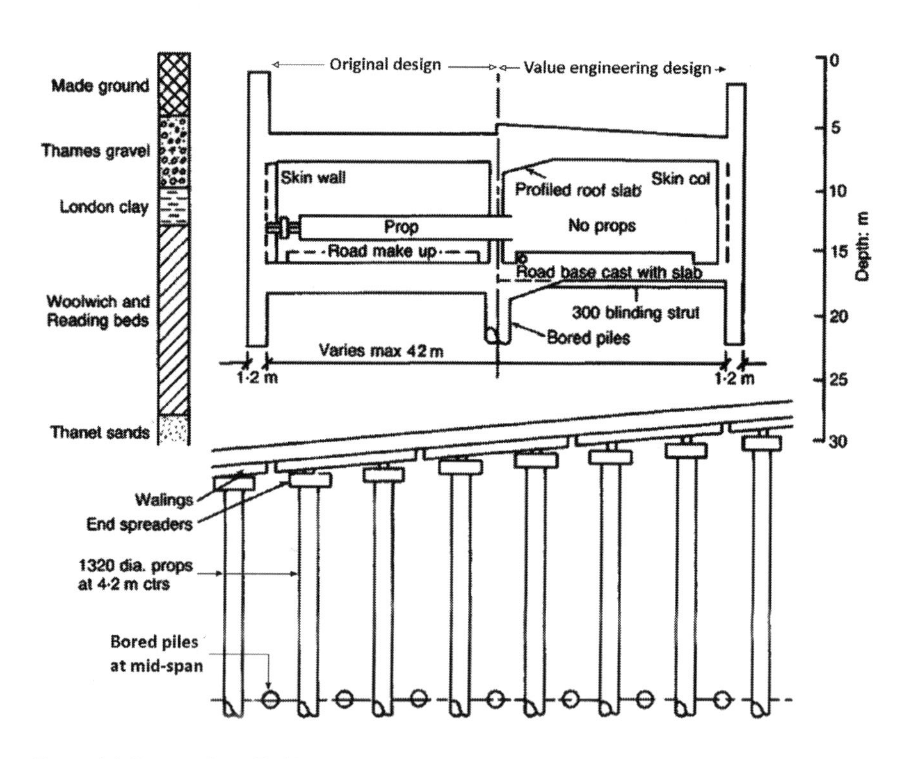

Figure 4.6 Ropemakers Fields: cross section and prop layout.

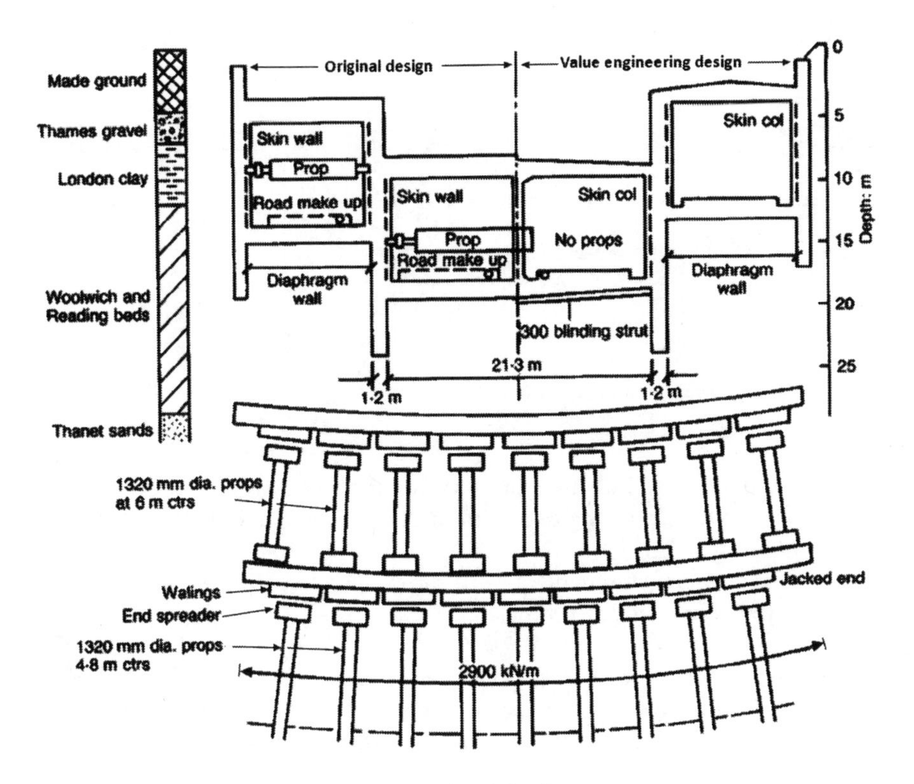

Figure 4.7 Dundee Wharf: cross section and multilevel prop arrangements.

movements. Trigger levels were accordingly set which related both to the amount and rate of movement (Figure 4.9). The first trigger level of 20 mm set the upper boundary for the green zone. Movements contained within this zone were not a cause of concern. The second trigger level of 25 mm marked the upper boundary for the amber zone. Measurements recorded in this zone would signify that wall movements were a potential cause for concern. This would initiate the contingency of increased frequency of monitoring. The focus here would be on assessing whether the rate of wall movements would mean exceeding the second trigger level before the blinding strut was cast at formation level. Such tracking would allow early identification of any adverse trends and so enable the timely implementation of contingency measures. If measured wall movements were consistently less than the first trigger level, then the design would be modified by initiating a construction sequence without mid-height props. Also, if consistently low wall movements were sustained,

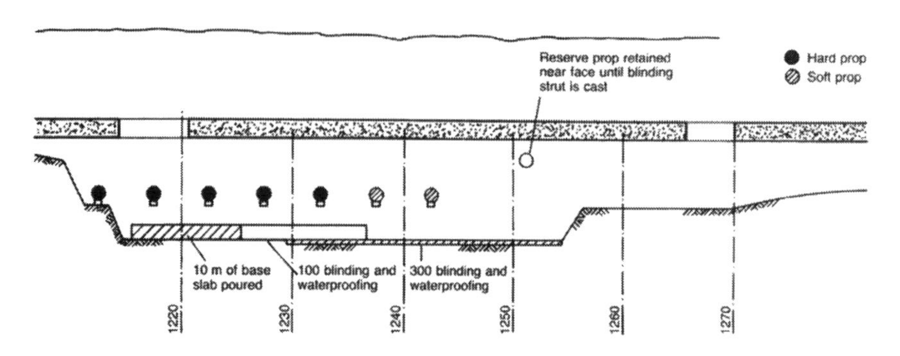

Figure 4.8 Initial construction sequence.

there would be potential for the additional progressive modification of increasing the length of the excavation bay and/or reducing the thickness of the blinding strut.

The contingencies initiated for wall movements entering or predicted to enter the red zone were as follows:

(1) If it were during the soft prop stage – that is, with a gap between the prop and the wall, the use of hard props would be resumed and possibilities for a reduced amount of temporary propping only would be assessed.
(2) If it were during construction without props, installation of the contingency prop would be triggered. Modification of the design would then have involved assessments to reintroduce more temporary support. This could have included, for example, shorter excavation bays or the reintroduction of hard props but at greater centres.

4.4.4 Measured Performance

The observed wall movements during the first propping trial strongly indicated that construction could generally proceed without the mid-height props. However, concerns regarding the actual performance, given the variable range of soils, placed constraints against proceeding on this basis. Thus, as noted, each construction front commenced with a propping trial before the viability of the OM design was confirmed by favourable observations and was accordingly approved and progressed.

The OM was introduced at the start of the first tunnel construction front in Ropemakers Fields (Figure 4.6). The first stage at a construction front started with the installation of hard props in accordance with the original design (Figure 4.8). These were nominally pre-stressed to 10% of the specified design load. Prop loads and temperatures were monitored as excavation progressed, the former being measured using vibrating wire strain gauges linked to an automated data logger. Actual prop loads were assessed allowing for temperature corrections. The loads were found to be considerably lower than was originally assumed and typically not greater than the nominal pre-load (Table 4.2).

The second stage repeated the first except that the 100-mm-thick blinding at the base slab formation, specified in the original design, was increased to 300 mm. This was intended to act as a robust blinding strut. When the blinding gained sufficient strength, the props were destressed and wall closure was monitored. The third stage was to install 'soft props' for the propping trial. These had a 25 mm gap between the diaphragm wall at one end of each prop compared to hard props installed with no gaps. This allowed easy monitoring but would acceptably limit wall movements before the prop started to act. Up to this stage, the propping system as originally designed was still being used.

As the observed wall movements remained comfortably acceptable, the next stage was to proceed without props. Excavation started at the designated prop level and was taken down the full width of the tunnel to the base slab formation. This was limited to a length of 5 m and the blinding strut was cast the same day. Measurements of wall movement were taken at designated positions and times during construction. This was done using tape extensometers and surveying techniques – the former to monitor convergence and the latter to establish absolute movement. These measurements correlated well with those obtained by monitoring the gap in the soft props. To enable prompt action to arrest any adverse trends towards unacceptably large wall movements, a contingency prop and a number of reserve props were kept in the tunnel beneath the roof slab near to each construction front (Figure 4.8). (The overall process developed for observations and actions is described in more detail in Sections 4.4.6 and 4.4.7.)

Observations were taken at two-hourly intervals during the initial stages. When a satisfactory trend of wall convergence had been established, this frequency was reduced to twice and subsequently to just once a day. All the measurements of the wall movements were plotted and compared with a chart of predetermined zones and trigger levels (Figure 4.9). The allowable magnitude and the rate of movement which determined the boundary of each zone were set and agreed between all parties before any props were eliminated. A maximum wall convergence of 70 mm (35 mm for each wall) was allowed at formation level. However, in over 19,000 observations, the maximum convergence recorded was 11 mm and was generally less than 7 mm. Given these results, it is interesting to note that the diaphragm walls actually deflected more overall when the mid-height propping of the original design was used. This sounds counter-intuitive but was because the wall deflections were not effectively arrested until the base slab was constructed. With the mid-height props in place, construction was much slower and during this extended period the walls still moved inwards below the props and of course compressive loads were also being generated in the props. So, when the props were finally removed, the walls moved inwards, thus adding to the displacement developed in the deflected profile during the time before the base slab was constructed.

4.4.5 Increasing Beneficial Design Changes

A further advantageous modification to the design involved assessment of the effectiveness of the blinding strut. The objective was to reduce its thickness from 300 back to 100 mm. This was the minimum thickness considered necessary to support the base slab reinforcement cage. However, in the original design, no beneficial strutting support was assumed from this blinding layer. Bays of soft blinding strut were created by casting them with a 15 mm gap formed by a compressible filler along the edge abutting one of the diaphragm walls. Monitoring of the closure of the gap showed that action of the blinding strut was not initiated before the base slab was concreted. Wall closures had essentially remained the same as in the sections of tunnel that had been constructed previously with no soft gaps in the blinding. It was therefore decided to reduce the blinding strut to a nominal thickness of 100 mm.

The OM was applied to all nine tunnel construction fronts for the top-down sections. In each case after the initial stages of installing hard and soft props, no further propping was required. Plots of wall convergence

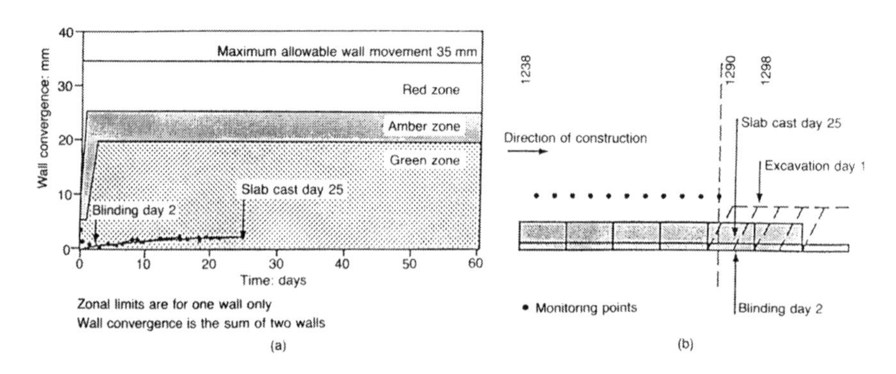

Figure 4.9 Typical convergence of diaphragm walls and base slab construction cycle.

against time were produced for over 300 monitoring locations during tunnel construction. Typical plots are shown in Figure 4.9(a) and movements in this range were consistently recorded at all the working fronts for the top-down construction as shown in Figure 4.10. (The profile and layout of the construction at North Quay, although not separately shown, is similar to that of Ropemakers Fields.)

4.4.6 Personnel and Communication

Application of the OM relied on highly motivated teamwork and complementary expertise between the contractor and his consultant. The support of the developer and the supervising officer was also essential. Clear and effective communication was essential. Before the start of each new construction front, a staff-briefing session attended by the contractor, his consultant and the supervising officer was held. A plan of action was predetermined and key personnel were selected and given the responsibility and authority to make decisions. Emphasis was placed on simplicity throughout the procedures established to operate the method. This embraced the lines of communication, the methods of monitoring and assessment and the actions to implement should any contingencies triggered by observed events.

Within the contractor's organisation a senior engineer was appointed to liaise between the construction monitoring teams and the consultant's design team. His role was to ensure that regular feedback and control of data from observations were maintained and to audit construction practice and contingency provision. The monitoring teams were given the trigger levels for movement limits so that, should the need arise, contingency

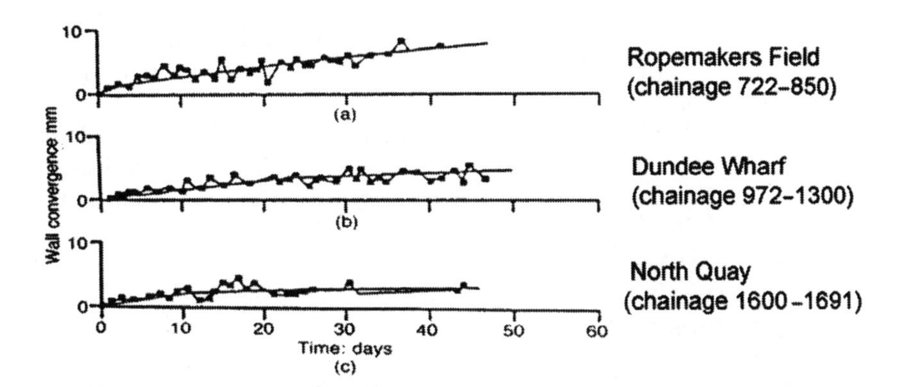

Figure 4.10 Typical diaphragm wall convergence.

measures could be implemented in a safe and timely fashion. The senior engineer was charged with briefing construction and monitoring personnel on methods and controls with particular emphasis on safety. Maintenance of all procedures to ensure adequate safety must be demonstrable and, for example, the installation of contingency props was undertaken at random to check the effectiveness of emergency provisions.

The amount of data to be assessed from the multiple working fronts created a substantial and ongoing requirement for analysis. This had to be processed quickly and effectively. It was therefore kept to simple plane strain analysis. Analysis was only a part of the design process. Design, particularly for the OM, has an important conceptual element. The designer needs to be adequately conversant with the specific construction methods and sequences and the progress on site. The consultant's design team leader acted as the point of communication between the design office and site. This required his regular presence on site during the construction.

4.4.7 Achieving Simple and Effective Monitoring

Simple and clear methods of monitoring were adopted. Wall convergence was established as the critical observation. This was monitored by using tape extensometers and taking measurements between pairs of eye bolts fixed at a set level on the face of the diaphragm wall. Readings on site were compared quickly with those previously recorded to monitor the actual movement trends against those allowable as set in the traffic light system. Additional measurements were taken using surveying techniques with a total station theodolite by sighting on to prisms fixed to the walls. Absolute movement was checked for each wall to compare relative movement against convergence measurements taken with tape extensometers. It soon became evident that wall movements were consistently low. As the plots in Figure 4.10 demonstrate, the overall wall convergence remained well within the green zone. It was therefore decided to adopt the simpler expedient of using wall convergence as the critical measurement for trigger levels. The theodolite survey continued to provide backup in a secondary role.

Accuracy and reliability of measurements was essential. To avoid the inevitable variances between different people using different tape extensometers, the same surveyor and extensometer were used at each of the monitoring points.

At the peak of construction activity four teams of surveyors were employed to cover nine construction fronts. Measurements were recorded manually on to spreadsheets. The data were issued to the section engineer at specified intervals and at least daily so that movement could be monitored and the necessary action taken to control construction. An advantage of direct measurement by extensometers was that convergence with previous readings could be compared immediately. Data transfer from the surveyor to the site management was therefore quick and efficient. Data were also passed to the monitoring engineer who transferred them to a database and spreadsheet format on a site-based PC. The data could then be converted into graphical format so that movement trends could be monitored and issued to the supervising officer. The monitoring engineer, in consultation with the contractor's consultant and the supervising officer, decided on any changes to monitoring frequency at regular weekly meetings.

4.4.8 Construction Plant and Resources

Increased speed, efficiency and safety of construction were paramount among the benefits sought. Specification and reliability of plant were important and each excavation stage used a Fiat Hitachi FH 200 hydraulic back-actor with spoil removal to each roof opening undertaken by a Leibherr 621 traxcavator. Before the OM was implemented, excavation below intermediate props was restricted and slow. Because of the limited headroom, a tracked 6 tonne mini-excavator was used to trim the base slab formation and excavate areas which could not be reached by the FH 200 back-actor. Elimination of the mid-height props allowed the more comprehensive use of much larger and more efficient machines. Each day after completion of excavation the blinding strut was cast. Pumped concrete was readily available from the site batching plant. Speed of reinforcement fixing and concreting of the base slab was ensured by maximising resources and supplying concrete at a rate of up to 90 m^3/h. Unhindered working without temporary propping helped considerably in this respect. Wall convergence was time dependent and so a reduction in the construction cycle directly helped to reduce the risk of unacceptable movements.

4.5 Limehouse Basin Construction

4.5.1 Reaching the Practical Limits of the OM

Stimulated by the success of the OM in the top-down construction, consideration was given to the potential for reducing the temporary works in Limehouse Basin. Confidence, trust and rapport within and between the teams were understandably high and the attempt to extend the success here was markedly ambitious. However, as related later, the attempt was constrained by reaching the practical limits of implementing the OM and, although some real benefits were realised, the prime objective of eliminating or at least substantially reducing the mid-height propping was not achieved.

The construction in the basin was programmed in three phases. These are shown in Figure 4.11 together with the form of construction and propping arrangement. It involved a bottom-up sequence within a temporary cofferdam. The OM was introduced during the second phase of the cofferdam construction. As with the top-down construction in the main section of the Link, the principal objective was to reduce or eliminate the need for mid-height props. But there the similarities ended.

The ground conditions in this section of the work were more uniform than along the rest of the tunnel. Within the depth of construction, they

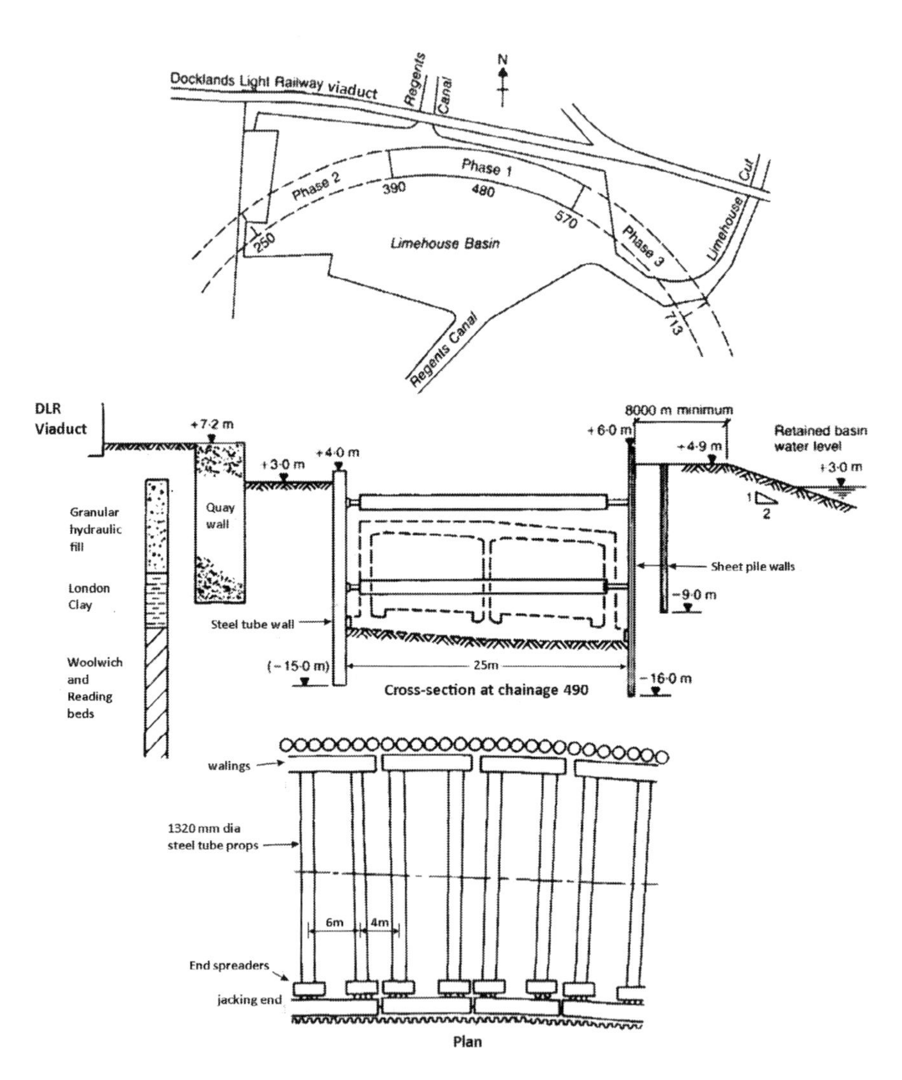

Figure 4.11 Limehouse Basin: location, construction form and phases and propping arrangement.

consisted of a reasonably homogeneous granular fill overlying London Clay. The more variable Woolwich and Reading Beds were not encountered. However, most of the lateral load was generated from the fill with the largest component coming from the high groundwater level (Figure 4.12). The hydrostatic loading could not be reduced easily. As before, the OM was introduced through a progressive modification approach. To ensure and demonstrate safety, a careful sequence of planned operations, related to observed behaviour, was developed. The design here was based on an assessment of detailed observations taken from the first phase of construction. These included prop loads as well as wall movements. It was therefore possible, and in this case necessary, to base the design on conditions that were known to be close to the most probable – that is, a design with inherently low factors of safety. It was appreciated that the likelihood of achieving full benefit from application of the method was relatively marginal.

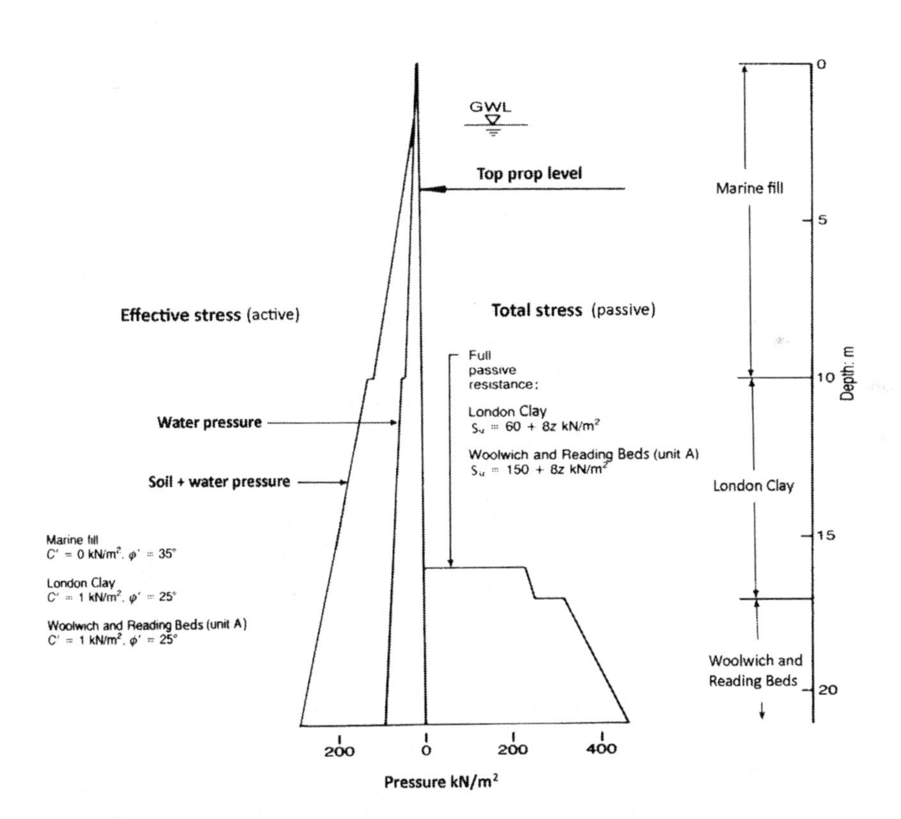

Figure 4.12 Limehouse Basin: OM design earth pressure parameters for cofferdam.

4.5.2 Monitoring Becomes Intensive

The design indicated that the likely failure trend would be in a bending mode in the southern wall of the cofferdam. This was constructed with heavy sheet piles but which are relatively flexible compared to the much stiffer tubes of the northern wall. The potential for such a failure in bending compared unfavourably with the worst-case scenario for the diaphragm walls in the main sections of tunnel using top-down construction. Failure there was a demonstrably remote possibility and would have been through a loss of passive support to the toe of the wall. For both the basin and the main tunnel construction, no loss of passive support was evident. In the basin, because of the sheet plied wall's potential vulnerability, a considerably more intensive monitoring regime was established. This comprised measurements of prop loads, wall convergence and absolute movement at several levels within the depth of the cofferdam and at every 5 m along its length.

Measurements of the hydrostatic pressures in groundwater level adjacent to the Limehouse Basin side of the cofferdam wall were also undertaken. This was achieved by a series of standpipe piezometers along the southern wall. Control of the maximum permissible hydrostatic head was essential and contingency pumping measures had to be pre-installed.

The southern wall formed with Larssen 6 sheet piles while the much stiffer northern wall utilised 1,220 mm diameter steel tubes. This higher stiffness was necessary to limit potential ground movements being imposed on the adjacent Docklands Light Railway viaduct. Consequently, with such disparity in wall stiffness, measurements of absolute wall movement were essential. Surveying techniques similar to those previously described for the top-down construction were used. However, the disadvantage of this method, compared with simple extensometer measurement, was the time needed to analyse and determine a value for absolute movement. The sheer quantity and frequency of readings and data management proved onerous in practice.

Monitoring of the overall deflected shape of the southern wall against a permitted envelope was achieved using inclinometers. These were installed in tubes fixed to the sheet piles before they were driven. Data from inclinometer readings served as a backup to the survey measurements.

4.5.3 Implementation of the OM in Limehouse Basin

As was initiated for the top-down tunnel construction, implementation of the OM in the Basin started with a propping trial. However, the objective of eliminating the mid-height props was not progressed beyond the soft prop stage. In practice, the deflected profile of the southern wall showed adverse trends, indicating that the limits set for acceptable movements would be exceeded. The gap in the soft props closed progressively within 7–10 days. This would have allowed insufficient time for construction of the base slab. The implications of monitoring wall flexure combined with the risks associated with the control of water levels were considered too onerous. Thus, for the Limehouse Basin, the point had been reached where the main benefit sought through the OM (i.e. the elimination of the mid-height propping) was outweighed by the demands of applying it (Figure 4.13).

4.5.4 Added Value Still Achieved

Despite failing in the original objective, the application of the method did lead to some very useful improvements through modifications to the cofferdam construction sequence. One of the main benefits achieved was a change in the propping sequence applied during the second phase. In this, the excavation to the base slab formation level proceeded in advance of the installation of the bottom props rather than afterwards as specified in the original sequence. This simple change of sequence made excavation easier, quicker and less restrictive – and consequently safer.

Figure 4.13 Limehouse Basin: Cofferdam construction with propping to tube wall (left) and sheet piles (right).

In the construction for the third phase, a different additional benefit was realised. The original propping sequence was reintroduced, but the embedment of the southern sheet pile walls below the base slab was reduced by nearly 90% – from 4 to 0.5 m. Considerable damage and deformation of the embedded length below the base slab had resulted from hard driving conditions in the first phase of the cofferdam construction. Potential reuse of these piles in the third phase was therefore severely limited. However, modification of the design using a dramatically reduced embedment, from the knowledge gained form implementing the OM, enabled extensive reuse and avoided the considerable expense in the procurement of additional sheet piles.

Back-analysis of the observations made for the Limehouse Basin was more definitive than for the diaphragm wall top-down construction. This is because it was possible to measure the loads in the top props as well as the wall deflections. This allowed a more accurate estimate of the soil pressures to be made. Water pressures were also known accurately. Back-analysis, on a simple plane strain basis, again suggested that initial in situ stresses were not high (typically K_0 = 0.5 in the gravels and K_0 = 1.0 in the clays). The OM design based on the assessment of the most probable conditions was based on the soil parameters given in Figure 4.12. It should be noted that these values relate to a very short-term and specific situation.

The other encouraging aspect with this particular application was that although the OM was taken to its practical limits requiring intensive monitoring and back-analysis and with wall movements consistently entering the red zone, the robust procedures developed and the close teamwork enabled construction safety to be maintained.

4.6 Risk and Insurance

Insurance for the works was provided by the developer's insurance under cover of an all risks policy established before the contract was started. The contractor had to provide standard insurance cover for employer's liability and for plant and equipment.

Full details of the proposals to use the OM were given to the developer and the contractor's insurers for information before the method was adopted. This included a technical presentation by his consultant to the insurers' representatives, followed by a site visit. No increase in premiums was necessary as a result of the proposals, thereby confirming that there was agreement to be no greater insurance risk. No allowance was therefore necessary for increased cost of insurance as part of the proposal.

4.7 Conclusions

The success of the OM for the Limehouse Link is best summarised by the cost and time savings achieved. As noted earlier, the principal benefit sought was the recovery of the construction programme. This objective was in fact exceeded since, not only were the early delays recovered, but the project was completed 5 months ahead of schedule.

The cost savings achieved were also impressive. The erection of 5.4 km of temporary propping and 2.7 km of waling, totalling 4,900 tonne of temporary steelwork, was eliminated (Table 4.3). Earlier introduction of the method could also have saved 500 tonnes of fabrication and the attendant handling and storage of materials. Although difficult to quantify, the use of the OM created safer conditions for construction. The constraints imposed by the temporary propping are highlighted in Figures 4.3 and 4.4, clearly indicating the dangers in handling and working around such substantial steelwork in confined spaces. The elimination of the vast majority of this steelwork consequently brought a substantial improvement in the safety of the construction environment.

Various views were expressed about the potential hazards and risks to the permanent works and adjacent properties through use of the method. However, in practice, these issues were shown to be negligible. This was achieved by the rigorous application of the method through progressive modification and the close collaboration and teamwork of all involved. The synergy of the OM with VE was clearly demonstrated. Both techniques share the objective of achieving savings in cost and time. Their implementation emphasised the major benefits

Table 4.3 Top-down construction: summary of savings in temporary propping.

	Top props (813 mm dia. and 1,320 mm dia.)	Intermediate props (1,320 mm dia.)		
	Original design	Original design	Observational method	Approximate saving
Total length (number of props)	2,300 m (115)	6,000 m (300)	600 m (25)	5,400 m (275)
Total length of walings and end spreaders (twin 914 UBs)	1,700 m	3,000 m	300 m	2700 m
Total weight of temporary steelwork to be handled	1,050 t	5,400 t	500 t	4,900 t

possible by strengthening the relationship between design and construction. Finally, the other key aspect was how the robustness and flexibility of the OM when implemented through progressive modification was comprehensively demonstrated. Despite the widely varying soil conditions and structural geometry, full success was consistently achieved across the nine individual working fronts for the top-down sections. Also, although significantly less successful, construction in the Limehouse Basin provided an example of how even when reaching the practical limits of an application of the OM it can be safely managed.

References

Dell'Isola, A. J. (1982). *Value Engineering in the Construction Industry*, Van Nostrand Reinhold, New York.

Egan, J. (1998). *Chairman, The Construction Task Force*. Rethinking Construction, p. 40, DETR, London.

Farley, K. R. and Glass, P. R. (1994). The construction of Limehouse Link tunnel. *Transp.*, **105**, August, 153–164, ICE, UK.

Glass, P. R. and Powderham, A. J. (1994). Application of the observational method at Limehouse Link. *Géotechnique*, **44**, No. 4, 665–679.

Institution of Civil Engineers (1996). *Creating Value in Engineering, ICE Design and Practice Guide*, Thomas Telford, London.

Latham, M. (1994). *Constructing the Team*, p. 130, HSMO, London, July.

New Engineering Contract (NEC). (1993). The Institution of Civil Engineers, London.

Padfield, C. J. and Mair, R. J. (1984). Design of retaining walls embedded in stiff clays, CIRIA Report 104. London, UK.

Peck, R. B. (1969). Advantages and limitations of the observational method in applied soil mechanics. *Géotechnique*, **19**, No. 2, 171–187.

Peck, R. B. (2001). Proc. Institution of Civil Engineers. *Geotechnical Engineering*, **149**, No. 2, 71–74.

Powderham, A. J. (1998). The observational method – application through progressive modification. *Proceedings Journal ASCE/BSCE*, **13**, No. 2, 87–110.

Powderham, A. J. and Rutty, P. (1994). The Observational Method in Value Engineering, Proc. 5th International Conference. DFI, Bruges, 5.7.1–5.7.12.

Stevenson, M. C. and de Moor, E. K. (1994). Limehouse link cut-and-cover tunnel: design and performance, Proc. 13th Int. Conf. Soil Mech. New Delhi.

Chapter 5

Heathrow Express Cofferdam (1994–1995)

5.1 Introduction

The Heathrow Express (HEX) Rail Link is an important new connection between Central London and Heathrow Airport. It provides a frequent, direct service between the airport terminals and London's Paddington Station (Figure 5.1) with a journey time of fifteen minutes. The line runs 19 km along existing track before branching off to continue for 6 km below ground to the airport's Central Terminal Area (CTA), where it involved substantial underground construction. In October, 1994, these works were in crisis following the major collapse of the large diameter station tunnels during their construction in the CTA. This case history describes how, through the single team approach, partnering, value and risk management and technical innovation were deployed to minimise the delay and cost overruns resulting from the collapse. The application of the observational method (OM) played a key role. The centrepiece of the recovery solution was the construction of a cofferdam measuring 60 m in diameter and 30 m in depth. Its design was specifically tailored to facilitate the implementation of the OM through progressive modification.

5.2 Key Aspects of Design and Construction

5.2.1 Overview

The original target date for the opening of the HEX was 1 December 1997. The bored tunnelling south of the M4 had commenced 3 years earlier. It included 3 km of running tunnels to the CTA, a series of station caverns underneath the CTA and a further length of running tunnel from the CTA to Terminal 4. A major collapse occurred when the station tunnels below the CTA, failed in October 1994 (HSE, 2000). The primary linings for these large diameter platform and concourse tunnels were then being constructed with sprayed concrete linings (SCL) (Figures 5.2–5.4). Tunnel invert is at a depth of around 30 m below ground surface. Fortunately,

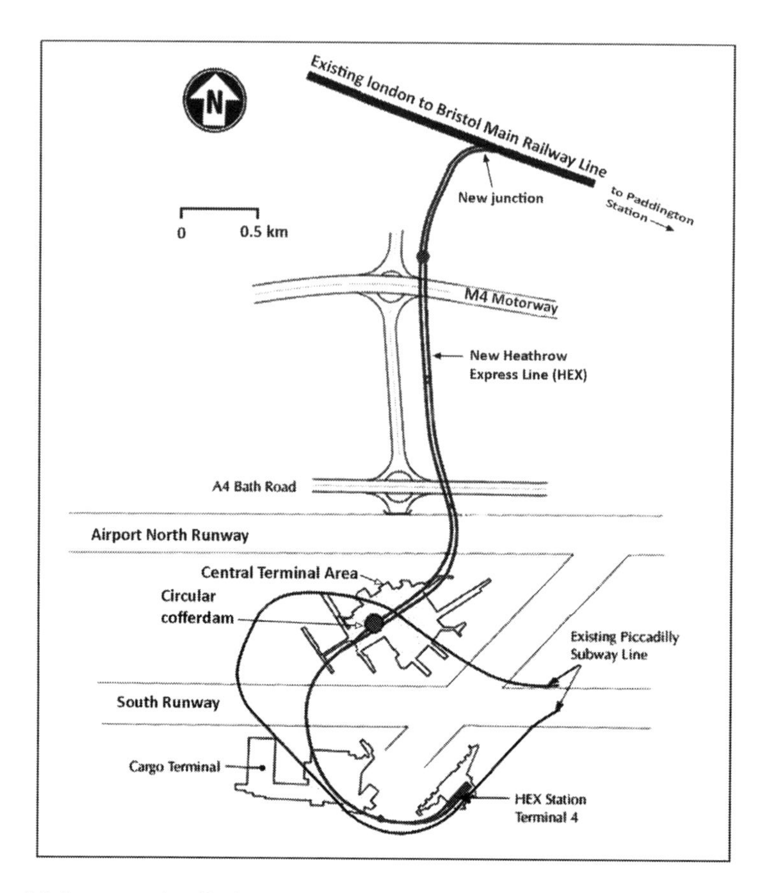

Figure 5.1 Location of cofferdam in CTA and HEX connection to mainline railway.

there was no loss of life or injuries but substantial damage to the works and adjacent structures occurred. At this stage, potential delay to the project resulting from the collapse was estimated to be about eighteen months. Fundamental to the recovery plan was the need to safely repair the damaged works and reconnect the tunnels and minimise delay to the opening date. The CTA cofferdam was the centrepiece in this strategy, and its successful construction enabled the line to be opened on 25 May 1998, recovering a full year of the lost programme (Thomas and Bone, 2000). An important early decision in the recovery was the formation of the Solutions Team, following from the single team approach developed by the owner BAA Airports. The Solutions Team members were selected from the main stakeholders of the project: BAA, main contractor, lead designer and the loss adjusters with their consultant.

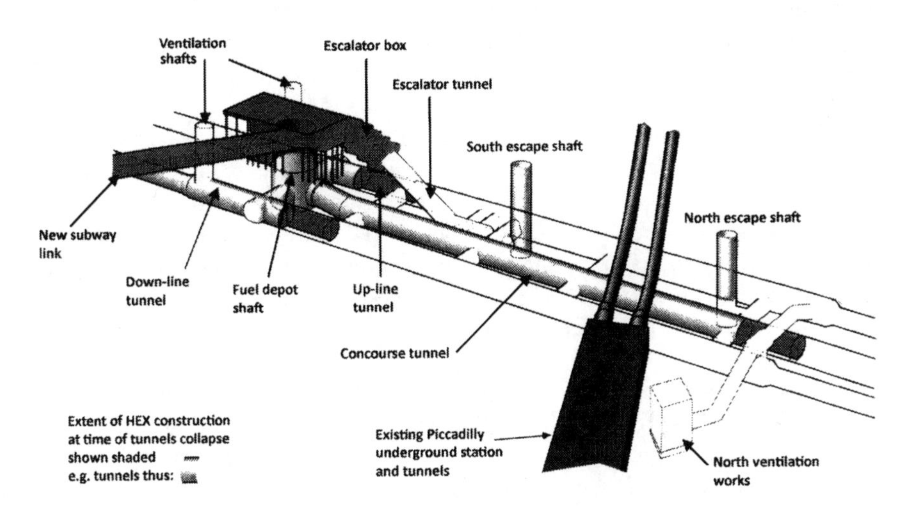

Figure 5.2 Extent of tunnelling construction at time of collapse.

5.2.2 Geology – Conditions Prior and Post Collapse

A site investigation, to evaluate the changed conditions, was initiated immediately after the collapse. The ground conditions in the CTA, prior to collapse, were relatively uniform with approximately 6 m of Terrace Gravels overlying the London Clay which has a thickness of around 60 m at this location. The London Clay overlies the Lambeth Group, which generally consists of heavily over-consolidated clays and sands. This in turn overlies the Chalk which is present at a depth of approximately 90 m below ground level.

The recovery strategy required early establishment of a ground model. This involved an iterative approach and, as new information became available, it enabled development and refinement of the model. Original ground horizon levels were carefully assessed and compared with those post failure. The focus was the top of the London Clay which had originally been about 6 m below ground surface and the new levels were mainly assessed from a series of shallow boreholes. A series of deep boreholes were also carried out to assess the condition in and adjacent to the collapsed tunnels. Detailed core logging was used with the emphasis placed on visual descriptions. Investigation and design development were proceeded in parallel, and it was important not to create unnecessary delay with a prolonged programme of laboratory testing. Two sets of data for the top of the London Clay, pre- and post-collapse, were collated. To achieve the best estimate of the contours for these two London Clay horizons, the data were statistically evaluated through a process known as kriging (Clark, 1979). The difference between the two kriged surfaces was plotted as contours of settlement as a result of the collapse and is shown in Figure 5.3.

This process indicated that there were four localised areas of highly disturbed ground. In view of the large collapsed volume, approximately 6,000 cubic metres, and the subsequent amount of excavation required, it was considered that there was likely to be significant time-dependent softening initiated as a result of the collapse. The excavation of the cofferdam would also create a further reduction in stresses leading to a prediction for soil strengths much lower than for typical conditions in London Clay.

On the basis of the site investigation and predictive numerical analysis, four zones were assigned within the London Clay (Figures 5.3 and 5.4). Zone 1 was undisturbed intact London Clay, but this was considered to lie beyond the active wedge of soil on the outside of the

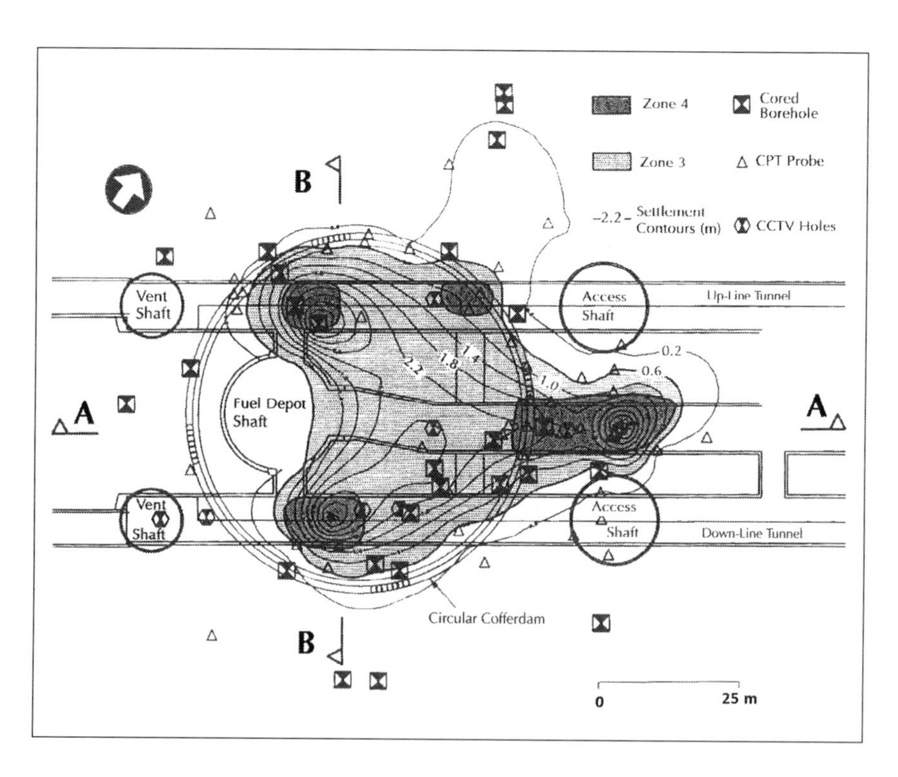

Figure 5.3 Settlement contours of London Clay with predicted zones of disturbance.

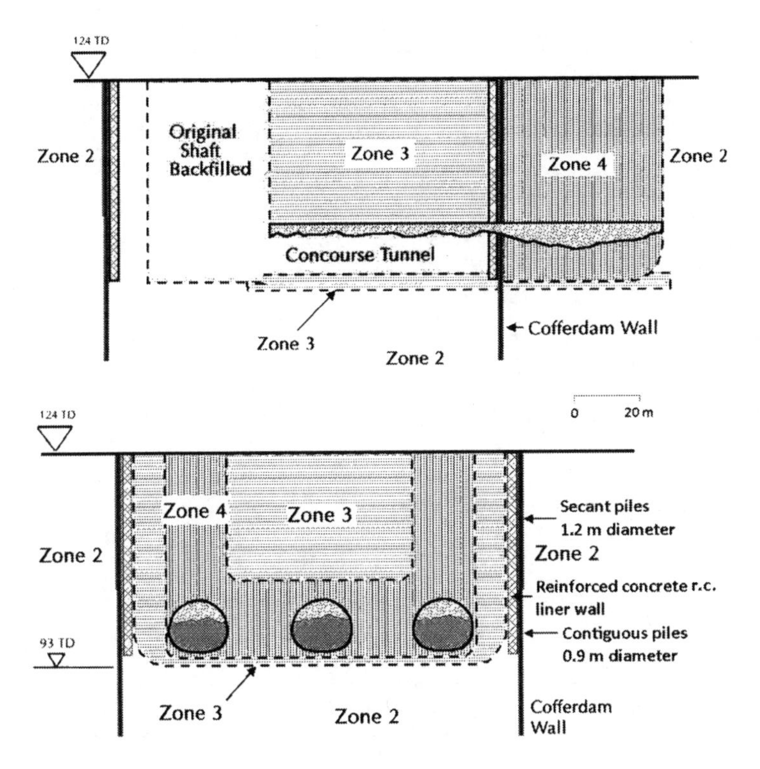

Figure 5.4 Cross sections showing predicted zones of disturbance.

cofferdam. So, the performance of the cofferdam would be principally influenced by Zones 2 to 4. Each of these was assigned two sets of bounding soil properties. The first represented moderately conservative (MC) parameters for the mass behaviour of that zone on the cofferdam as a whole. The subsidiary set was assessed as worst credible (WC) values, representing local influences that might have occurred where pockets of the most severely disturbed soil in that zone could result in adverse loadings on the cofferdam ring. These properties are summarised in Table 5.1.

5.2.3 Contractual Conditions

The New Engineering Contract (NEC) (ICE, 1993) for the project was specifically adopted by BAA project to facilitate a less confrontational approach to construction. The NEC emphasises the principles of trust and co-operation within a contractual framework. The report,

Table 5.1 London Clay soil parameters at Heathrow cofferdam site.

		Zone 2		*Zone 3*		*Zone 4*	
	Level	*MC*	*WC*	*MC*	*WC*	*MC*	*WC*
Su (kPa)	118–108 TD	50 + 7 d	30 + 7 d	30 + 7 d	0 + 7 d	0 + 7 d	10+ 1.5 d
	108 – 93 TD	105+ 3.5 d	85+ 3.5 d	85 + 3.5 d	55+ 3.5 d	55+ 3.5 d	(=0.25 σ'_v)
γ_B (kPa)		19.5	19.5	19.5	19	19	16
$\varnothing'$ degree		25	25	25	25	25	21
c' (kPa)		10	5	5	0	0	0
Strain (%)		<0.1	0.2	0.2	0.5	1	N/A
E_u/S_u		700	500	500	350	150	150
k_h (m/sec)		$(1 \times 10^{-8}$ to $1 \times 10^{-10})$		$(1 \times 10^{-7}$ to $1 \times 10^{-9})$		$(1 \times 10^{-3}$ to $1 \times 10^{-7})$	
k_v (m/sec)		$k_h \times 10^{-1}$		$k_h \times 10^{-1}$		$k_h \times 1$	
k_o (1)		1.0	0.8	0.8	0.6	0.6	0.6

Note: TD = tunnel datum (= ordnance datum+100 m), d = depth below ground level, MC = moderately conservative, WC = worst credible. Zone 4 extends to +95 TD and Zone 3 extends to +93 TD. Below +93 TD, Zone 2 MC apply.

'Constructing the Team' (Latham, 1994), had set out the parameters for enhanced performance to promote a healthier UK construction industry through teamwork and undertaking projects *'in a spirit of mutual trust and co-operation, trading fairly and nurturing the supply chain'*. This was followed by the report 'Rethinking Construction' (The Construction Task Force, 1998). Chaired by Sir John Egan, it provided the stimulus for the initiatives of the Construction Best Practice Programme and Movement for Innovation. The key emphasis on teamwork and creating a 'win–win' approach also led to the process of partnering (Brook, 1997):

> *"Partnering has emerged in a number of forms, partly to reverse the suicidal fall into institutionalised conflict with appalling relationships between contracting parties in the construction industry, and more recently as a means of securing more work by creating a competitive advantage"*.

5.2.4 The Cofferdam

The shape eventually adopted for the cofferdam was circular, 60 m in diameter and 30 m deep. As discussed in Section 5.4.2, it provided a very optimal creation of space compared to other options such as a square layout (Figure 5.5). It utilised 182 large diameter stepped secant/contiguous piles for the outer wall and 255 large diameter bored piles for the base slab. The design and construction had to deal with disturbed and unstable ground (including gravel beds and London Clay), water filled voids and major subsurface obstructions. These included mass and reinforced concrete and large items of buried construction plant. Being right in the middle of the CTA, there were also severe spatial limitations and environmental issues to address. The application of the OM through progressive modification was key to the overall design and construction strategy to manage risk and maximise opportunities to recover time (Powderham, 1998). Key issues were the control of ground movements and the associated effects on the cofferdam itself and protection of adjacent structures while achieving fast track design and construction.

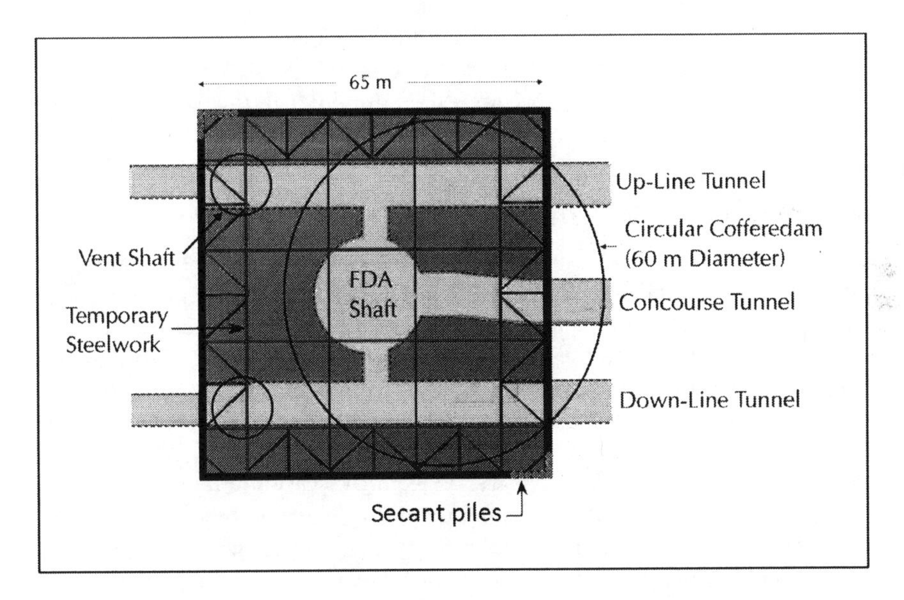

Figure 5.5 Square cofferdam option (not implemented) compared to circular form.

5.3 Achieving Agreement to Use the OM

5.3.1 Safety and Innovation

The HEX tunnel collapse presented a major challenge to develop appropriate measures to rescue the project. However, in practice, it was unacceptable to adopt experimental or untested methods and risk further delay. The key elements – the concept of the cofferdam combined with the application of the OM – were both established technologies. The innovative features related to the form and the circumstances under which they were applied (Powderham, 2009). The key innovations, as illustrated in Figures 5.5–5.8, were as follows:

1. The scale and geometry of the cofferdam and its means of construction in a tightly constrained space.
2. The unique structural features of the 40 m deep piled wall construction for the cofferdam (see Figure 5.6) and the varying and difficult ground conditions through which it had to be installed.
3. The design was specifically tailored to facilitate the application of the OM through progressive modification. This bespoke approach to intimately match the design with construction methods maximised potential benefits while concurrently enhancing risk management.
4. Development of a comprehensive data management system to monitor the performance of the whole cofferdam in real-time. Particularly with adoption of the electro-levels for primary instrumentation, this was considered 'leading-edge' at the time and was featured in a BBC documentary for the Open University entitled 'Talking Buildings'.

5.3.2 Progressive Modification Offers the Way Forward

Although the client BAA was very disposed towards the use of the OM, given the critical environment created by the collapse, achieving agreement demanded clear and compelling advocacy. A series of presentations by the OM team were made to the client's executive boards. The key role of the progressive modification approach and how it was intimately related to the bespoke design and construction of the cofferdam were highlighted in these presentations. A moment of truth came in the Q&A following the presentation when the OM team were asked: 'All very impressive but what do you regard as the critical point in the construction sequence – i.e. at what stage will we be able to collectively breathe a sigh of relief?' The response could have been 'when excavation has past the stage of maximum bending moment in the wall' or 'when the base of the cofferdam is stabilised'. Such responses would not have satisfied the concern behind the question but simply stimulated the need for further explanation and perhaps a lessening of confidence. The actual answer given was: "*Thank you – your question touches the very heart of the matter. The whole thrust of the process we advocate is specifically directed to eliminate all such critical hurdles. Throughout construction of the cofferdam the performance will be measured and all trends assessed. Any adverse trends will be addressed and resolved in a timely way and well before they approach critical conditions".* Agreement to implement the OM promptly followed this presentation.

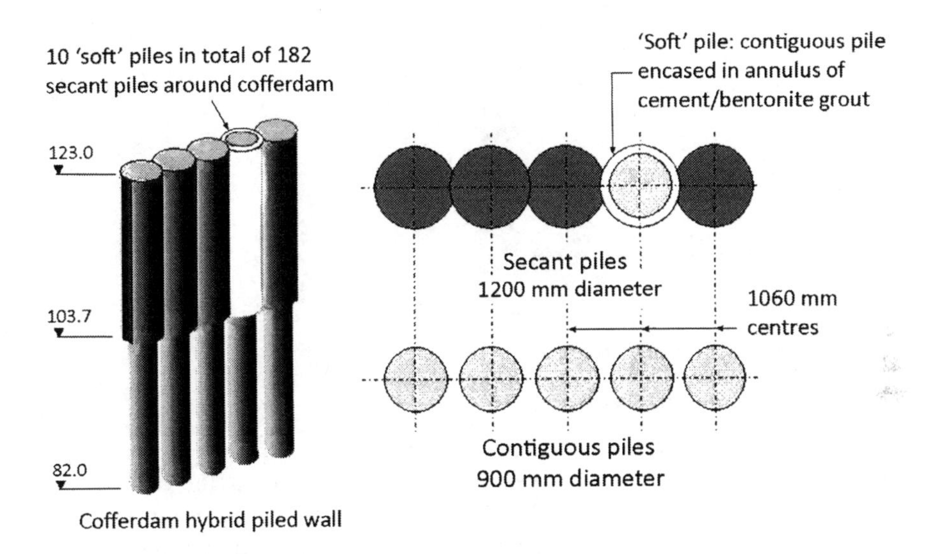

Figure 5.6 Details of bespoke hybrid piled wall of cofferdam.

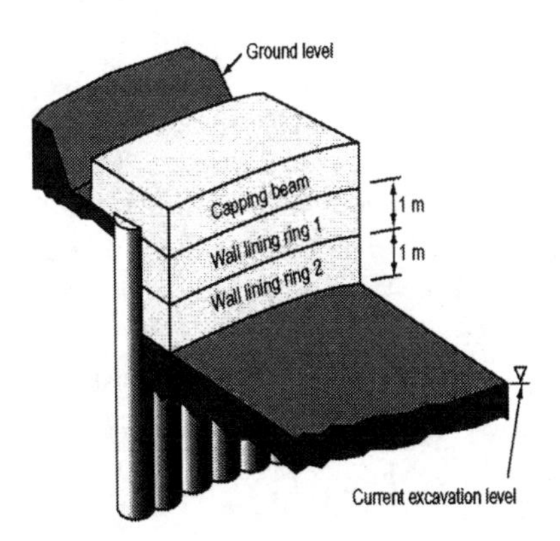

Figure 5.7 Construction sequence for lining wall.

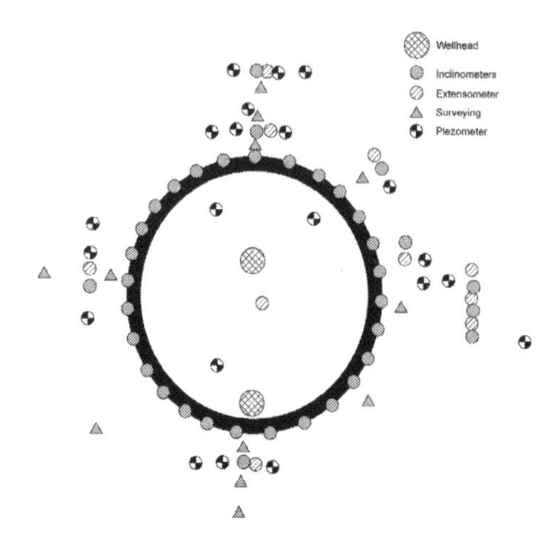

Figure 5.8 Cofferdam instrumentation.

5.4 Implementation of the OM

5.4.1 The Recovery Solution

The application of the OM formed an integral part of the risk management strategy for the overall design and construction of the cofferdam. The potential offered by the OM was identified right from the start of the design concept. Careful consideration was given throughout the design development to enhance compatibility with construction methods. The scale involved is conveyed in Figures 5.9 and 5.16.

Early brainstorming workshops produced a wide range of schemes including micro-tunnelling and jacked tunnels, but the team quickly came to the conclusion that a large cofferdam would be the key feature of the recovery solution. There was a basic requirement to encompass most of the disturbed ground and the majority of the damaged subsurface structures. These structures included reinforced concrete piled surface slabs and escalator box, the original 20 m diameter 30 m deep fuel depot shaft and the three partially constructed large diameter SCL platform and concourse tunnels. The shaft and tunnels had been filled with around 13,000 cubic metres of concrete as an emergency measure for short-term stabilisation of the collapse. There was clearly a need for a robust design for the cofferdam which would be built in highly disturbed ground containing major obstructions. Initial options favoured a top-down construction

Figure 5.9 Construction progress by mid-May 1996 with 12 rings completed.

sequence based on a square shaped layout. However, top-down construction had the following major disadvantages:

(1) It required early decisions for the detailed design of the final internal structure. This applied particularly to floor layout and levels and internal columns and piles. Being on the critical path for design development and approval, these design requirements risked causing substantial delay.
(2) Although any permanent slabs may have been able to provide a measure of temporary lateral support during construction, a substantial amount of additional temporary strutting was still likely.
(3) A prime reason for such a large cofferdam was to enable safe access and removal of major obstructions. Top-down construction could severely inhibit such access. The presence of major obstructions throughout the 30 m depth would conflict with the construction of the permanent lateral supporting slabs.

For these reasons, although top-down construction is often the preferred method for large excavations in urban environments, it was rejected here in favour of a bottom-up sequence. However, with a rectilinear shape, bottom-up construction would maximise the amount of temporary strutting required. For example, the amount estimated for a

Figure 5.10 Buried plant in concourse tunnel presenting obstruction to piling.

65 m square cofferdam was about 5,500 tonnes of structural steelwork. Apart from the cost and programme implications, this would also present significant restrictions to access and working space. Ongoing concept development therefore concentrated on ways to reduce the need for such strutting. The type of construction for the outer wall of the cofferdam needed to be decided early in the concept development since the whole process for the way forward demanded close integration of design with construction methods.

5.4.2 Cofferdam Configuration

As noted range of options for the cofferdam were considered including a 65 m square (Figure 5.5). However, in December 1994, a circular cofferdam was selected as the preferred option. At a diameter of 60 m and depth of 30 m, it offered a dramatically simple solution. Larger circular cofferdams had been constructed (e.g. as reported by Kirmani and Highfill, 1996) but not in such disturbed and variable ground conditions or utilising a unique bored piled wall (Figure 5.6). The circular cofferdam brought the following major advantages:

1. Complete elimination of temporary cross strutting maximising available space for construction operations.

2. It minimised the total volume of excavation. This was because it was possible to arrange the two permanent ventilation shafts to the south close but external to the cofferdam rather than being contained within a rectangular arrangement. Since these two shafts are in the relatively undisturbed Zone 2, the circular cofferdam still encompasses the majority of the disturbed ground including most of the areas of greatest settlement. In comparison with the square cofferdam option, it required about 20,000 cubic metres less bulk excavation. Apart from major cost savings, this brought important environmental and programme benefits, particularly since construction in the centre of a busy airport could significantly affect airport operations. The ground was also likely to be contaminated with aviation fuel which had been stored in this location. Bio-remediation was undertaken to mitigate this risk.
3. The symmetry of the solution allowed a uniformly progressive step-by -step sequence of construction for the cycles of excavation and the casting of the inner 1 m thick reinforced concrete liner supporting the piles. This rhythm and symmetry greatly facilitated the progressive monitoring of ground and structural movements so that the associated trends, and in particular any adverse ones, could be detected at an early stage. This latter aspect was also highly compatible with the application of the OM through progressive modification and was key to the overall risk management strategy for construction of the cofferdam and central to the potential for additional cost and time savings.

5.4.3 Innovations in Piled Wall Design and Construction

The site conditions created particular requirements for construction methods. A fundamental aspect of the cofferdam was the design and construction of the outer wall. Diaphragm (or slurry) walls are typically adopted for this type of construction, but the combination of major obstructions (Figures 5.10 and 5.12) and potentially extensive voids were critical in eliminating this form of wall with its dependence on bentonite slurry support for panel excavation. A primary consideration was the need to provide a good cut-off to the groundwater in the Terrace Gravels above the London Clay particularly in view of its potential contamination. So, groundwater inflow had to be controlled along with the need to fully retain any loose, disturbed ground caused by the collapse. The outer wall also needed to be reasonably stiff and robust. Large diameter piles secanted to an appropriate depth offered the potential to satisfy all the criteria. Secant piles were therefore adopted as the basis for development of the wall design. Depth, diameter, spacing and construction tolerances now needed careful consideration in conjunction with construction methods and sequences.

The disturbed ground and obstructions which, apart from reinforced concrete structures, included large items of buried construction plant entombed in mass concrete would clearly present challenges to the construction of any wall. This initiated detailed discussions between the OM team and the specialist piling contractor. Maintaining tight alignment tolerances and structural integrity were key issues. To facilitate this, emphasis was placed on keeping the bored piling plant well within its operating range. The typical maximum operating depth to ensure good quality construction for secant piles of this diameter was stated to be 25 m. Thus, it was determined to limit the 1,200-mm diameter secant piles for the top 20 m. The piled wall then stepped in below this level to continue as 900 mm diameter contiguous piles for the next 20 m (Figure 5.6). Contiguous piles are much easier to construct since they do not involve overlap cutting into adjacent piles as in the case of secant piles. The base case for the OM is shown in Figure 5.11.

Both primary and secondary piles were reinforced with bar reinforcement except above the tunnels where structural steel sections were used. It is essential to relate design to construction, and this was particularly relevant to the piled wall where considerable interactive effort and support were provided to the team by the specialist piling subcontractor. The piles were installed using an oscillator and casing through an accurately constructed reinforced concrete guide wall. The specified minimum vertical

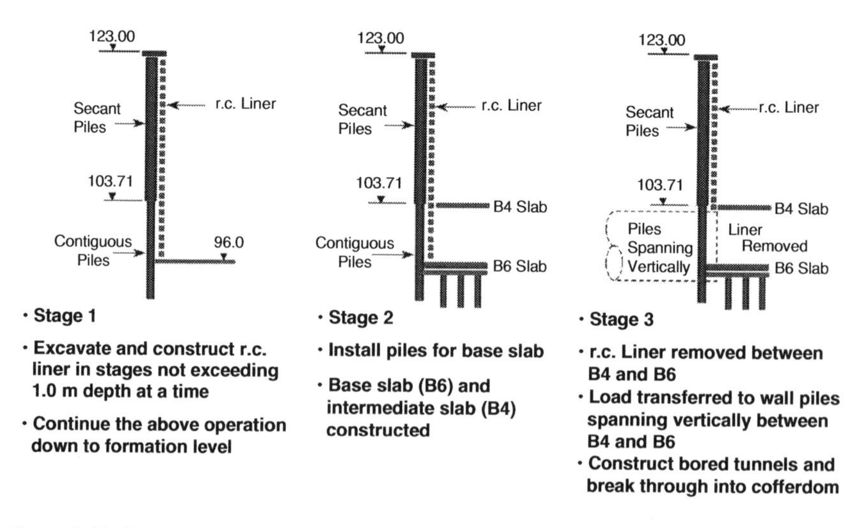

Figure 5.11 Base case construction sequence for OM design.

Figure 5.12 Inspection of the collapsed SCL lining of the concourse tunnel at the base of the cofferdam.

tolerance for these piles of 1 in 150 was satisfied and was generally achieved close to 1 in 200.

As noted, the principal function of the secant piles was to provide a barrier to groundwater and continuous support to any zones of weakened soil. At a depth of 20 m, the secant piles would reach the original level of the crown of the tunnels. It was judged that the vast majority of the disturbed ground would be encountered over this depth and the secant piles would thus form an effective barrier. It was accepted that some limited zones of ground treatment could be necessary to complete

Figure 5.13 Inspecting the annulus of a 'soft' pile.

Figure 5.14 Early tunnel breakthrough.

this cut-off. Pre-treatment of the ground prior to piling focussed on the highly disturbed zones above and around the collapsed tunnels. A program of permeation grouting was initiated to infill any remaining voids in these zones and so minimise risks and potential delays arising from the need for additional contingency ground treatment during excavation. In practice, the secant piles performed extremely well overall, uniformly controlling ground movements and maintaining a very effective cut-off to groundwater.

The inclusion of the so-called 'soft' piles was another unusual feature. As shown in Figure 5.6, there was a total of ten 'soft' piles installed at approximately 19 m centres within the ring of secant piles, below which the piles continued as the rest of the wall as 900 mm diameter reinforced concrete contiguous piles. The objective of these 'soft' piles was to address a concern expressed by the independent checking team for the loss adjusters regarding the build-up of hoop compression around the ring of secant piles. The concern was that it presented a risk of compressive failure at the overlap interface in the secant piles. The OM team judged that the combination of the secant piles and the reinforced concrete liner wall had adequate redundancy in which at least one alternative load path was available. It was therefore considered as satisfying the criteria of the 'lower bound' or 'safe' theorem (Heyman, 1995). Thus, given the construction method, with each reinforced concrete liner ring being promptly cast after each excavation sequence, compressive failure in the secant piles would not be an issue. Any unacceptable build-up of compressive force and associated strain would be avoided through an alternative load path transferring the compressive force vertically upwards along the overlap and into the liner wall. However, the OM design had to be acceptable to all parties and, in this context, the provision of 'soft' piles was approved. Equally spaced around it, their function was to ensure an acceptable limitation in the hoop force by the crushing of the cement/bentonite annulus. These ten piles were constructed by using a temporary steel liner, with the same internal diameter as the contiguous piles, placed concentrically within the secant pile liner over the whole 20 m depth. The gap between the two liners was then filled with a cement/bentonite grout. Thus, the 'soft' piles were formed as a 40 m long contiguous pile with an annulus 150 mm thick over the top 20 m. Such a grout was considerably weaker than the reinforced concrete of the secant piles. It is also quite brittle when fully cured. Consequently, the validity of the judgement of the effectiveness of the alternative load path would be definitively tested. The state of the 'soft' annuli of cement/bentonite was comprehensively inspected after completion of the bulk excavation to formation level (Figure 5.13). No compression failure was evident.

5.4.4 Design Development

Given the importance of recovering construction schedule, a fast track design and construction approach was required. This was greatly aided by the formation of an integrated single team including the establishment of site-based designers during construction.

Following approval of the design concept for the cofferdam which included the location, diameter and wall type, design effort focussed on detailed design development linked to the needs of construction.

As noted, a key advantage of the circular cofferdam was its inherent integrity and simplicity of form. It provided a robust base scheme consisting of an outer ring of piles, reinforced concrete liner, piled base slab and an intermediate slab at level B4 (Figure 5.11). This allowed the design and construction to be independent from whatever internal structure above the B4 level was eventually developed thus avoiding delay from the associated design development and approval process. This benefit was mainly due to the circular geometry which eliminated the need for any permanent cross strutting above this level but also because the piled base slab would allow a wide range of options for vertical loading. The secant section of the piled wall essentially had a temporary function, as the inner reinforced concrete liner was designed to take all of the permanent ground loading. The secant section stopped at the B4 level, just above the crown of the bored tunnels. The contiguous piles extending below extending below this level were designed to span between the B4 and B6 slabs.

Figure 5.15 VE incorporation of FD shaft into base slab.

Figure 5.16 BBC film unit for Open University.

5.4.5 A Bespoke Design for Progressive Modification

The scale of the cofferdam within the CTA is illustrated in Figure 5.9. It was necessary to address the potential risks from the range of newly weakened ground conditions, particularly the WC criteria which required establishing appropriate contingency plans. Upon completion of the ground treatment, 182 secant piles were installed to form the outer ring. The 255 bored piles for the base slab served both to support the slab structurally and minimise medium to long-term ground heave – the latter issue being critical to maintain acceptable alignment for the permanent trackwork. Rather than installing these piles as low cut-off from a higher level, they were installed from the base of the excavation during July 1996. This meant that all of the piling rigs, equipment and materials had to be lowered (and assembled as required) some 30 m below the ground surface. However, since the piling teams then had complete and unimpeded access over the whole area at formation level, the piles were installed in record time and construction of the base slab completed by September 1996.

The design was specifically developed and tailored to maximise its compatibility with the OM and thus enhance the acceptable performance of the cofferdam throughout construction. The critical observations were the lateral deflections of the piled wall. These deflections had to be assessed with regard to the trends in overall ground movements and the potential effects on adjacent structures. Emphasis was placed on simplicity and ease of monitoring. The importance of such aspects as construction rhythm and symmetry has already been noted. The particular conditions at Heathrow demanded a demonstrably robust design and one that could sustain, with appropriate pre-planned contingency measures, the WC ground conditions should they be encountered. However, given the key need to safely recover the delay caused by the collapse, the OM also offered major opportunities. This progressive modification facilitated further time savings by introducing advantageous design changes on the basis of feedback from measured performance during construction.

The application of the OM addressed three main aspects:

1. The principal objective to control the risks associated with such a major excavation.
2. Contingency measures. The method needed to allow timely implementation of any of the pre-planned contingency measures to maintain and control safety. However, given the robustness of the design coupled with the OM, it was expected that it would be possible to convincingly demonstrate that contingencies were not necessary, or at least, to minimise them and thus mitigate their effect on time and cost.
3. The potential for additional benefits by introducing design changes that would create extra time savings. Examples here were the increased liner depth, the early tunnel breakthroughs and the incorporation of the base slab of the fuel depot shaft (Figures 5.14 and 5.15).

The primary instrumentation comprised vertical arrays of inclinometers installed in the piles and adjacent ground along with precise levelling. The inclinometers were beam mounted electro-levels and, in the piles, were installed in every sixth one around the cofferdam. This enabled real-time monitoring of its overall performance. Secondary instrumentation involved piezometers, extensometers and spatial survey (Figure 5.8).

5.4.6 Observations, Trigger Levels and Contingencies

The critical quantities to be measured were deflection of the piled walls and the associated ground movements. These two factors relate to the flexibility of the structure and the global movements generated by the unloading created by the bulk excavation within the cofferdam. Parametric studies had indicated that, under the WC conditions, the maximum design bending moments would occur in the contiguous piled section at a depth of 22 m generating a deflection of around 75 mm. This was established as the limiting condition for acceptable performance of the cofferdam wall – i.e. the red limit. The computed value at this depth using assumed MC parameters was around 60 mm as shown in Figure 5.17. The red trigger level for this depth was accordingly set for a deflection of 60 mm. In terms of the traffic light system, described in the previous case histories, the amber trigger level for this depth was set at 45 mm thus setting the green zone for deflections up to this value. These applied specifically to the depth of 22 m, so it was imperative to be able to closely monitor the increasing deflections in concert with the cycles of excavation and, in particular, the identification of any adverse trends. In the event of such trends, two principal contingency measures were planned in advance. These were to introduce thicker stiffer reinforced concrete rings in the cofferdam lining and to excavate down the sides only thus creating a substantial time lag between the main central excavation. Construction of the reinforced concrete liner rings would then progress significantly ahead of the bulk excavation thus providing early support and limiting wall movement. To successfully implement such a process, if deemed necessary, would need early and reliable identification of movement trends. While the performance of the cofferdam was continuously monitored throughout the construction process, detailed reviews were planned for when the excavation depth reached 7 m and 15 m. This was to fully assess what trends were evident by these stages. Detection of any significantly adverse trends developing would then have led to implementation of contingency measures. In practice, no such trends developed. On the contrary, it proved possible to modify the design advantageously to enhance construction progress as discussed in Section 5.4.7.

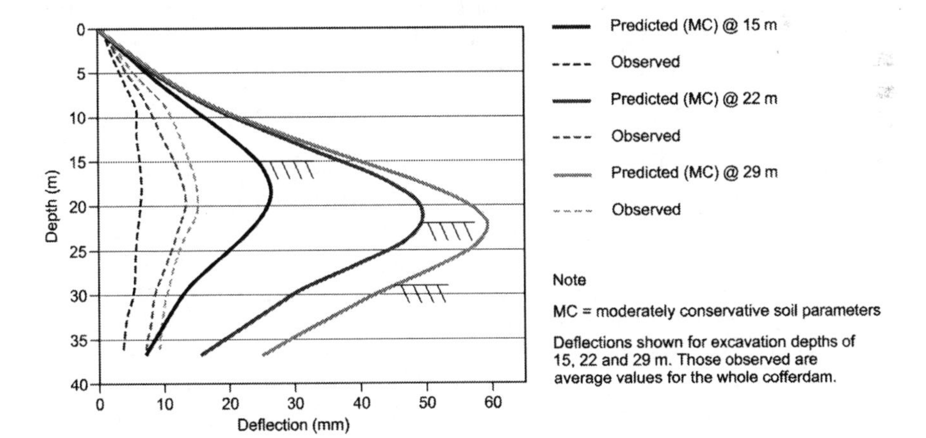

Figure 5.17 Measured vs computed wall deflections.

5.4.7 Measured Performance and Design Improvements

In practice, the cofferdam significantly out-performed all of the predictions computed from a wide range of analyses. As discussed below, the measured deflections were more than three times less than those predicted by analysis using the assumptions for MC conditions. The associated movement trend was already very evident at the 7 m depth review and enabled a variety of advantageous design changes to be implemented. The first change was to increase the depth of excavation and liner ring construction from 1 m to 1.2 m. In fact, it was evident from a design perspective, on the basis of the observed performance, that this increase could have been at least 50%. However, the depth of 1.2 m was adopted. At this value, the volume of bulk excavation reached the practical limit for the weekly cycle, noting that these substantial works were being undertaken at the centre of a busy airport which naturally imposed constraints on spoil removal. This change, allowing a faster rate of construction, was undertaken after completion of liner ring 9, all those thereafter being of the increased depth. This reduced the total number of rings from 29 to 22.

A comparison of the measured performance with the wall deflections determined for MC soil properties is shown in Figure 5.17. The observed values shown are highest maxima recorded and occurred adjacent to the substantial spoil heap situated close to the cofferdam. As discussed below, the associated surcharge would have significantly increased the lateral loading and hence the resulting deflection of the wall. So, the performance can be seen to be very much better than the range of predictions generated for MC soil parameters – especially since the analyses did not include the effects from the spoil heap.

Although the cofferdam offered an attractive structural simplicity, the variable soil/structure interaction together with the construction sequence presented great complexity. It was a challenge to produce an appropriate model for numerical analysis. Even assessing the performance of the cofferdam in hindsight, that is, using a 'Class C1' prediction (Lambe, 1973), replication of the measured profiles proved elusive even for state of the art software. Various finite element analyses were undertaken, in advance of construction, by the parties involved which included the consultants for the loss adjuster, the main designer and the independent checker. The results of these analyses, being based on estimates ranging from the most probable to WC soil properties, varied widely. They produced a range of maximum deflections, at the depth of 22 m, ranging from 45 to 115 mm. In practice, the typical maximum deflection of the piled walls around the cofferdam was only in the order of 15 mm. This compared with the analytic prediction of 60 mm based on MC parameters which thus over-predicted the measured values by a factor of 4. Interestingly, the measured wall deflections did not reflect the assumed effects of ground weakening arising from the collapse. There was no apparent correlation between the observed wall deflections and the areas of the most disturbed compared to the relatively undisturbed ground. In fact, the greatest deflection of the wall, measured around 24 mm, occurred adjacent to the temporary spoil heap. This created an additional lateral loading from the surcharge. It was understandably located in an area well away from collapse zones. Although the circular cofferdam brought a simple symmetry, the soil–structure interaction overall was very complex. The effectiveness of the ground treatment, for example, would have also added to this complexity. However, on a very basic and somewhat simplistic assessment of the observations, it did seem surprising that the wall deflections did not readily relate to the wide range of assumed soil strengths. The control of the lateral ground movements achieved compared very favourably with other case histories of deep excavations in London Clay (e.g. Burland and Hancock, 1977; Marchand, 1993).

Another major design change, based on the observed performance, was the introduction of early tunnel breakthroughs (Figure 5.14). The original design plan was to take the lining sequence completely down to base slab level thus maintaining the rhythm of construction and the ease of monitoring (Figure 5.11). However, with the performance so demonstrably robust, it was decided to break through into the cofferdam from the adjacent shafts at a much earlier stage than originally planned. The effects of the breakthroughs were carefully monitored by implementing each of them progressively in defined stages. The pilot tunnels were sequentially enlarged to full size and temporarily plugged with mass concrete to maintain ring action around the wall of the cofferdam. Early tunnel breakthroughs were thus safely completed substantially ahead of excavation within the cofferdam. Apart from advancing tunnel construction adjacent to the cofferdam, this allowed early progress for track work.

5.5 Conclusions

The cofferdam marked a comprehensive success in an integrated approach to design and construction on a high profile project. BAA merits particular recognition as the initiator for such a positive team environment and one in which safely managed innovation was strongly encouraged and supported.

The New Engineering Contract (1993) facilitated the introduction of design changes during construction. This combined with the creation of the single team culture made the conditions very conducive for application of the OM. 'Rethinking Construction' emphasises the need to measure performance. Teamwork and good communication are key factors for an improved and innovative construction industry. The inherent complexity of ground conditions will inevitably present uncertainties and risks.

Measuring actual performance through the OM can significantly improve the management of such risks. There is also a strong link between innovation and improved business opportunities. In this context, the HEX cofferdam was evaluated against the performance targets set by Egan. These were to reduce capital cost and construction time by 10%; increase predictability of outcome by 20%; reduce defects and accidents by 20% and increase productivity, turnover and profits by 10%. The HEX cofferdam scored well in all of these categories. The initiatives set out in the reports by Latham and Egan created an environment that was particularly conducive to the use of the OM.

The circular shape adopted for the cofferdam created simplicity, symmetry and rhythm. This in turn led to efficiency of function and ease of construction and its monitoring. Recovery of the works was achieved in a demonstrably safe manner with major savings being achieved in time and cost. Overall delay to the project was reduced by a whole year. The success provided a further strong example of the synergy between value engineering (VE) (ICE, 1996) and the OM (Powderham and Rutty, 1994). One of the VE alternatives introduced was the incorporation of the original 19 m diameter base slab for the fuel depot shaft into that for the cofferdam. The shaft itself was irretrievably damaged during the collapse but by exploiting the benefits enabled through the OM, its base slab proved salvageable as shown in Figure 5.15.

Progressive modification brought additional comfort and control in addressing the variable ground conditions and the uncertainties in soil/structure interaction. Contingencies were avoided and a range of design improvements were introduced during construction delivering substantial time savings. The main outcomes of the single team approach, for the Heathrow Express as a whole, are described by Powderham and Rust D'Eye (2003).

References

Brook, M. D., (1997). Partnering – are clients getting a fair deal? Construction Best Practice Programme Procurement Committee, April.

Burland, J. B. and Hancock, R. J. R. (1977). Geotechnical aspects of the design for the underground car park at the Palace of Westminster, London. *Structural Engineer*, 55, No. 2, 87–100.

Clark, I. (1979) *Practical Geostatistics*, Elsevier Applied Science, London, UK.

Egan, J. (1998). *Chairman, The construction task force*, Rethinking Construction, 40 pp, DETR, London.

Heyman, J. (1995). *The Stone Skeleton*, Cambridge University Press, Cambridge, UK.

Institution of Civil Engineers (1996), *Creating Value Engineering, ICE Design and Practice Guide*, Thomas Telford, London.

Kirmani, M. and Highfill, S. C. (1996). Design and construction of the circular cofferdam for ventilation building No.6 at the Ted Williams Tunnel. *BSCE/ASCE*, **II**, No. 3, 31–50.
Latham, M. (1994), *Constructing the Team*, pp. 130, HSMO, London, July.
Marchand, S. P. (1993) A deep basement in Aldersgate Street, London, Proc. ICE, Feb, 19–26.
New Engineering Contract (NEC). (1993). *The Institution of Civil Engineers*, The Institution of Civil Engineers, London.
Powderham, A. J. (1998). The OM – application through progressive modification. *BSCE/ASCE*, **13**, No. 2, 87–110.
Powderham, A. J. (2009). The Vienna Terzaghi Lecture, The Observational Method – using safety as a driver for innovation.
Powderham, A. J. and Rust D'Eye, C. (2003). Heathrow Express Cofferdam: innovation and delivery through the single team approach. *Proc BSCE/ASCE, Civil Engineering Practice, Spring/ Summer*, **18**, No. No.1, 25–50.
Powderham, A. J. and Rutty, P. C. (1994). The OM in value engineering, Proc. 5th. Int. Conf. Piling and Deep Foundations, Bruges.
Thomas, G. and Bone, W. (2000). Innovation at the cutting edge, CIRIA Research Report RP588, September.

Chapter 6

Heathrow Airport Multi-Storey Car Park 1A (1995–1996)

6.1 Introduction

The Heathrow Express (HEX) Rail Link provides a fast direct connection between the airport and central London. A principal component was the new underground station in the airport's Central Terminal Area (CTA). This involved three large diameter bored tunnels, constructed in close proximity (Figure 6.1). Within the CTA, the observational method (OM) was being applied on a global basis to a range of structures that were exposed to tunnelling-induced settlement. These included three multi-storey car parks (MSCP) 1, 1A and 3 (Figure 6.1). Only MSCP 1A suffered any visible damage. This building is effectively two structures separated by a central movement joint. The south western half sits on pad footings where it is located above the London Underground Piccadilly Line tunnels, while the north-eastern half of the building is founded on bored piles. It provides a unique case history of a very adverse building response to tunnelling-induced settlement in which relatively small ground movements caused substantial local damage. This applied to both types of foundation. The unexpected sensitivity also highlighted an important limitation in the use of the OM which was further compounded by an outlier beyond the general approach to managing risks of damage to buildings. All this arose despite the OM being implemented by an experienced team and in an 'OM aware' environment. The case history provides a dramatic comparison with that of the Mansion House in Chapter 3.

6.2 Key Aspects of Design and Construction

6.2.1 Overview

MSCP 1A is a five-storey structure of reinforced concrete (r.c.) construction and measuring 60 m by 142 m overall in plan (Figures 6.1– 6.3). The spatial relationship between the CTA station tunnels, the Piccadilly Line and the building is shown in Figures 6.1 and 6.4. Although the building

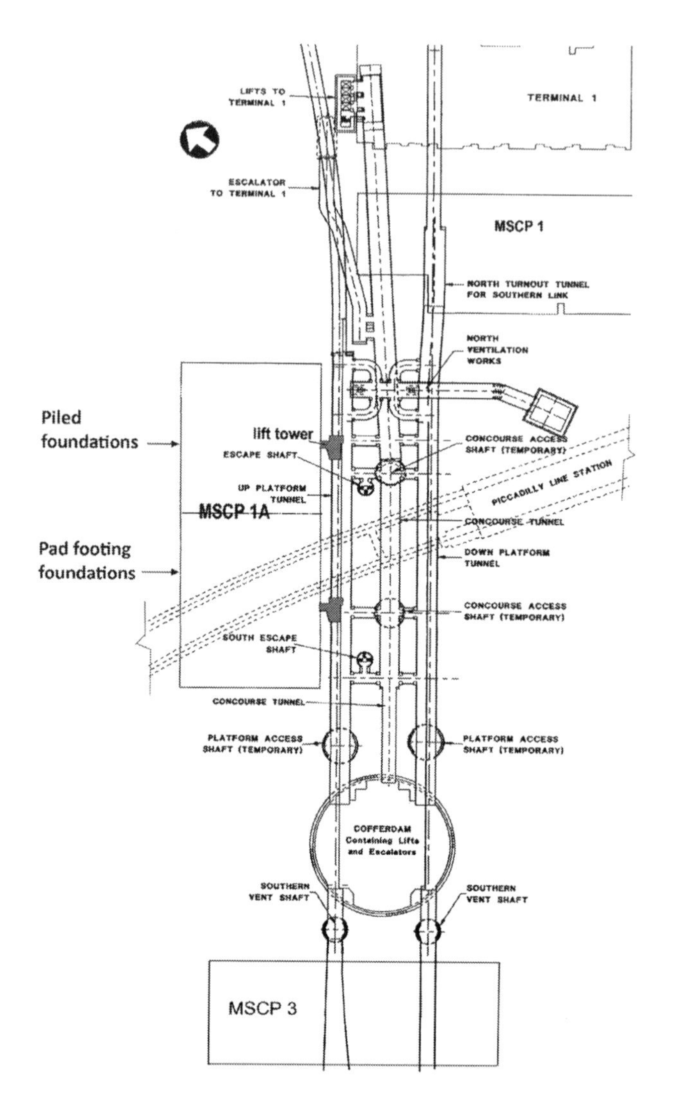

Figure 6.1 Layout of HEX CTA Station showing multi-storey car parks MSCP 1, 1A and 3.

has the appearance of a single structure, the foundations are different in each half. The south-western half of the car park together with its lift tower is founded on pad footings and was completed in October 1964. The north-eastern half of the car park and its lift tower have piled foundations and were completed in April 1965. Both lift towers are separated structurally from the car park.

The London Underground Piccadilly Line tunnels run beneath the south-western half of the building as shown in Figure 6.1 and were constructed in this area between 1971 and 1976 (Cooper *et al.*, 2002). These tunnels have a diameter of 3.8 m and are at about 13 m below existing ground level (Figure 6.4). The HEX CTA station comprises two 230 m long platform tunnels and a central 300 m long concourse tunnel. The HEX tunnels pass beneath MSCP 1 and MSCP 3, which are r.c. frame structures on pad footings being two and five stories, respectively (Figure 6.1).

6.2.2 Geology

Ground conditions across the site are relatively uniform consisting approximately 6 m of Terrace Gravel over London Clay – although there is some doubt regarding the depth and presence of the gravels below the north-eastern half of MSCP 1A, which may account for the use of piles. Ground water level is approximately 1 m above the top of the London Clay that, in this area, has a thickness of around 60 m (Powderham and Rankin, 1997). Except for the upper levels of the shafts and escalators, the HEX construction is within the clay that, at the depth of the running tunnels, has a shear strength of around 100 kN/m^2.

6.2.3 Tunnel Construction

The original tunnel construction for the CTA station was undertaken using sprayed concrete linings (SCL) as the primary support. This system was adopted following its successful use in tunnels around the world and on a trial tunnel for this project. However, in October 1994, some eight months after commencing the work, a major collapse of the tunnels at the CTA occurred. Some settlement was suffered by the Piccadilly Line tunnels, although the effects on MSCP 1A were minor with settlements in the range of 2–3 mm.

As part of the recovery operation, a 30 m deep, 60 m diameter cofferdam was constructed enclosing the majority of the collapsed area (Powderham and Rust D'Eye, 2003). The CTA Station tunnels were constructed from access shafts just north of the cofferdam. The 150 metres of partially collapsed concourse tunnel that had been filled with concrete following the incident were re-excavated using pneumatic breakers and lined with 8.0 m diameter cast iron segments. The remainder of the CTA Station tunnels was constructed with precast segmental linings. The 9 m diameter shields incorporated hydraulic forepoling, face gates and tables in the top half of the face to reduce settlement and a backhoe excavator in the lower half of the face. Pilot tunnels with a diameter of 3.3 m were also driven ahead of the main shields to further aid control of settlement (Deane *et al.*, 1997).

With regard to the induced settlement of car park MSCP 1A, the most crucial dates concern the construction of the upline tunnel which was closest to the car park. The pilot tunnel for this commenced 9 August 1995 and was completed beyond the zone of influence of the car park by 4 October 1995. The tunnel enlargement progressed past the car park on 22 December 1995 and was completed on 15 March 1996. Comparative dates for the downline tunnels were for the pilot: 3 November 1995 to 8 December 1995 and the enlargement: 8 March 1996 to 31 May 1996. Breaking out of the concrete for the concourse tunnel started from the southern-most shaft. Tunnelling commenced in both directions from this shaft from mid-January 1996. The concourse tunnel was completed past the car park in August 1996.

Compensation grouting was used to limit settlement during construction of the tunnels. The grouting was carried out from the south escape shaft to limit tilting of the southern lift tower, both pre collapse during 1994 and during the second half of 1995. Compensation grouting was also carried out from the north shaft following similar concerns regarding the northern lift tower. This commenced in February 1996. The resulting heave is particularly evident around chainage 110 (see Section 6.4.4).

6.2.4 Car Park Structure

The main structure of MSCP 1A is an in situ r.c. frame with flat slabs spanning monolithically between the beam and column construction. The columns are generally set out on a grid measuring 7.85 m by 5.49 m, with the longer dimension parallel to the tunnels. However, the outer line of columns on rows A and J is set 3.67 m from the edge of the building and is 9.68 m from the adjacent rows B and H. The columns are similar in both halves of the building and at the ground floor level are typically 0.46 m square.

As noted, the building foundations comprise two distinct types. In the south-western half of the building above the Piccadilly Line tunnels, the structure sits on discrete r.c. pads, which are founded just below ground level on the gravels. The pads are typically 2.9 m square and 0.84 m thick. The other half of the building is founded on piles. The r.c. pile caps, which are constructed at the same level as the pad footings, are rather more variable in size but the dimensions are typically 2.31 m square and 1.07 m thick with an arrangement of five piles beneath the pile caps.

While several of the pile caps are structurally tied together with r.c. ground beams running below the ground floor slabs, these connections are mostly in the central zones of each half of the building and, more particularly, where such r.c. ground beams are present they are aligned parallel to the CTA tunnels (see Figure 6.2). The pile caps of the north-eastern lift tower are also structurally tied together with r.c. ground beams aligned at right angles to the HEX tunnels. The significance of these structural details is discussed in Section 6.5.5.

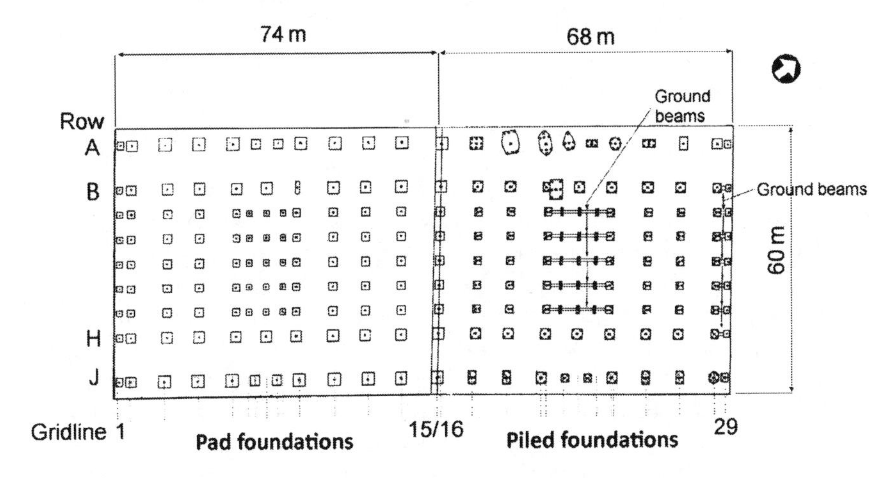

Figure 6.2 Plan of MSCP 1A showing foundation layout and limited locations of ground beams.

6.3 Achieving Agreement to Use the OM

As noted in Section 6.1, the OM was being extensively implemented in the CTA. This applied to the construction both before and after the collapse of the SCL station tunnels. Many hard lessons were quickly learnt in the detailed investigation that followed and prominent amongst these were the vital issues of effective communication and ownership. Correcting these failures was paramount. It is a real testament to the vision of the owner, BAA, that, following the collapse, a retreat into onerous conservatism was overtly eschewed and instead the enlightened development of the 'Single Team' approach was fostered (Powderham and Rust D'Eye, 2003). As described in the HEX cofferdam case history, this engendered a very conducive environment to introduce and implement the OM. Apart from aspects intrinsic to the tunnels, the OM was applied to the structures within their zones of influence and the associated construction including the HEX cofferdam and access shafts. Given the continued and even enhanced positive disposition by BAA towards the use of the OM, it was relatively straightforward to achieve agreement to apply it to the MSCPs.

6.4 Implementation of the OM

6.4.1 *Traffic Lights, Trigger Levels and Contingencies*

As with other applications for structures in the CTA, including the other MSCPs and the cofferdam, the OM was implemented using a traffic light system with the trigger levels defining the green, amber and red zones set on a reasonably conservative basis in relation the predicted performance. The actual performance within these parameters was measured and assessed progressively in line with the tunnelling construction. As was used for the OM at Mansion House, the primary system of instrumentation was based on strings of beam-mounted electro-levels and precise levelling. Similarly, the focus was on assessment of the angular distortion developed and the maintenance of this within an acceptable level of risk. These measurements were supplemented with condition surveys in the overall assessment. In the event of an adverse trend developing, this would be addressed in advance of reaching critical conditions by implementing contingency measures such as reducing the volume loss from tunnelling and/or installing an appropriate form of structural strengthening. At least this was the plan. Such plans are fully feasible when the generated ground movements and the response of the structure are mutually progressive. If there is a sudden or brittle event in either, then it is not generally possible to avoid the associated damage. That, quite unexpected, response proved to be the case for MSCP 1A. As described below, such sensitivity was not indicated in advance by the detailed risk assessments. In fact, of the three MSCPs in the CTA, MSCP 1A appeared to be exposed to the lowest risk. The predicted and measured values for the induced ground settlement for this building were quite close with the angular distortion being recorded at 1/1,000.

6.4.2 Risk Assessments

The risk assessments for Stage 1, based on Rankin (1988), were undertaken in early 1994 to identify which buildings might be adversely affected by the HEX construction at Heathrow. These involved several MSCPs, including MSCP 1, 1A and 3 (Figure 6.1).

Stage 2 assessments were then developed to provide more specific evaluations for individual structures. These assessments produced maximum predicted settlements, ground slopes and maximum tensile strains from calculations of ground movements based on expected tunnelling volume losses. They were undertaken in the same year but before the collapse of the CTA Station tunnels in October 1994 (HSE, 2000).

The Stage 2 assessments were based on the 'greenfield' settlement troughs. This conservatively assumes no beneficial effects from the modification of the trough from the soil/structure interaction with the affected buildings. Accordingly, the maximum ground slopes from the 'greenfield' settlement trough beneath the buildings were used to estimate the angular distortion. This evaluation was undertaken on a bay-by-bay basis. The maximum tensile strain was determined from the sum of the maximum bending strain in the building and the average tensile strain in the ground across the estimated width of the settlement trough. For MSCP 1A, this

Figure 6.3 MSCP 1A looking north, showing five storey structure and lift towers.

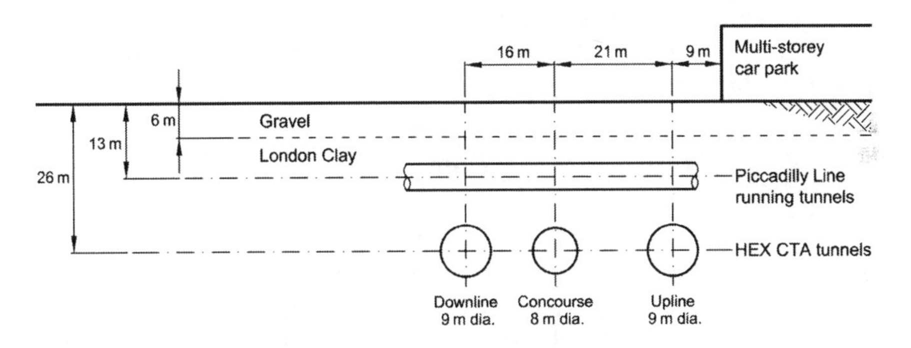

Figure 6.4 CTA Station tunnels in relation to car park MSCP 1A.

width was determined as 30 m. A check was also made for the combination of diagonal shear and horizontal strain. These are not directly additive but combined using a Mohr's circle. It is important to note, in this context, that the process of summing strains is based on the idealisation of the building (or specific parts of it such as a flank wall) as a simple elastic beam (Burland and Wroth, 1974). The depth of this notional beam would generally be taken as the full height of the main structure. Consequently, such an idealisation does not directly model the individual components of a framed structure but represents a generalised risk. Thus, in such a global assessment, diagonal tensile strains, for example, would typically relate to shear walls or infill panels. Such features were absent in the zone of the car park under consideration and so did not feature in the risk assessment. However, in practice, shear and bending failures were indeed a critical issue.

The combination of angular distortion and horizontal strain from the 'greenfield' computations was related to levels of risk of damage as set out by Boscardin and Cording (1989). Thus, although no direct measurements of horizontal strain were made, it was taken into account in the risk assessment using this approach. This essentially followed that developed for the Mansion House as described in Chapter 3 (see Figure 3.8 in that case history).

These assessments, which, as noted above, were derived on a conservative basis, indicated that MSCP 1A had the lowest predicted risk of damage for the three car parks while MSCP 1 had the highest. For MSCP 1, the predictions produced maxima of 33 mm settlement, 1:310 ground slope and tensile strain of 0.16%. Correlation with Boscardin and Cording (1989), placed it well into the risk category of 'moderate' damage. The comparative values for MSCP 1A were 15 mm, 1:790 and 0.06%, respectively, placing this building in the risk category of 'very slight' damage. However, with the proximity of the Piccadilly Line tunnels and the potential for some differential movement that might lead to some separation of the towers, it was decided to upgrade the risk of damage for MSCP 1A to 'slight'.

6.4.3 Predicted Ground Movements

Two key questions were:

a) How would the differing foundations between pad footings and piles influence the ground movements imposed on the buildings?
b) How would the sub-surface trough differ from that at the surface and thus potentially further influence the soil–structure interaction with the piles?

Surface 'greenfield' settlement, S, was estimated using the procedure outlined by O'Reilly and New (1982) using the following relationships: $S_{max} = (0.313 V_L D^2 / K_T\ z_o) = \delta_1$ and for the trough, $S = S_{max} \exp\ (-y^2 / 2 K_T^{\ 2} z_o^{\ 2})$, where z_o is the depth to the tunnel centreline, y is the transverse distance from it and D is the tunnel diameter. The width of the settlement profile is defined by the key parameter i $(=K_T z_0)$. This is the distance from the tunnel centreline to the point of inflexion of the trough. The total half-width of the settlement trough is effectively approximates to $2.5i$ (see Figure 6.5).

A volume loss (V_L) of 1.5% was assumed, with the trough width parameter, K_T of 0.5, being judged appropriate for stiff clay. To account for multiple tunnels, it was assumed that the ground movements for each tunnel were generated independently, and hence could be superimposed. To assess these effects on the piled structure, a sub-surface trough was

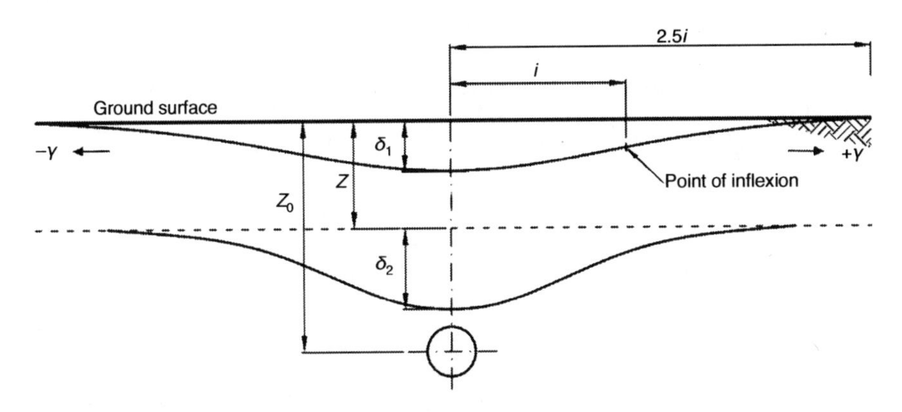

Figure 6.5 Forms of surface and sub-surface settlement profiles (after Mair *et al.*, 1993).

established for a depth of 15 m. This depth was the estimated pile length based on design loads found in archival material.

The sub-surface trough was based on Mair *et al.* (1993), who proposed that the trough width parameter, K_T, increases with depth such that $K_T = (0.175+0.325(1-z/z_o))/(1-z/z_o)$. The comparative forms of surface and sub-surface settlement profiles are shown in Figure 6.5 with their respective maxima S_{max} of δ_1 and δ_2. As can be seen, the sub-surface trough is the narrower but with a greater magnitude of settlement than the surface profile.

Field measurement during construction of the Jubilee Line Extension (JLE) gave good correlation with this relationship (Standing and Selman, 2001). The sub-surface trough is computed using $S = S_{max}/S.\exp(-y^2/2K_T{}^2(z_o-z)^2)$ where $S_{max}/r = 1.25V_L \times 1/\{0.175+0.325(1-z/z_o)\}\times(r/z_o)$.

A comparison of the computed 'greenfield' surface and sub-surface settlement troughs is given in Figure 6.6. The narrower sub-surface trough is shown with a greater magnitude of settlement at the upline tunnel axis (zero on the x-axis). However, it is interesting to note that while the sub-surface settlement is greater at the tunnel axis, at Gridline J, the magnitude of settlement is less for the sub-surface trough. The 'greenfield' estimates for horizontal ground movement are shown in Figure 6.7.

6.4.4 Measured Settlements

Precise levelling was undertaken from June 1994 until August 1996. The levelling points are shown in Figure 6.8. Plots of the settlement recorded along Gridline J for certain key dates relating to tunnel construction are shown in Figure 6.9.

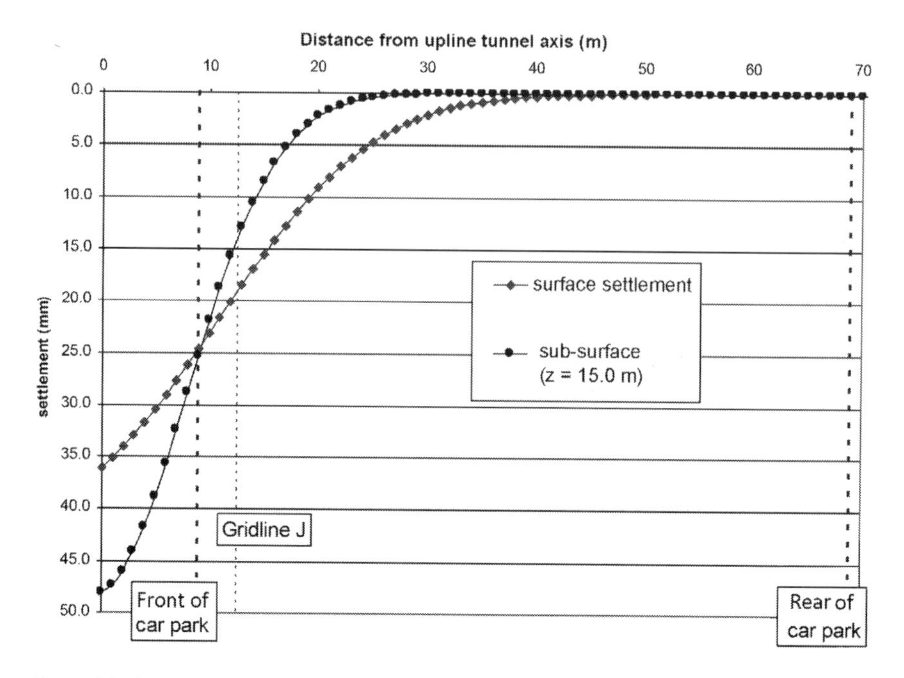

Figure 6.6 Comparison of 'greenfield' surface and sub-surface settlement profiles.

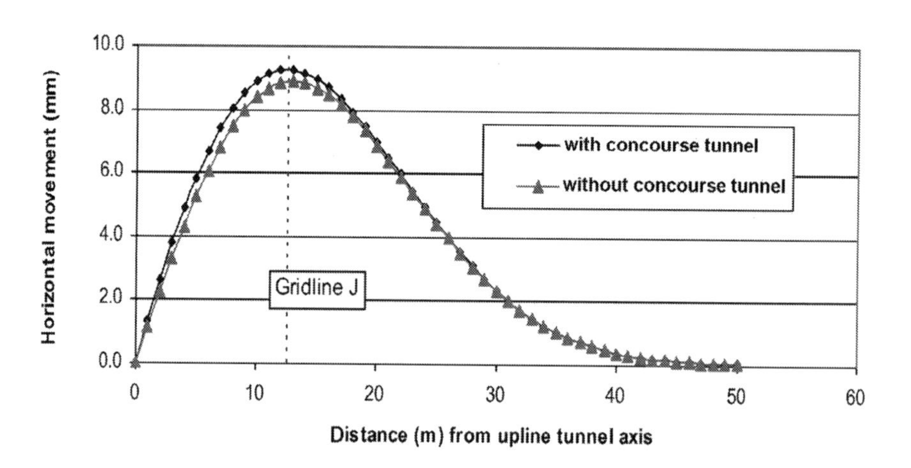

Figure 6.7 Horizontal movement computed for 'greenfield' conditions.

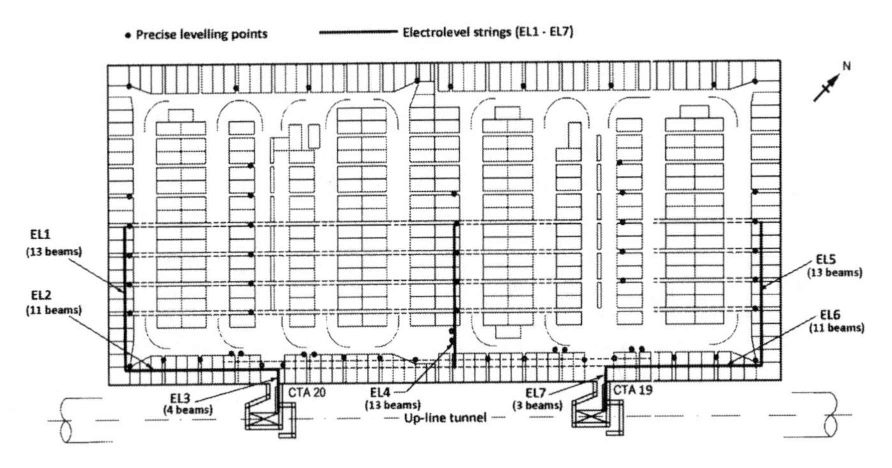

Figure 6.8 Electro-level strings and precise levelling points.

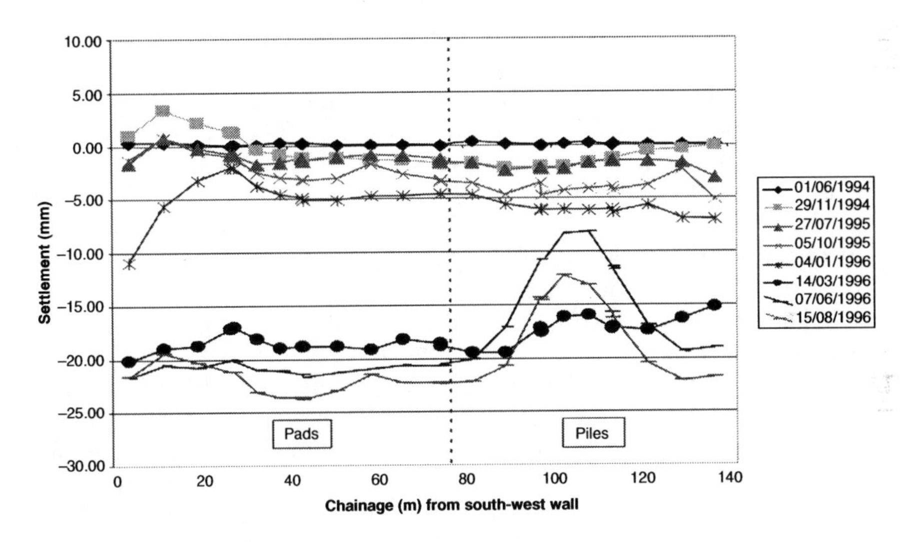

Figure 6.9 Measured settlements along Gridline J.

The line for 1 June 1994 indicates that at this time there had been no movement recorded along Gridline J. By 29 November 1994, the collapse of the HEX station tunnels in the CTA had occurred, and settlements in the range of 2–3 mm were recorded as shown. These plots also indicate that localised uplift from compensation grouting was evident at the south-western end of the building. As the concourse tunnel had not been constructed over the full length of the building at the time of the collapse, there was little to no settlement then evident at the north-eastern end.

The lines for 27 July 1995 and 5 October 1995 correspond to pre and post construction of the upline pilot tunnel. Settlements in the order of 3 mm were generally evident between these dates. The exception is the south-western end where virtually no settlement was recorded over this period. This is likely to be the result of compensation grouting.

Upline tunnel enlargement started on 22 December 1995, but the first profile obtained for this period is the line shown for 4 January 1996. The 'dip' in the line at the south-western end is due to the upline tunnel enlargement having already started to proceed past these points. The line dated 14 March 1996 shows the measured settlements by completion of the upline enlargement. Settlement for this period was around 16 mm, with no big differences apparent between the piled and pad footings. Settlements at the overall completion of tunnelling within the zone of influence, concluding with the concourse tunnel, are shown by the line dated 7 June 1996, and the post tunnelling situation by the line of 15 August 1996. The effect of the compensation grouting is evident for the latter three profiles, with some significant heave recorded close to the northern lift tower. Some compensation grouting was carried out from the southern escape shaft up to this period, but the last time that grouting had been performed from the northern escape shaft was 7 June 1996.

The profiles indicate that the pad foundations suffered slightly more settlement than the piles. Ignoring the more obvious effects of compensation grouting, this difference in vertical settlement appears to be in the region of 3 mm. Figure 6.10 shows a comparison of the measured settlement profiles of 15 August 1996 and the predicted 'greenfield' troughs for surface and sub-surface. It can be seen that the actual settlement for both pad and piled foundations is somewhat greater than the 'greenfield' prediction. However, it should be noted that the 'greenfield' prediction is based on immediate settlement, whereas the measured troughs in Figure 6.10 represent the medium term where some consolidation would have occurred. It can be seen that the measured values in the period immediately following the upline tunnel enlargement closely matched those computed for the 'greenfield' surface prediction (Huggins, 2002). This provides significant evidence that the inherent structural flexibility of the building frame as a whole presented a general inability to modify the settlement trough through soil/structure interaction.

Generally, horizontal ground strains and the degree of their transmission into the foundations of a building are challenging to measure

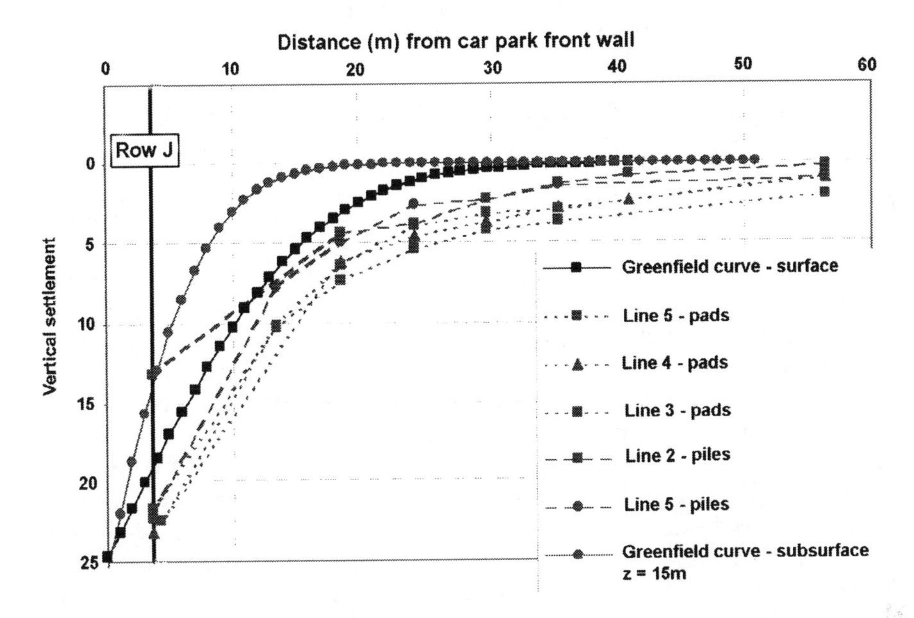

Figure 6.10 Comparison of 'greenfield' prediction with actual vertical settlement.

reliably. Given that the initial predicted level of risk for MSCP 1A was very slight, no direct measurement of horizontal movement was considered necessary prior to tunnel construction. This approach accorded with established practice for managing the risk of damage to buildings from induced settlement – especially for such low levels of perceived risk. However, in hindsight, given the unusual sensitivity of the building that was revealed, this proved to be an error. A detailed assessment of the nature and locations of the cracking in the ground floor columns post construction was undertaken. It strongly indicated, contrary to case history evidence, that most of the horizontal tensile strain generated in the ground was transmitted to the foundations. Under such conditions, given the high structural stiffness of the affected columns, very high bending moments and shear forces would be generated from quite small movements. An optical plumb survey of the columns was performed after completion of tunnelling during October 1996. However, the results were inconclusive since the verticality of the columns was not known prior to the construction of the tunnels. A survey of the current verticality of columns outside the zone of influence did provide a potential reference base and lateral movements in the order of 5 to 10 mm of the column bases in row J were broadly indicated. This degree of horizontal movement is generally in line with that computed for 'greenfield' conditions (Figure 6.7).

Thus, it does not intrinsically invalidate the usual assumptions which, conventionally, would be considered conservative.

6.5 Effects on the Building and Remedial Measures

6.5.1 Damage Assessment

As part of the damage assessment performed in October 1996, individual columns along row J were surveyed for cracking. Where cracks occurred they generally exhibit a pattern similar to that indicated in Figure 6.11. Table 6.1 provides a summary of crack details.

As noted, the columns along row J suffered a combination of horizontal and diagonal cracks. These are consistent with an imposed sway caused by the lateral movement of the footings. The orientation of the diagonal cracks was consistent throughout occurring in the column faces at right angles to the alignment of the tunnels. The columns most severely affected were those located at the south-west and north-east ends of the building with the columns on piles being, on average, more severely affected than those on pad footings. However, it should be noted that some variation in the condition of the structure could be relevant, including the quality of original construction and subsequent deterioration by de-icing salts. The factors of safety of the worst affected columns were assessed to have been reduced to around slightly over 1.3 under dead-load conditions and less than 1.2 for full live loading. A normal value would be in the range of 2–2.5.

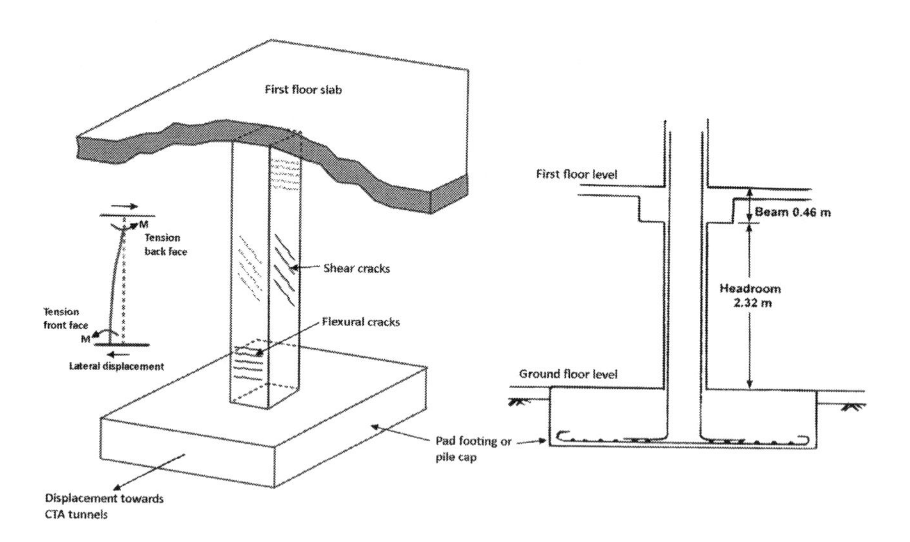

Figure 6.11 Typical pattern of cracking in row J columns and cross section.

Table 6.1 Summary of column crack details along row J.

Column	Flexural Cracking		'Shear' cracking		Total cumulative crack width (mm)
	South	North	East	West	
PADS					
J1	x	x	x	x	0.55
J2	x		x	x	0.95
J3	x	x	x	x	0.75
J4	x	x	x	x	0.65
J5	x	x			0.20
J6					0.60
J7/8	x				0.45
J8/9	x	x	x	x	0.50
J10	x		x		0.15
J11					0.00
J12	x				0.00
J13	x				0.00
J14	x	x			0.20
J15	x				0.10
J16		x			0.15
PILES					
J17	x	x	x	x	0.85
J18	x		x	x	0.60
J19	x				0.00
J20	x				0.00
J21/22	x		x		0.00
J22/23			x	x	0.40
J24	x				0.00
J25		x			0.15
J26	x	x	x	x	3.10
J27	x	x	x	x	2.55
J28	x	x	x	x	2.30
J29	x	x	x	x	1.35

PADS:

Flexural Cracking	
South	North
80%	47%

'Shear' cracking	
East	West
40%	33%

Mean crack width
0.35 mm

PILES:

Flexural Cracking	
South	North
83%	50%

'Shear' cracking	
East	West
67%	58%

Mean crack width
0.94 mm

6.5.2 Remedial Measures

During this period, the overall strategy for the airport car parking was under review. This included a combination of maintenance and re-build. It was not fully established at the time what design life requirements would be assigned to specific car parks particularly those in the CTA. For MSCP 1A, following the completion of the CTA station tunnels, it was difficult to quantify what effects any ongoing settlement trends might have. There were still also further adjacent tunnelling works to be completed. Consequently, it was decided in the interim to maintain the car park operational and encase the five worst affected columns in steel jackets. The comprehensive programme of monitoring included strain gauges on the steel jackets and a detailed log of all cracks remaining visible together with vertical and horizontal movements of the structure.

6.5.3 Structural Response: Actual vs Predicted

As previously noted, the level of potential damage predicted did not raise a cause for concern even with the more detailed Stage 2 assessments.

These indicated a risk category of very slight damage. However, although damage was confined to the columns in row J, it was locally severe. The basic process described here for assessing the risk of building damage from tunnelling adopted for the HEX was subsequently refined and developed on a more global basis for the JLE with an extensive research programme (Burland, 2001). For the JLE, a consistent pattern was evident. Almost all of the damage reported for the buildings studied by the research team was in the category of slight or less. Most of the reported damage related to discontinuities in structural integrity – notably at connections between adjacent structures particularly where they had mutually different founding conditions. Consequently, this process of predicting the risk and categories of damage is considered to be reasonably conservative. Furthermore, it was noted that for buildings founded on rafts or strip footings, there was overwhelming evidence that little or no horizontal ground strains are transmitted into the building. However, it was also concluded that further measurements were required for buildings founded on isolated pad footings or piles (Burland, 2001).

The risk assessment for MSCP 1A, based conservatively on the 'greenfield' settlement trough, effectively placed it in the category of very slight damage. This makes the actual response a far outlier. While in hindsight, the probable causes are evident, this case history presents an exceptional case of building response to excavation-induced movement. The results merit careful consideration and are discussed below.

6.5.4 Background to the Risk Assessment

a) **Precedent.** It is quite normal for buildings and other structures to be exposed to the combined effects of settlement and tensile ground strain from adjacent excavation. However, for a wide range of structures exposed to the degree of ground movement in this case history, little or no damage has been experienced. Powderham *et al.*, (2004) were aware of no similar instance – either in the nature and severity of the damage or the marked disparity between the expected level of damage and the results. Indeed, the damage categories are based on the data from field evidence.
b) **Local experience.** Many similar structures at Heathrow were exposed to higher levels of apparent risk from similar but more onerous ground movements. No damage was noted to these structures. The tunnelling had already been completed under MSCP 3 before MSCP 1A with no adverse effects apparent and both MSCP 1 and MSCP 3 had notably higher levels of potential damage predicted. MSCP 1A itself would have been exposed to settlement and tensile ground strain

from the construction of the Piccadilly Line tunnels in the late 1970s but no damage relating to this was evident.

c) **Inherent robustness.** The codes of practice require an appropriate degree of robustness that, apart from the usual factors of safety on individual elements, includes resistance to accidental lateral loading. For car parks, vehicular impact is an issue and foundations should be designed to provide sufficient lateral resistance to such events. This is achieved, for example, by the use of ground beams to tie adjacent discrete foundations together. By whatever means this robustness is provided, it could normally be expected to provide adequate resistance to low levels of tensile strain. Even if, in the case of MSCP 1A, all of the predicted tensile ground strain is concentrated within the bay between rows J and H, and thus averaged over 10 m rather than 30 m, the maximum combined tensile strain would have only amounted to around 0.07%. This would still place the potential for damage in the very slight category (Mair *et al.*, 1996), indicating no real change in level of risk. Reinforced concrete framed structures are inherently more flexible than masonry structures and are consequently able to accommodate small movements without compromising their integrity.

6.5.5 Evaluation of Actual Response

With the benefit of hindsight, aided by the data obtained from the OM, the reasons behind the observed response of the building can be evaluated. Typically, we do not have complete information and the actual behaviour would have derived from the combination of several factors, some of them quite complex. These shortfalls would include the present state of the building, its stress history and the soil/structure interaction – particularly that of the piled foundations. Despite these limitations to providing definitive answers, it is considered that a reasonably consistent picture emerges. The immediate questions are as follows: 'Why did the damage occur and why was it confined to the particular columns of this structure?'

Taking the various aspects in turn:

a) The columns on row J. It is evident from the recorded damage that local lateral movement of the isolated footings on row J produced exceptionally high bending moments and shears in the columns spanning between the ground and first floors. On a simple basis, it can be shown that a small amount of sway concentrated in such a short, stiff column would cause such effects. Rotation of the footings along row J was not evident from inspection of the structure after tunnelling. It is relevant to note in this context that the ground slope and the deflection of the first floor would have tended to be additive to the moments generated from sway of the columns. However, the lateral force required to cause sway is considerable. For a sway of only 5 mm, the shear force generated would be around 500 kN. Although the tensile ground strain was low, the predicted horizontal movement along row J was around 9 mm and the measured values were generally indicated in the range of 5–9 mm. If we consider the case of the pad footings, the implications are that: (i) the frictional force between the footing and the ground was sufficiently higher than the combined restraining force of the structure at this location and (ii) there was hardly any shearing movement required in the ground to mobilise this frictional force.

The first criterion can be met if we consider each column individually with no benefit from any restraining force in the rest of the structure. There are no ground beams providing structural ties between rows J and H. The combined dead load of the column and the footing would be around 2800 kN and a coefficient of friction 0.3 would provide enough force to sway the column.

The soil/structure interaction for the piled footings is inherently more complex but a similar case could be made. It is relevant to note, for both types of footing, that the induced lateral forces would not have been sufficient to move two pad footings simultaneously. In other words, if just the footings of rows J and H had been adequately tied together, the tensile strain would have been largely dissipated in the ground with little or no effect on the structure. This is what normally happens since, in the vast majority of cases, the foundations of buildings are significantly stiffer than the ground. This has been noted even in potentially sensitive structures such as old masonry buildings where they have been located on the limbs of settlement troughs and subjected to the combined effects of hogging deformation and tensile ground strain. For example, no adverse effects were recorded for the Mansion House from the construction of the deep tunnels for the Docklands Light Railway Extension to Bank in 1990 (See Chapter 3).

b) Other structural elements of MSCP 1A. It is relevant to question why the damage was wholly confined to the columns on row J. While the predicted movement from tensile ground strain was less at row H, it was still over 6 mm and apparently enough, on the scenario above, to cause some damage to this line of columns. No damage was evident. This may be partly explained by consideration of the spatial development of the settlement trough with time. For the footings of row H to move laterally towards the CTA tunnels, the ground floor slab that abuts them must also move. Apart from self-weight friction with the ground, this slab would be restrained by those footings on row J that the progressing wave of the settlement trough had not yet fully reached. Such considerations could also explain why there was no damage apparent in the columns in the south-western half of the car park from the construction of the Piccadilly Line tunnels. The other potentially vulnerable elements of the structure would have been the lift towers. As noted, these were already receiving extra focus because of the risk of differential settlement. The preventative measure of compensation grouting had been implemented to address this. Perhaps more relevant to the issue of tensile ground strain, however, was that the discrete footings of the towers were robustly tied together with ground beams. This could also explain why the columns in row J were least affected near the towers. At these locations, with the tensile ground strain being absorbed in the ground beneath the towers, much reduced effects would have been transmitted to the adjacent columns in row J. The tied foundations of the towers would have produced a shielding effect as well as a possible abutting reaction for the footings on row J.

6.5.6 Key Lessons Learned

It is a frequently voiced truism that we learn more from failures than successes – as noted, for example, by Petroski (1985). Such lessons are understandably harder and typically create more publicity than those from success – especially if the failure is close to home. However, lessons learned from success are considerably more agreeable. As the vast majority of case histories in this volume clearly demonstrate, the successes of the varied applications of the OM have, apart from the prime objectives, yielded the by-product of a rich trove of important lessons.

The failure of the application of the OM for MSCP 1A was partial in that the building did not suffer catastrophic damage but, with some precautionary strengthening, remained functional for the rest of its service life. This had already been downgraded because of the existing deterioration of the building structure from corrosion plus the overall plan of refurbishment and rebuild for the Heathrow terminals in general. This, in turn, had mitigated the owner's usual concerns about minimising the risk of potential damage from the tunnelling. Some limited damage, such as a few new cracks or widening of existing ones, was considered acceptable compared to intrusive and expensive preventative works – as long as there was no compromise to safety. It was felt that some minor damage of increased incidence of cracking would only marginally affect an already compromised serviceability limit state in the r.c. structure. In practice, the actual damage, though far more severe than expected, did not compromise the overall plan of staged renewal. Nevertheless, the failure was dramatic and, as noted, required some local strengthening. And, the failure did deliver important lessons – both directly to the application of the OM and also highlighted a general disparity in the approach to managing risk of building damage. These lessons are summarised as follows:

(a) Regarding the application of the OM. In practice, the actual response of the structure was one impractical to address effectively through the OM. Two key errors were made. Firstly, the most critical observation was not adequately identified. This was the imposed horizontal movement. While it had been notionally catered for through the combination of angular distortion with horizontal tensile strain – as successfully adopted for the Mansion House – its transmission into the structure was not measured directly. Secondly, the response in the structure was sudden and brittle involving extensive cracking in both bending and shear. While not presented as excuses, it is instructive to consider the understandable mitigating reasons behind these errors. The overall approach to managing the risks of building damage in these circumstances is broadly conceptual. The soil/structure interaction is too complex to be addressed purely by analysis, and it is accepted that there will always be inevitable unknowns. Success requires good engineering judgement. Such judgement is distilled from theory, observation and experience. Here, the theory was based on an established conservative approach. Detailed observations of the contemporary performance of identical structures in the CTA and the experience derived from an extensive range of previous case histories did not indicate any issues relating to imposed horizontal movements. The response of MSCP 1A was unique at the time and consequently very challenging to identify in advance. To the best of the authors' knowledge from their international networks, it remains a unique outlier today. This case history certainly constitutes a failure in the application of the OM. However, it may be characterised as a partial one since the building, though subjected to severe damage locally, remained safely functional. Even though the application of the OM was critically compromised, it still managed to deliver significant benefits. Granted it was not possible to prevent the onset and development of visible damage. However, the data from the detailed monitoring and condition surveys for the OM enabled continual close assessment of the building response. It thus provided the possibility of taking emergency measures if the trends were judged to be moving towards initiating unsafe conditions. Such measures would have involved a temporary halt to tunnelling and the installation of ties between column bases and/or temporary structural steel columns. The other benefit from the OM application was that the same data facilitated a detailed post assessment of the building response thus providing a much more informed understanding of the underlying causes.

(b) In relation to managing risk of building damage, historically, the development of the approach to protecting buildings from the risk of damage from settlement induced by excavation has focussed primarily on stone masonry and brick structures. This is understandable since they generally comprise the predominant stock of buildings in our cities which are most likely to be exposed to such settlement. They also constitute some of the most important and sensitive buildings of national heritage. In the context of a r.c. structure such as MSCP 1A, there are a range of issues to consider: (1) The established classification for categories of severity is based on ease of repair of visible damage. It specifically applies to masonry and not to other structures such as those of r.c. and structural steel – the common form being frame buildings. (2) This differentiation is particularly important. It is essential to appreciate that there are profound differences between the structural performance of masonry compared to these other structural forms (Heyman, 1995). They follow very different criteria. These may be summarised as follows:

- The assumptions for masonry are that: (i) it essentially acts in compression and has negligible tensile strength; (ii) its performance is principally governed by stability acting through lines of thrust in which the correct proportions of geometry are fundamental, and (iii) cracking is an inherent feature. Its presence is inevitable and where it does not conflict with the criteria in (ii) above has no deleterious effect on structural performance. Cosmetic repairs for such cracks, which may be centimetres wide, can be easily addressed by filling them with mortar.
- Reinforced concrete and structural steel design comply to very distinct criteria. In particular: (i) structurally, they are governed by two limit states – one for serviceability (the SLS) and the other for the ultimate conditions; (ii) such buildings are typically framed structures constructed with an assemblage of beams and columns. These have both significant tensile and compressive strength and can thus sustain substantial bending moments; (iii) crack widths in r.c. are strictly limited under SLS – typically to 0.3 mm or less – to comply with durability criteria. Cracks significantly wider than these limits would at least compromise durability and would probably indicate unacceptable structural overstress. Such conditions, in contrast to masonry, cannot be simply addressed by filling them with mortar; (iv) Compared to masonry, such frame structures are inherently flexible.

Despite the range of marked differences listed above, the same approach to manage the risk of damage to buildings is generally adopted for these different structural forms – certainly, at least, in the initial risk assessments. In fact, although fundamental, these differences are often ignored or discounted since they generally do not become overtly manifest even with assessments of actual performance. This is because the inherent flexibility of framed structures allows them to accommodate the relatively small movements without compromising their design limit states. Where elements within these framed buildings are relatively inflexible, such as shear walls, they can be treated as structural masonry. Nevertheless, these differences are striking and they became very evident in the case history of MSCP 1A.

Another potential irony may be seen here in that the actual movements induced in this structure were very close to those computed for 'greenfield' conditions. Even the measured crack widths, as indicated in Table 6.1, generally ranged up to 1 mm, with a mean width of 0.94 mm. Thus, apart from some limited local cracking in excess of this in columns in the north-eastern corner, the crack widths were broadly in compliance with the initially assigned category of very slight risk. Thus, it may simplistically be assumed that, since the conservatively assumed level of risk had not been exceeded, then any associated damage would have been quite acceptable. Such a conclusion may seem temptingly tenable. However, for MSCP 1A, due to very high local bending stiffness being combined with an atypical lack of horizontal restraint, these differences were dramatically emphasised. This resulted in the mean crack widths exceeding the SLS value of 0.3 mm by a factor of three, thus seriously compromising this limit state. Such cracking would be structurally admissible in a masonry structure as the associated maximum horizontal deflection of around 10 mm would have not affected stability. However, it represented a level of severe damage in the r.c. column and which could not be repaired by simply filling in the cracks.

Thus, a striking disparity, starkly highlighted in hindsight by this case history, is the inherent vulnerability of r.c. frame buildings founded on isolated footings when these are not secured to resist horizontal tensile strains imposed from the ground. Counter-intuitively, this vulnerability is greatly enhanced if the structural members involved are very stiff. This is essentially the opposite of the structural performance of masonry buildings. They are typically more massive, which intrinsically tends to reduce horizontal movement. Moreover, masonry can typically accommodate the relatively small movements involved without compromising the key criterion of stability. More generally, this case history also underlines the importance of nurturing good communication and integration between areas of specialist expertise – for example between geotechnical and structural engineering. Further commentary on this subject is provided by Burland (2006).

A key issue arising from this review is the particular importance of structural types and how realistically they are represented by the idealised model being used to analyse the building. This applies to both simple and complex models. In simpler models, the gap between the idealisation and reality is generally addressed by a higher degree of conservatism in the overall approach. However, in all models, a disparity with the nature of the real structure may be significant and potentially critical. This particularly applies in the assessment and management of the risk of damage.

With MSCP 1A, the combination of short stiff columns constructed monolithically with isolated footings produced a local but very high sensitivity to horizontal ground movement. This effect was recorded in columns founded on both pad and piled footings. There was some indication that those founded on piles suffered more damage on average than those with pad footings, but there was no evidence that the vertical settlements in either case played a significant role. Although the piles were sited well above the tunnel horizon, the vertical settlements were similar to those of the pad footings. While the shear and flexural cracking to the outer row of columns subsequently led to greater sensitivity to seasonal temperature variation, the damage as a direct result from the construction of the CTA tunnels was essentially confined to the short term with no significant effects being noted from the longer-term settlement.

6.6 Conclusions

6.6.1 A Triple Irony

It is a curious combination of ironies that:

(a) Using a well-established approach based on 'greenfield' settlements and one generally agreed to be comfortably conservative, MSCP 1A was predicted to be exposed to the least risk the three car parks. And yet it was the only one that suffered any observed damage. This damage, while local to the ground floor columns, was both extensive and severe.
(b) The predicted level of risk was so low at 'very slight' that it was not expected to require any preventative measures to maintain the risk of damage within acceptable levels – especially for a utilitarian structure such as a MSCP. This view was reinforced since almost identical structures had been exposed to significantly higher predicted levels of risk from the tunnelling for HEX. These other structures had all survived unscathed. Such results were also assessed to apply to the construction of the Piccadilly Line tunnels in the late 1970s beneath MSCP 1A itself. Here, no damage to the building was evident from condition surveys.
(c) Finally, even though the induced settlement trough closely matched the presumed conservative prediction for 'greenfield' conditions, the

severity of the actual response was unexpected, sudden and brittle. In this, it breached one of the cardinal limitations of the OM thus rendering its prime objective ineffective.

6.6.2 Beware the Oddball – Reflections on Peck (1998)

This case history has highlighted the potential for an unusual but critical response to tunnelling-induced ground movement. It was and still remains a risk that would not be routinely identified in an initial evaluation of risk using the generalised assumptions for the structural integrity of a typical building. The MSCPs in the CTA appeared to be as typical construction of their kind and all of them, including MSCP 1A, had performed as expected. Despite tunnelling within their zones of influence, no induced damage was evident. That was the observed situation up to the commencement of the CTA station tunnels adjacent to MSCP 1A. It was then dramatically demonstrated that MSCP 1A did not respond in a typical way. It had also been expected that, because of the modifying benefits from the stiffness of the structure, the actual damage would be less than that indicated from the category assigned on a 'greenfield' basis. Moreover, based on the evidence from the large body of field data, detailed evaluation beyond the Stage 1 assessments would not normally be undertaken for such utilitarian structures with an assigned category of slight risk or less.

Certainly, flexible buildings, such as those with r.c. or structural steel frames, will tend to impose less modification to the vertical components of induced settlement in comparison to the very stiff response of a monolithic masonry building. In general, any such modification tends to be beneficial as it flattens out the settlement trough and thus the imposed distortion on the building. Similarly, with respect to the horizontal components of ground strain, the foundations of most types of building typically resist the associated movements. This usually substantially reduces or even effectively eliminates any significant contribution from horizontal tensile ground strains to cause damage. This includes flexible buildings since they are typically founded on rafts or tied footings. Paradoxically, in the particular case of MSCP 1A, it was the high local stiffness of the ground floor columns that proved a dominant factor in the problem. Other elements of the building suffered greater deformation – such as the first floor slabs vertically – but being adequately flexible, no damage was evident.

References

Boscardin, M. D. and Cording, E. J. (1989). Buildings response to excavation-induced settlement ASCE. *Journal of Geotechnical Engineering, American Society of Civil Engineers*, **115**, No. 1, January, 1–21.

Burland, J. B., (2001), Assessment methods used in design, Building response to tunnelling – case studies from construction of the Jubilee line extension, Volume 1. Thomas Telford, London.

Burland, J. B., (2006), Interaction between structural and geotechnical engineers, The Structural Engineer, 18, April, Presented at IStructE/ICE Annual Joint Meeting 26, April 2006, London, UK.

Cooper, M. L., Chapman, D. N., Rogers, C. D. F. and Chan, A. H. C. (2002). Movements in the piccadilly line tunnels due to the heathrow express construction. *Géotechnique*, **52**, No. 4, 243–257.

Deane, A. P., Myers, A. G. and Tipper, G. C. (1997). Tunnels for heathrow express, Proceedings Tunnelling '97, London.

Heyman, J. (1995). *The Stone Skeleton*, Cambridge University Press, Cambridge, UK.

Huggins, M. F., (2002), A comparison of tunnel induced subsidence of two identical structures on pad and pile foundations, MSc Thesis, Imperial College.

Mair, R. J., Taylor, R. N. and Bracegirdle, A. (1993). Subsurface settlement profiles above tunnels in clays. *Géotechnique*, **43**, No. 2, 315–320.

Mair, R. J., Taylor, R. N. and Burland, J. B. (1996). Prediction of ground movements and assessment of risk of building damage due to bored tunnelling. *Geotechnical Aspects of Underground Construction in Soft Ground.* Mair and Taylor (editor), Balkelma, Rotterdam, pp. 713–718.

O'Reilly, M. P. and New, B. M. (1982). Settlements above tunnels in the United Kingdom, their magnitude and prediction, Proc. Tunnelling '82, Brighton, pp. 173–181.

Peck, R. B. (1998). Beware the oddball, Urban geotechnology and rehabilitation, Seminar, ASCE, Publication No. 230, Metropolitan Section, Geotech Group, New York.

Petroski, H. (1985). *To Engineer Is Human – The Role of Failure in Successful Design*, St Martin's Press, New York.

Powderham, A. J., Huggins, M. F. and Burland, J. B. (2004), Induced subsidence on a multi-storey car park, The Skempton Conference, 29–31 March, London, UK.

Powderham, A. J. and Rankin, W. J. (1997), Heathrow collapse recovery solution cofferdam – planning design and implementation, Proc, International Conference on Foundation Failures, Institute of Engineers, Singapore, 12–13 May.

Powderham, A. J. and Rust D'Eye, C. (2003). Heathrow express cofferdam: innovation and delivery through the single team approach. *Proceedngs of BSCE/ASCE, Civil Engineering Practice, Spring/Summer*, **18**, No.1, 25–50.

Standing, J. R. and Selman, R. (2001). The response to tunnelling of existing tunnels at Waterloo and Westminster. *Building Response to Tunnelling – Case Studies from Construction of the Jubilee Line Extension*, Volume 2, Burland, J. B., Standing, J. R. and Jardine, F. M. (editors), Thomas Telford, London, pp. 509–546.

Chapter 7

Boston Central Artery Tunnel Jacking (1991–2001)

7.1 Introduction

The Central Artery Tunnels (CA/T) in Boston, Massachusetts, constructed between 1991 and 2007, is a landmark project in the history of urban infrastructure redevelopment. It has rejuvenated the city which had become increasingly choked with traffic, establishing an enhanced environment and sustainable future for its citizens. A vital need was to minimise the disruption during construction – which, for this project, has been likened to prolonged urban open-heart surgery (Wheeler, 1997; Soudain, 1999). Known locally as the 'Big Dig', it was hailed as the single largest civil engineering project in American history (The Economist, 2003).

One of the largest sections, Contract 9A, required constructing three interstate highway tunnels beneath an operating railway (Figures 7.1, 7.2 and 7.13). The tunnel jacking, which provided the key solution to solving this huge construction challenge, was by far the largest, most complex application of its kind in the world. It needed a quantum leap in scale, with each jacked tunnel being well over ten times the size of any attempted in the US before and involved a wide range of innovation (Powderham, 2004, 2009). The construction delivered a low maintenance, robust product while bringing important environmental benefits and contributed to over US$300 million in construction savings (Angelo, 1996). From initial concept to construction in 2001, it took over ten years of sustained effort and teamwork. Contract 9A was officially opened on 17 January 2003. This case history provides an overview of the tunnel jacking within the wider project and describes the range of innovations and the key role of the observational method (OM).

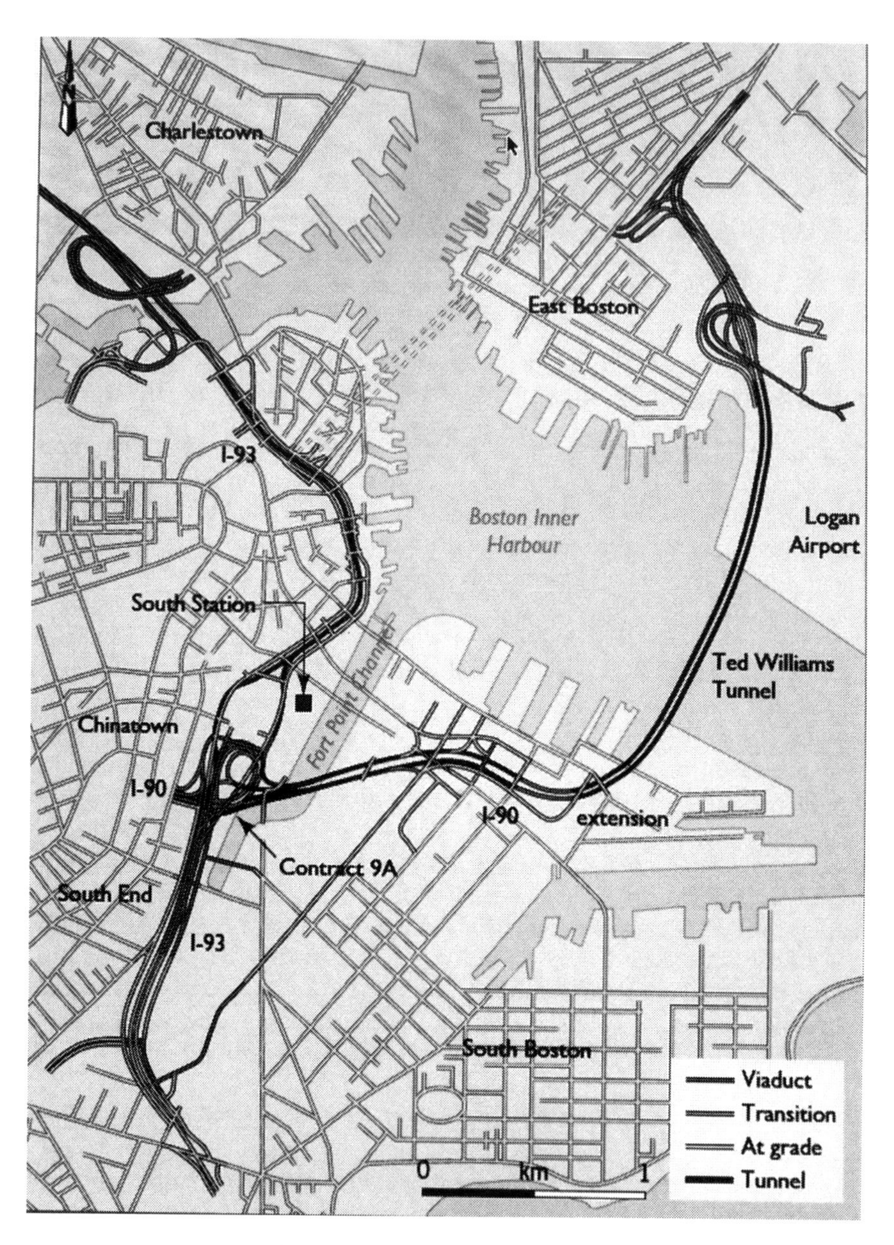

Figure 7.1 Contract 9A of the Boston Central Artery project takes both interstate highways, I-90 and I-93, intersection underground.

Figure 7.2 Looking north before construction along the I-93 towards the I-90 intersection. Photo credit www.bigdig.com.

7.2 Key Aspects of Design and Construction

7.2.1 The Project

In its entirety, the CA/T project was the largest ever US Federal Works engineering undertaking and took over twenty years to plan, design and construct. It involved the replacement of an ageing steel elevated highway structure, built in the 1950s, which carried the Interstate 93 Highway through downtown Boston, with a new highway underground. Much of these tunnels were built directly beneath the existing elevated structure. In addition, a new highway, also mainly in tunnel, was constructed from the original termination of Interstate 90 (I-90) in downtown Boston, to Logan International Airport in the east in Boston Harbour. The new CA/T now takes much of the city's interstate highway network underground thus bringing substantial relief from heavy traffic congestion and pollution. Overall, the project entailed construction of 7.5 miles of new highway with approximately half running underground (Figure 7.1). The final cost of the project in 2012 was estimated at $24.3 billion, with generally 70% of the funding from the Federal Government (through the Federal Highways Administration), and 30% from the Massachusetts State

Government. The Massachusetts Turnpike Authority (MTA) became the eventual owner and operator of the new highway system.

7.2.2 Geology

The ground conditions beneath the railway presented an exceptionally challenging array of mixed face conditions for tunnelling. This included contaminated fill and buried structures overlying compressible strata, which became major obstructions during construction (Figures 7.3–7.5 and 7.16). The geological profile in Figure 7.16 shows the alignment of I-90 Eastbound (I-90 EB) tunnel, which was longest of the three jacked tunnels, and the confliction with the obstructions.

The upper part of the site's geology consists of 6.1–7.6 m of miscellaneous fill material. This is primarily granular and contains numerous large obstructions including boulders and segments of concrete as well as an abandoned, low-level reinforced concrete track-way, old masonry foundations, reinforced concrete and hundreds of timber piles – in short, the various artefacts of two centuries of waterfront development (Figures 7.3, 7.4 and 7.17).

Groundwater levels at the site are typically 1.8–3.0 m below existing grade. Underlying the fill are extensive, variable deposits of organic materials in the range of 3.0–4.6 m thick, which consist largely of organic silt with fine sand and some peat. Below these are local lenses of relatively dense sand and inorganic silt of alluvial origin, generally less than 1.5 m thick. The thickest soil deposit at the tunnel alignments is the

Figure 7.3 Construction of buried trackway base in 1890's.

Figure 7.4 Timber piles showing at low tide.

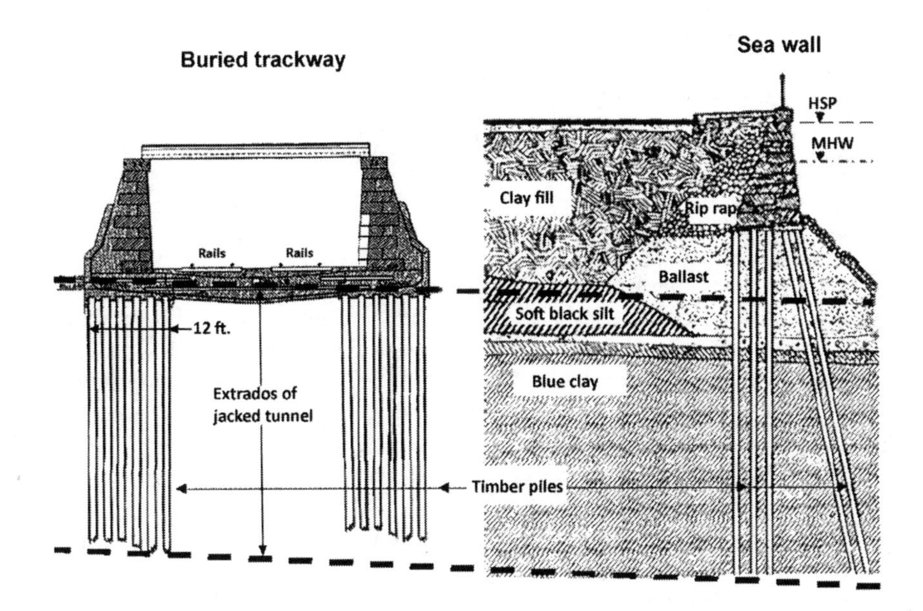

Figure 7.5 Vertical alignment of jacked tunnel through buried trackway and sea wall.

Boston Blue Clay, consisting of clay and silt. The deposit is stronger and less compressible at the top, over a thickness of approximately 4–5 m, where it has been over-consolidated by desiccation. The lower section of the marine clay is normally or lightly consolidated and considerably softer.

7.2.3 Original Design Concept

Before the CA/T, the east-west Interstate I-90 Highway terminated in south Boston forming a very complex surface intersection with the I-93 which runs north-south through the heart of the city (Figure 7.1). Creating the new I-90/I-93 interchange and extending the I-90 to Boston's International Logan Airport required multi-lane highway tunnels to be constructed under the approach to South Station (Figure 7.2). This is a complex network of seven interconnecting rail tracks which carried over 40,000 commuters and 400 train movements daily.

Original design concepts for the CA/T tunnels were based around traditional cut and cover construction techniques. However, for the unique challenges presented in Contract 9A, this required five phased relocations of the railway tracks and associated infrastructure. Such an approach was unacceptable to the railway authorities. Apart from moving the tracks which included complex crossovers and switches, each phase would have involved re-establishing the extensive control systems with their sensitive buried fibre optics. Moreover, it would have been necessary to sequentially construct the elements of each tunnel in deep and narrow isolated trenches between the temporarily relocated tracks (Figure 7.6). This would have been very time consuming with major access and operational challenges to safely manage the multiple interfaces.

7.2.4 Innovations in Tunnel Jacking

The unique combination of challenges that the site presented led to the introduction and development of a wide range of innovation (Powderham *et al.*, 2003, 2004). Some innovations were specific solutions to the various challenges while others created additional benefits in a process of continuous improvement. The temporary 'parking' of the first two units of the I-90 EB jacked tunnel is an example of the latter as discussed in Section 7.4.2.

The key innovative features were as follows:

(a) Scale: The most prominent innovation was the huge leap in scale (Figure 7.7). Each of the three jacked tunnels was well over ten times the size of any constructed in the USA and, at the time of bid, nearly seven times that of any built in the UK (Ropkins, 1998). When this scale was combined with the complex geometry of the site, the adjacent infrastructure and the difficult geology, it led to the development of further innovations that minimised risk and enhanced delivery. As shown in Figure 7.7, with the dramatic juxtaposition of the jacked

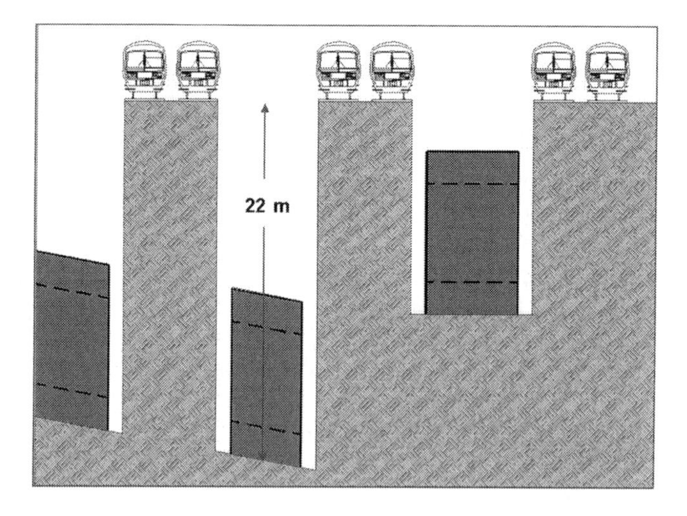

Figure 7.6 Illustrative phase of construction in original design concept.

Figure 7.7 The 24 m wide, 12 m high I-90 EB tunnel during jacking. As noted, it had the least clearance between the tunnel and the railway above.

Photo credit Jason Rodwell.

box and the train passing above, the clearance between them for I-90 EB was minimal, being less than 2 m. The clearance for the other jacked tunnels was in excess of 6 m but which, of course, is still a very low clearance for such large tunnels. There was a trade-off in this context since, with greater depth of the tunnel, the less obstructions were encountered but the ground loading and associated jacking forces were proportionately higher. The structural details of the jacked tunnels are provided in Table 7.1.

(b) Retaining walls: The severe spatial constraints on this site and the geometrical complexity led to the development of an innovative range of retaining wall systems (Figures 7.7, 7.15 and 7.24). The dimensions of the three thrust pits are given in Table 7.2. The headwalls for the three tunnels (Ramp D, and the I-90 WB and EB) were constructed with soldier pile tremied concrete for ease of control during shield entry to commence tunnel jacking (Figures 7.8 and 7.9). Structurally efficient pre-stressed diaphragm walls were utilised for side walls and which were some of the largest of this type ever constructed. In other locations providing lateral support to the top of the walls was impracticable – for example, the back walls of the I-90 WB and the sidewalls of

Table 7.1 Structural details of the three jacked tunnels.

Jacked Tunnel Units	*Unit*	*Ramp D*	*I-90 WB*	*I-90 EB*
Width	m	23.77	23.77	24.08
Height	m	11.58	11.58	10.82
Roof and base slab thickness	m	1.8	1.8	1.8
Wall thickness	m	1.8	1.8	1.5
Tunnel length in final position	m	48.15	75.9	112.78
Shield length	m	2.74	2.74	2.74
Number of tunnel units	No.	2	3	3
Total tunnel concrete volume	m^3	6,200	9,800	11,200
Total reinforcement weight	t	900	1,470	1,680
Total tunnel weight	t	15,500	24,500	28,000
Maximum jacking capacity	t	25,850	25,850	25,850
Normal jacking capacity	t	15,500	15,500	15,500
Actual jacking load on lead unit	t	9,600	15,400	9,900

Table 7.2 Thrust pit elements and dimensions.

Thrust Pits	*Unit*	*Ramp D*	*I-90 WBD/I-90 EBD*
			Combined thrust pit
Average width	m	30	80
Average length	m	60	80
Retained height	m	25	25 (max)
Diaphragm wall	m	1.2	1.2
Cantilever T-section rear wall	m	Not used	0.9 (with 4 m deep web section)
Base slab thickness – front	m	0.9	0.9
Base slab thickness – rear 6 m	m	1.5	1.5
Jet grouted strut below base slab	m	6	6

the I-90 EB thrust pits. For these, 3.5 m deep T-panel cantilever diaphragm walls were used (Figures 7.15). While the ground freezing (described below) robustly stabilised the ground beneath the railway, the ground movements it generated (locally up to 300 mm both laterally and vertically) had major implications for the railway alignment and for retaining the walls of the thrust pits (e.g. see Figures 7.25 and 7.26).

(c) Ground freezing: The challenging and variable ground conditions demanded close attention to stability and comprehensive ground movement control. While ground freezing beneath an operating railway

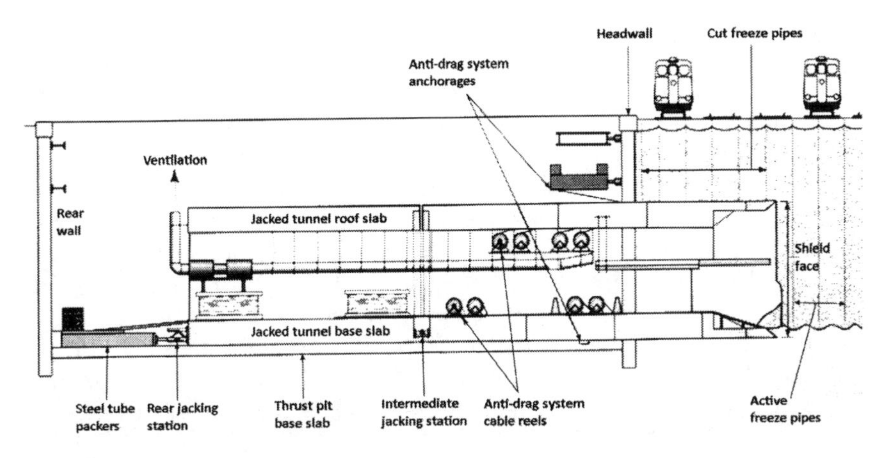

Figure 7.8 Long section showing overall system for tunnel jacking.

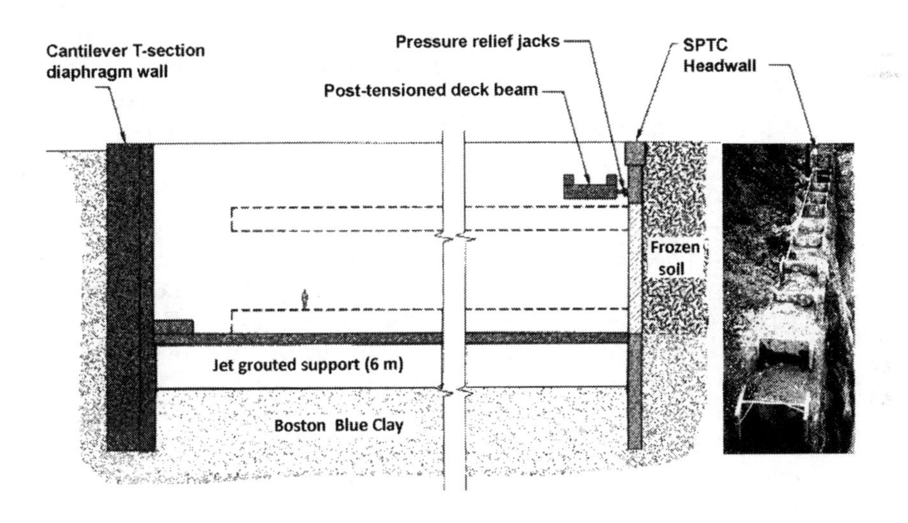

Figure 7.9 Thrust pit section showing support from jet grout and headwall beams.

system was not unique, its application here was on an unprecedented scale when viewed in context with the ground conditions and, in particular, the organic deposits and the clay. The low permeability of these deposits leads to high expansion during the freezing process with the potential to cause large ground movements both laterally and vertically. The organics would also create high volume changes during the thawing process. Some 1,800 freeze pipes to depths of 20 m were installed using a track mounted sonic drilling rig mostly during night-time possessions. The ground freezing was phased over a period of two years in sequence with the jacked tunnel construction. Liquid ammonia was used as the primary refrigerant to cool the calcium chloride brine to an entry temperature of –30°C. Apart from stabilising the ground and creating a much safer working environment at the tunnel face, ground freezing greatly simplified the shield entry through the headwalls of the thrust pits. It also allowed the unique temporary 'parking' of two 10,000 tonne tunnel box units (the first two elements of the I-90 EB jacked tunnel) beneath the operating railway. This was another 'first' that successfully addressed the spatial constraints imposed on the combined thrust pits and delivered extra programme savings by allowing the optimum sequence of tunnel installation (Figure 7.15). The numerous obstructions encountered in the tunnel face were also considerably easier and safer to deal with in the frozen ground.

(d) Shield design and excavation: Another key benefit enabled by the ground freezing was the simplification of the shield design. The stable faces allowed much larger cells to be used so that the tunnel could be fully excavated using roadheaders (Figure 7.10). All of these factors contributed to mitigating the risk of interruption either to or from railway operations. However, in creating these major benefits, the ground freezing also brought some additional challenges. While the excavation at the tunnel face was safer, the ground had been converted from a soft deposit to a rock – and a rock that was essentially without joints. Efficient excavation required development of enhanced performance for the British Webster roadheaders. Further innovations were introduced to cope with the large ground movements and potentially high pressures created by the freezing. These included pressure relief systems and a heating control system in the ground and tunnel walls to prevent the tunnels becoming frozen in the ground – particularly during the 5 months of temporary parking of the I-90 EB units (See Sections 7.4.3 and 7.7.3).

(e) Anti-drag system: The special anti-drag system (ADS) developed for this project had the highest capacity and the most extensive ever installed. A total of 900 closely spaced 19 mm steel cables were used on the roof (Figures 7.8, 7.11 and 7.12). Steel cables were also used under the base slabs of each tunnel. As shown in Figure 7.8, the cables were fed from reels housed in the roof and base of the lead section of each jacked tunnel. The cables were continuous over the full length of each tunnel running from the anchorages in the headwall and thrust base for the roof and base slab sets, respectively. They were sacrificial being cut on completion of the tunnel jacking and left in place. The primary role of the ADS was to decouple the

Figure 7.10 Breakthrough of I-90 WB tunnel.

tunnel boxes from the ground and infrastructure above. This prevents the ground moving laterally with the tunnel as it is jacked forward. The ADS also helps to reduce jacking loads and, when used with a lower set of cables, provides improved alignment control. Trains safely moved uninterrupted overhead as the tunnel sections were installed below the tracks at a rate between 1 and 2 m per day.

7.3 Achieving Agreement to Use the OM

7.3.1 A Dramatic Simplicity

Closing the railway, except for very limited periods for possessions during non-peak operations, was not an option. Neither were the multiple relocations of

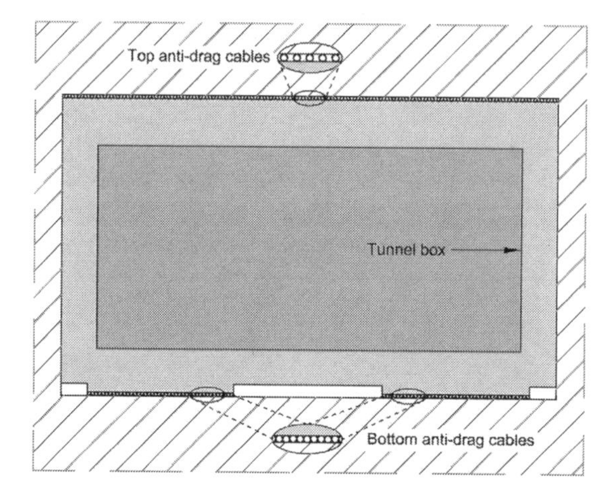

Figure 7.11 Arrangement of ADS cables for roof and base slabs.

Figure 7.12 View inside jacked tunnel showing ADS cables fed through roof.

the tracks acceptable as proposed in the original base case design. Tunnel jacking beneath the railway as a potential alternative to cut and cover construction was presented at the bid stage for the design contract in February 1991. Tunnel jacking provided a radical solution but its underlying concept was dramatically simple: instead of moving the railway, move the structures.

The method was relatively well known in the UK as a non-intrusive soft ground technique particularly suited for large tunnels with shallow cover

Figure 7.13 Jacked tunnels under construction looking east towards Logan airport. Photo credit www.bigdig.com.

(Ropkins, 1998). However, for the Boston project, it was unique and, given the scale, a dramatic proposal. It immediately generated much interest but, predictably, its acceptance took much longer. While the potential benefits were huge, it was a big step to take and there were many issues to resolve. Credibility was an immediate factor which depended on the experience, track record and confidence of the team and the ability to establish the case for the scheme's viability. Convincing the client to adopt such a radical alternative required a fully developed design concept for the specific site conditions – particularly in relation to the spatial constraints, existing infrastructure and geology. The level of detail required by the MTA was understandably demanding. Beyond providing detailed reports on constructability and risk assessments backed up with calculations for predicted jacking loads and ground movements, this process entailed a series of presentations and workshops with representatives from the main stakeholders – the MTA with their management consultant, the Federal Highway Administration (FHWA), and the three railway authorities – the Federal Railroad Administration (FRA), Massachusetts Bay Transportation Authority and Amtrak. It took over 18 months of sustained advocacy and detailed concept development before the MTA was prepared for it to be presented for final approval to the FHWA. It is interesting to note that the consultant team who proposed and developed the scheme was not then tasked with giving the presentation to the FHWA. Instead, it was the MTA that took the lead and presented the tunnel jacking alternative to the FHWA. This was a good team-building initiative. With approval secured, tunnel jacking was formally incorporated into the design development for the 9A Contract.

It is important to appreciate that it would have been impossible to introduce such a fundamental change after the award of the construction contract without huge and unacceptable delays. This was because the complex and inter-related components of the 9A Contract and the interfaces with adjacent contracts demanded very specific design requirements, construction methods and sequences. It would have also been very unlikely that the short-listed contractors bidding for the contract, if the underground works had been based on traditional and local methods, would have had the necessary expertise and experience in the design and construction of large jacked tunnels.

7.3.2 'A Rose by Any Other Name'

As noted, achieving formal approval for the alternative design of tunnel jacking in its entirety required an extended period of advocacy. Gaining approval to use the OM presented an extra hurdle. While the specialist tunnel subconsultant viewed the implementation of the OM as essential in managing the risks pertaining to ground movement control and the potential effects on the railway, the MTA were less than enthusiastic for its use. This reluctance arose from a temporary works failure during the construction of a retaining wall in another contract on the project. The cause of this failure was viewed, in part, to have derived from a flawed application of the OM. This led to an understandable sensitivity to its general use on the project. Thus, some discretion was needed and so, rather than proclaiming the use of the OM overtly, it essentially progressed in borrowed clothes under an alias of instrumentation data review. For this, a committee, comprising representatives of all stakeholders, was formed and recognised under the somewhat unwieldy but descriptive acronym SCIDRAT (Standing Committee on Instrumentation Data Review Attributed to Tunnelling). This acted effectively to coordinate the roles and responsibilities of all the relevant parties in the implementation of the OM.

7.4 Implementation of the OM

7.4.1 The Primary Objective

The principal need was to maintain the safety of the railway within the specified tolerances for the tracks. The process, using the OM traffic light system, is described in Section 7.6.1.

It is always important to avoid viewing applications of the OM in isolation. They must be considered and integrated within each specific process for the design and construction. The OM is typically focussed on the control of risks associated with various forms of movement. The successful delivery of Contract 9A was in many ways all about movement – keeping the trains moving and moving the tunnels but not the tracks. More specifically, it was about ground movements and their control and these arose from multiple sources as described below. Consequently, the use of the OM was central to the success of the tunnel jacking overall and the wide range of innovations that were involved. Its implementation was effectively guided through the SCIDRAT – an integration that was also a convenient discretion as mentioned in Section 7.3.2.

7.4.2 Solving the Spatial Challenges

The site is located at the intersection of the I-90 and I-93 Interstate Highways (Figure 7.1). Apart from the notably difficult ground conditions, there were severe spatial and operational constraints. The space available for construction was limited to an area of 300 by 700 m enclosed between the interstate highways, the railway system and a marine waterway – the Fort Point Channel. The key requirement was to avoid any interruption to the adjacent surface traffic – particularly that of the railway. The spatial limitations were significantly eased by the introduction of tunnel jacking since the phased relocations of the railway were no longer needed (Figure 7.6).

However, available construction space was still considerably less than the normal needs for this method of tunnelling. This was overcome by the unique use of combined thrust pits and global ground freezing (Figures 7.14, 7.15 and 7.18).

7.4.3 The Process for Tunnel Jacking

Jacked tunnels, particularly the longer ones, are typically constructed using several separate units. These tunnel elements are linked together by intermediate jacking stations and the tunnel is installed employing a 'caterpillar'-like action (Figure 7.8). This helps to minimise the overall jacking force since it is then only dependent on the maximum required to move the units individually. For this project, based on a comfortable factor of jacking capacity, there were two jacked box units for Ramp D and three each for the I-90 WB and I-90 EB tunnels. The units are cast on a thrust base that, apart from being able to withstand the jacking forces, must be constructed to a high tolerance since it controls the initial alignment of the tunnel. A shield is used to support the face of the excavation at the front of the tunnel. In soft ground, the shield is divided into a series of small compartments across the face, and the leading edges of

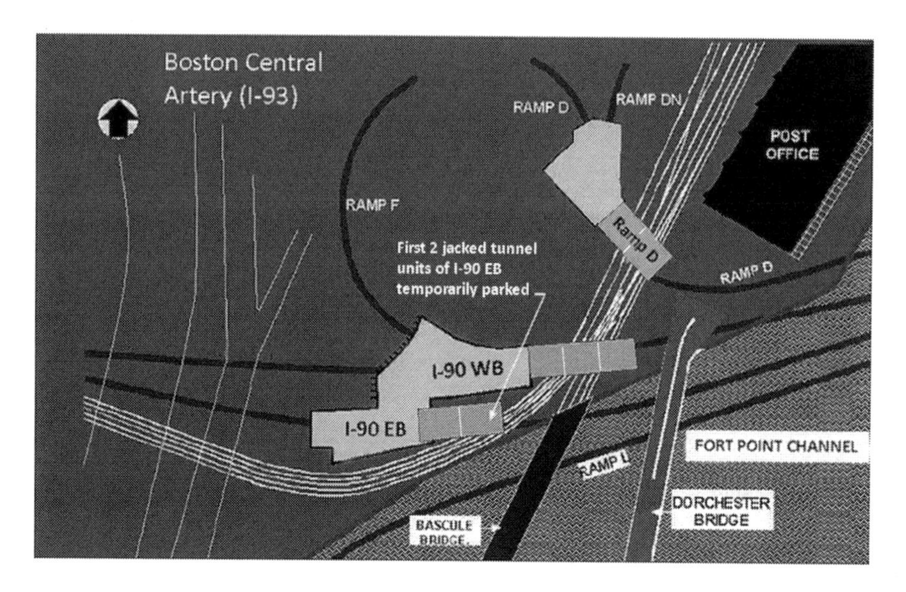

Figure 7.14 Plan showing the three jacked tunnels and associated thrust pits.

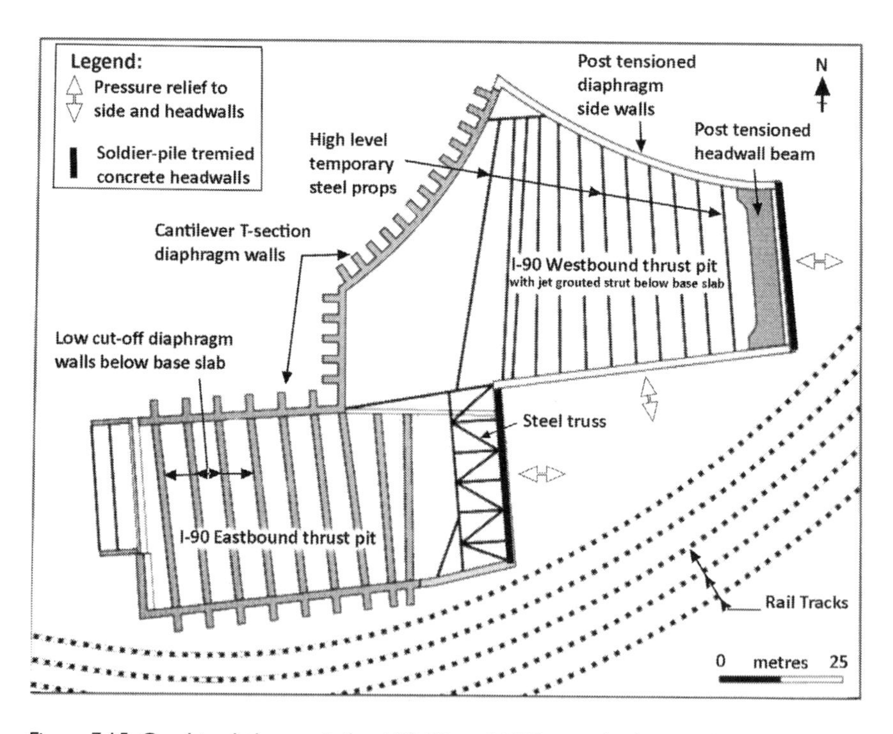

Figure 7.15 Combined thrust pit for I-90 EB and WB tunnels showing bespoke designs for wall types and support systems.

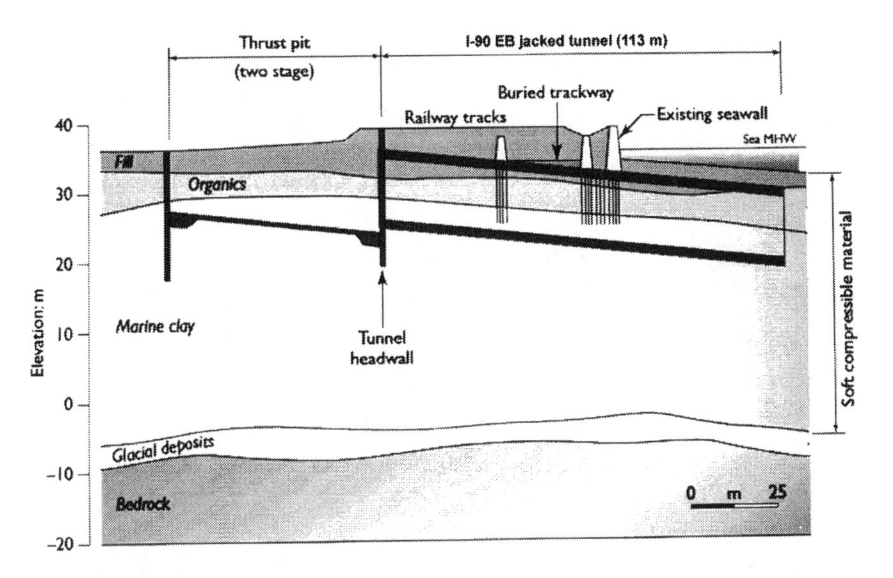

Figure 7.16 Cross section for EB jacked tunnel.

Figure 7.17 Two centuries of waterfront development left an extensive legacy of obstructions.

the shield are generally maintained embedded in the soil to ensure its adequate support. Excavation is progressed in each compartment using hand tools or small machines. Usually, all units for a jacked tunnel are cast and aligned together on the thrust base prior to jacking. Once jacking is started, it is a continuous process until the whole tunnel has reached its final position. This is to avoid the risk of the build-up of ground pressures around a stationary section of tunnel during an extended hiatus in jacking which could lock the unit in place before it has reached its correct position or cause undue ground movement because of the build-up of friction on the external faces of the tunnel.

However, for Boston, the limitations imposed by existing transport infrastructure did not allow enough space to extend the I-90 EB thrust pit to accommodate all of its three units simultaneously prior to commencing the jacking process. One way to address this constraint was to utilise the whole combined thrust pit to cast the I-90 EB tunnel units. The last eastbound unit could then be slid laterally across into line from the westbound pit to the eastbound pit so that continuity of jacking could be maintained. The downside would be a substantial delay to the westbound jacked tunnel since it could not commence until its area of the combined thrust pit was cleared of the third eastbound unit. This clash in construction sequence was solved by installing the I-90 EB jacked tunnels in two time-separated stages. This is illustrated in Figure 7.14, showing an intermediate phase of construction where the jacking of the units for both Ramp D and the I-90 WB has been completed, and the first two units of I-90 EB have been jacked into their temporary parked positions. After this stage, the I-90 WB thrust pit thus became available to construct the third unit of the I-90 EB. It could then be moved laterally into position into the I-90EB thrust pit and thus in line behind its two parked companions. The set of three units could then be jacked into their final location. As noted, this was a unique approach to solving a spatial challenge and involved an additional application of the OM. This related to assuring that the first two units of I-90 EB did not become seriously locked into the ground during the period during which they were parked awaiting the arrival of the third unit. The jacking of the first two units of I-90 EB started on 1 March 2000 and was completed on 10 May 2000. When the jacking of the WB tunnel units was completed, construction of the third unit of the EB tunnel could commence in the vacated WB thrust pit. The jacking of the WB tunnel started on 6 June 2000 and was completed on 16 August 2000. The jacking of the second stage of the EB tunnel commenced on 4 October 2000. Thus, the first two units of I-90 EB were parked for 5 months. The OM was applied through monitoring the jacking forces required to mobilise small incremental movements of these two units at regular intervals throughout the parked period. This enabled the potential development of any significant trends in locking forces to be identified. It was predicted that the ground being maintained in its frozen condition would enable it, over this 5 month interval, to effectively span around the external perimeter of each tunnel unit thus preventing full closure of the over-cut gap. The back-up contingency measure to address locking of the units was to lubricate the affected zones. To achieve this, heating systems were incorporated in the roof and walls of the units and also installed in the surrounding ground. To enhance the lubrication, ports to enable the injection of bentonite around the periphery of the units were also provided. In practice, these contingencies were not required as established through the regular incremental movements.

7.5 Construction Activities Causing Ground Movements

The OM was focussed on the control of ground movements and their effects on the alignment of the railway. For the tunnel jacking operations at this site, there were a total of seven distinct but interrelated activities, each of which had the potential to cause ground movements affecting the rail tracks. Each is discussed below.

7.5.1 Installation of Diaphragm Walls Forming the Thrust Pits

As described in Section 7.2.4, various types of diaphragm walling were used to form the thrust pits depending on geometry and depth. All of these walls were installed with traditional equipment with clamshell buckets and using either bentonite or synthetic slurry. All obstructions in the fill layer were removed by pre-trenching with slurry providing the support. In sensitive areas, close to the rail tracks or services, sheet piling was occasionally used to prevent over-excavation or the risk of localized ground collapse. Ground movements from the diaphragm wall installations were very small, and no significant effects in the adjacent rail tracks or services were recorded.

7.5.2 Jet Grouting of the Low Level In-Situ Strut

Base stability in the Boston Blue Clay was marginal for the untreated deeper excavations. Jet grouting was generally adopted to provide pre-installed lateral support below formation level prior to excavation. A 6 m deep zone of improved ground with a compressive strength of 1.0 N/mm^2 was specified for this. The jet grouting included the complete treatment for the Ramp D and I-90 WB thrust pits, and partial treatment for the I-90 EB thrust pit which was about 4 m shallower. It was installed from existing ground level within the area bounded by the diaphragm walls. The installation equipment first drilled the 20 m or so down to formation level and then commenced jetting. The soil at this elevation was always clay. To provide 100% coverage, 1.8 m diameter jet grout columns were installed on a 1.35 m grid. A three-fluid system was employed using pressures of up to 415 bars.

Substantial heave was evident adjacent to the drill probes. This was estimated to be in excess of a metre locally over a 2 m diameter area, although it was difficult to measure accurately as the area was awash with returned grout/soil spoil and there was a continuous process of clearing the spoil. Such local heave presented no difficulty, as it was remote from the railway or other infrastructure. However, the jet grouting also caused the ground to 'boil' at a distance of up to 50 m from the drill probe. It appeared that the high-pressure grout was finding paths of relatively low resistance within the ground at depth and permeating along those paths to the ground surface. This indicated that relatively large volumes of the ground were being pressurized at depth by the jet grouting operation. Close to the diaphragm walls, the walls effectively cut-off the paths of low resistance and prevented grout travelling up to the surface. However, in doing so, the walls formed a barrier against which pressure built up to such an extent that the walls were pushed outwards. Inclinometer readings indicated that the walls were moved by as much as 100 mm which caused heave in the adjacent ground. The rail tracks heaved by up to 50 mm as a result of this effect. Controlling the ground movement from jet grouting was managed by several methods:

(1) Test results from core samples indicated that the grout strength was higher than required. An evaluation of this allowed the 100% coverage to be reduced to approximately 85%, leaving small zones of soft ground in place. These soft spots provided areas which enabled the high pressures to be relieved.
(2) Installation of the jet grout columns was sequenced to allow dissipation of pressure in any given area before re-pressurization with more grout.
(3) Pressure relief pipes were drilled down into the clay layer. These were positioned adjacent to the diaphragm walls which were close to the rail tracks and provided a path of low resistance to the high pressures (Figure 7.15).

These measures enabled the pressure on the diaphragm walls to be considerably reduced resulting in a substantial reduction of heave and lateral movement to the nearby rail tracks.

7.5.3 Bulk Excavation of the Thrust Pits

Bulk excavation of the thrust pits proceeded smoothly. The Ramp D thrust pit and the combined pit for both I-90 tunnels were excavated over a periods of approximately 2 and 4 months, respectively. Wall movements were within the ranges predicted and well within the specified trigger levels. Maximum deflections were recorded around 12 mm for the walls with a high level strut and around 40 mm for the cantilever walls. Both horizontal and vertical ground movements affecting the rail tracks during bulk excavation of the thrust pits were significantly less than the wall movements and were no greater that 15 mm.

7.5.4 Installation of Vertical Freeze Pipes

Tunnel jacking, in common with all tunnelling, requires adequate control of the tunnel face to prevent ground movement into the tunnel creating volume loss leading to ground settlement and lateral movement. On this project, the contractor chose to control the tunnel face by stabilizing and strengthening the ground by freezing it. As noted, the existing ground conditions were very challenging having high potential instability and crowded with a wide variety of obstructions. This presented a wide range of uncertainties and potential risks. Dealing with these in untreated ground during tunnelling would have proved time consuming at best. Globally treating the ground with conventional methods prior to tunnelling and then on an ad hoc basis during excavation at the face would have also risked delays and further uncertainties. Effectively converting the variable deposits from a soft ground to a weak rock by freezing resolved these difficulties and also made it far easier to safely remove obstructions. Approximately 1,800 freeze pipes were installed vertically at spacing of 2.0 to 2.5 m to freeze the ground over the entire plan area of the tunnels. This extended from 1 m below ground surface to a depth of 1 m above the tunnel base (Figures 7.9 and 7.18). The freeze pipes comprised 110 mm diameter closed ended steel sections which were installed using sonic drilling methods. The sections were vibrated while being rotated and were installed quite rapidly. A 20 m long pipe would typically be installed in less than 1 hour. All pipe installation above the rail tracks were undertaken during night-time possessions to avoid disruption to the rail services.

Ground movements during freeze pipe installation were very limited. Generally, the vibration of the pipe created a cone of ground settlement less than 300 mm diameter around the pipe, over a depth of 500 mm or less. This was immediately backfilled with ballast after drilling had been completed.

7.5.5 Heave and Lateral Movement of the Ground during Freezing

Water expands by approximately 9% when it freezes. The amount of expansion of the ground is dependent on its permeability, the speed of the freezing process, the geometry of the ground being frozen and the sequencing of the freezing operation. After an assessment of the ground characteristics and laboratory testing of the soils, it was established that the expansion would occur in two orthogonal horizontal directions from the centre of each frozen mass and vertically upward from its base. The magnitude was estimated to

Figure 7.18 View showing heads of freeze pipes vertically installed between the rail tracks.

Photo credit www.bigdig.com.

be roughly equal in the three directions. The majority of the expansion was anticipated to be caused by freezing of the organic and the clay strata. The fill was expected to have sufficient permeability to allow the water to disperse away before it fully froze. Figure 7.19 shows heave contours generated by the ground freezing for the I-90 WB tunnel.

Maximum vertical and lateral heave were each predicted to be approximately 175 mm for Ramp D and I-90 WB tunnels. The I-90 EB tunnel, which was the shallowest as shown in Figure 7.7, involved less freezing in the clay strata and consequently caused less heave.

7.5.6 Ground Movements during Tunnelling Operations

During tunnelling, ground movement towards the face combined with overcut around the external periphery causes volume loss which produces settlement troughs at the ground surface. This is most evident in soft ground. Here, however, the frozen ground behaved like a soft rock with no significant movement towards the face. Thus, one of the original concerns about potentially high volume loss in the soft ground was effectively eliminated with minimal effect on the alignment of the railway. The ground freezing

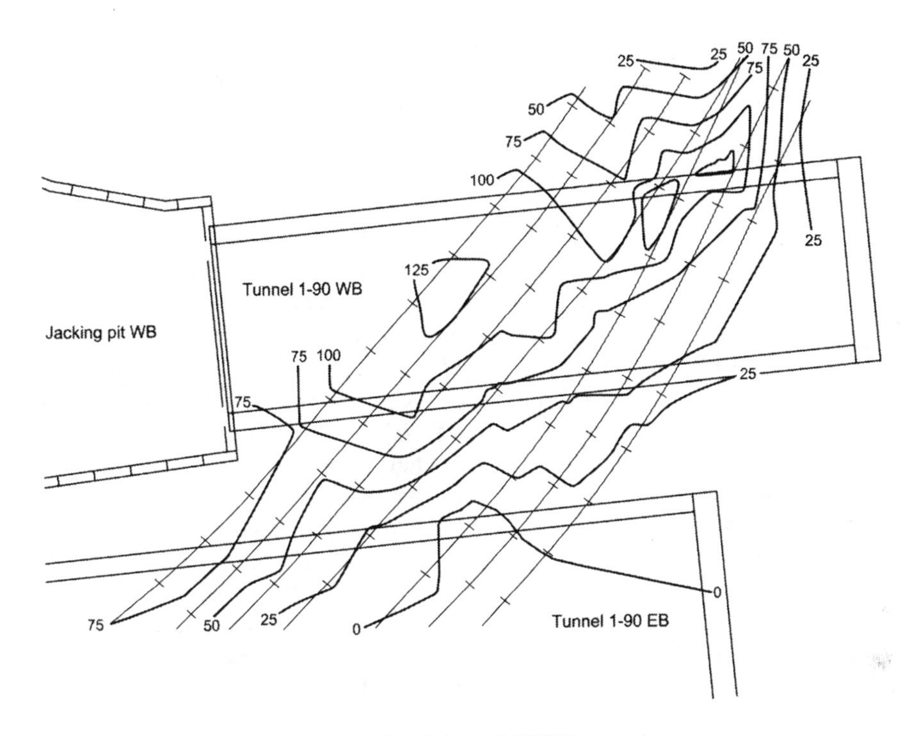

Figure 7.19 **Surface heave contours (mm) above I-90 WB tunnel.**

also resolved the concerns regarding the removal of the many obstructions and the attendant risk of causing additional instability and ground movements. This benefit particularly applied to the extensive presence of timber piles which can be difficult to remove in a safe and timely way in soft and potentially unstable ground. An example of the concentration of the timber piles encountered in the tunnel faces is shown in Figure 7.17. In their frozen condition, they were simply excavated with the rest of the ground using the road headers. Even the more obdurate obstructions such as masonry were much easier to deal with such stable tunnel faces.

The presence of numerous obstructions at the upper levels did require some over excavation of the perimeter of the tunnel face. This was in addition to the pre-determined shield over-cut of 85 mm for the walls and 35 mm for the roof to allow the ground to relax and thus reduce resistance to jacking. Backfilling the over excavation was not always fully successful. The frozen ground which remained above the tunnels, as they were jacked forward, was able to arch over considerable distances. Later, as the ground slowly thawed and lost strength, there was a tendency of it to settle into any voids created above the tunnel roof. Typically, settlement occurred above the tunnels producing contours similar (although in the opposite direction) to those produced by the freezing operation. In general, the time-dependent settlement progressed in slow, well-defined trends enabling the track alignment to be progressively maintained within safe operational tolerances.

7.5.7 Longer-Term Ground Movements during Thawing and Consolidation

A comparison between the predicted and measured ground heave is shown in Figure 7.20. Settlement continued to occur above the tunnels for several months after completion of the jacking. Generally, the movements created settlement contours, which centred around the mid-point of the tunnel and the area, which had been frozen. Once the primary settlement due to the closing of the over-cut had ceased, secondary settlement due to thawing and ground consolidation continued at a reducing rate locally reaching maxima of around 150 mm. During the period post-jacking, several localized non-typical settlements occurred above and around the jacked tunnels. This was attributed to voids left in the ground from the tunnel excavation operations resulting either from over-excavation, obstruction removal or from the delayed closure of the planned over-cut for the shield along the roof and walls. In all cases, the

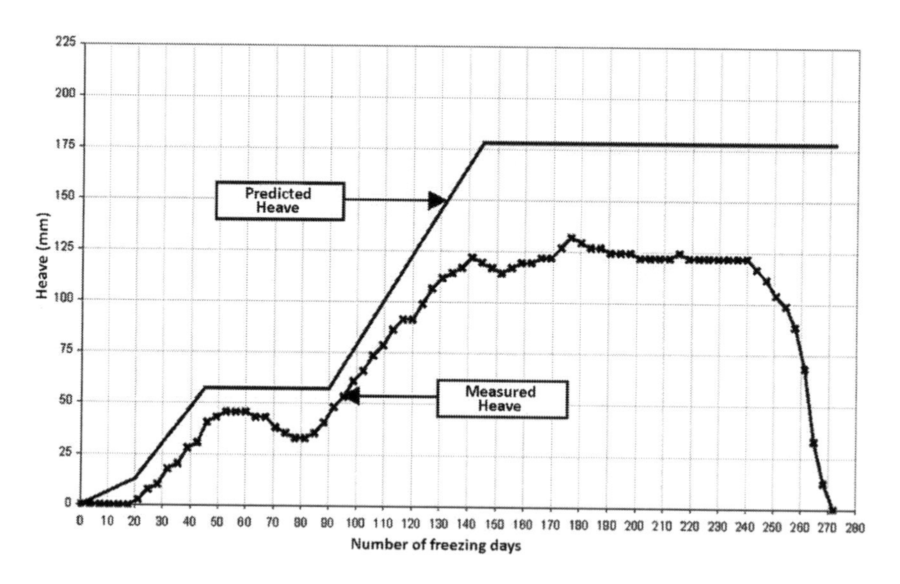

Figure 7.20 Typical example of the predicted and actual ground heave.

settlements were identified early and any necessary corrective measures taken to maintain the rail tracks within tolerance.

7.6 Instrumentation and Monitoring

7.6.1 Railway Track Monitoring

During design development, the allowable rail track movements were established in accordance with the tolerances to FRA standards. Trigger levels for track movements were set on the basis of the OM traffic light system and were considerably less than the accumulative ranges of movement generated during and after the construction.

The selection of an acceptable track monitoring system was the subject of much investigation and debate. Instrumentation systems evaluated for monitoring the horizontal alignment and vertical movement of the tracks included the use of robotic optical survey instruments, beam-mounted electro-levels and a portable geometry trolley. The preferred system selected was a relatively simple and conventional one. This was based on manual survey techniques and was one with which the railway authorities were both comfortable and confident. Readings were taken every 4.7 m, a spacing equating to a quarter of a standard track chord length of 18.9 m and enabled easy correlation of data to railway design criteria. Cross level, warp and other alignment criteria were checked against the allowable limits set by the FRA. Trigger levels and limits were set for each criterion. Horizontal alignment was monitored by manually read surveys using a total station optical instrument and nylon tape. Non-metallic tape was used to avoid potential effects to the track signalling system.

Due to the importance of maintaining normal railway operations, the tracks were monitored at least once per day, every day of the week, during the ground freezing operations. This was increased to twice daily during the tunnel jacking activities. Horizontal alignment was checked once per week. Railroad authority personnel also maintained continual visual inspections. These visual checks provided early warning of any problem areas and were essential to ensure the safe operation of the railroad. In addition, there was an imposed speed limit of 10 mph on all trains entering and leaving South Station during the freezing and tunnelling operations.

7.6.2 Diaphragm Wall Monitoring

Twenty-nine inclinometer casings were cast into or placed just behind the diaphragm walls of the thrust pits. These were installed to monitor lateral wall movements due to jet grouting, excavation and ground freezing. Each inclinometer casing was installed into bedrock to ensure base fixity.

In most cases, this required drilling down over 30 m through the clay and glacial till. The frequency of monitoring was coordinated with the progress of construction activities, varying from daily readings where developing movements were evident, to monthly intervals where construction was substantially complete and no significant movement had been recorded over a prolonged period. In general, inclinometers were read twice weekly during active construction.

7.6.3 Ground Movement and Groundwater Monitoring

Extensive instrumentation was installed for the monitoring of both the ground and structures. Figure 7.22 shows the location of the geotechnical instrumentation. For clarity, all other forms of monitoring, such as the precise levelling monitoring points and temperature sensors, are not included in this diagram. Some 25 inclinometers and 15 magnetic probe extensometers were installed at various locations to monitor horizontal and vertical ground movements associated with the thrust pit and tunnel construction. They were used to monitor the effects of jet grouting, thrust pit excavation, ground freezing and tunnel jacking. These were read generally once per week, with increased frequency for key events such as the breaching of the headwalls of the thrust pits to commence the jacking process. Precise levelling points were also established to monitor ground

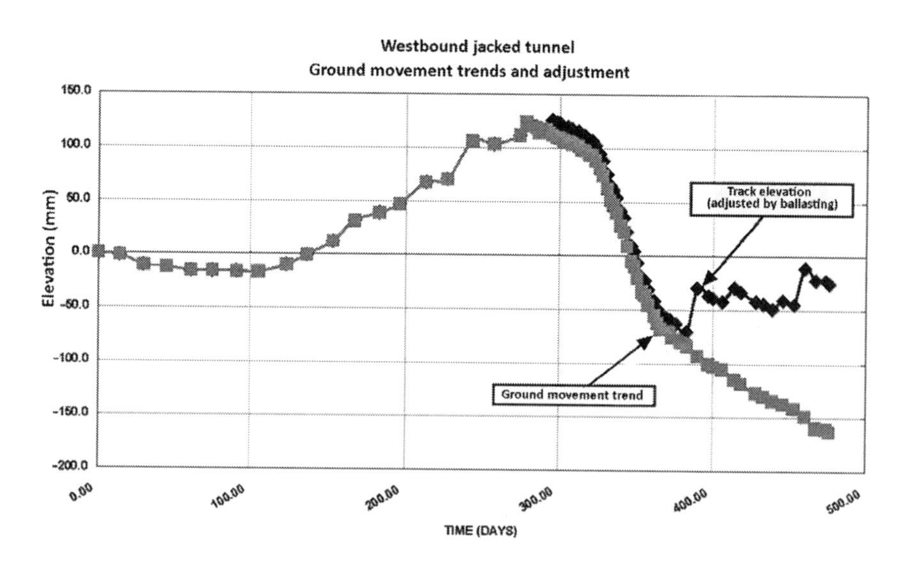

Figure 7.21 Settlement/heave plot for a typical point of maximum movement on the rail tracks.

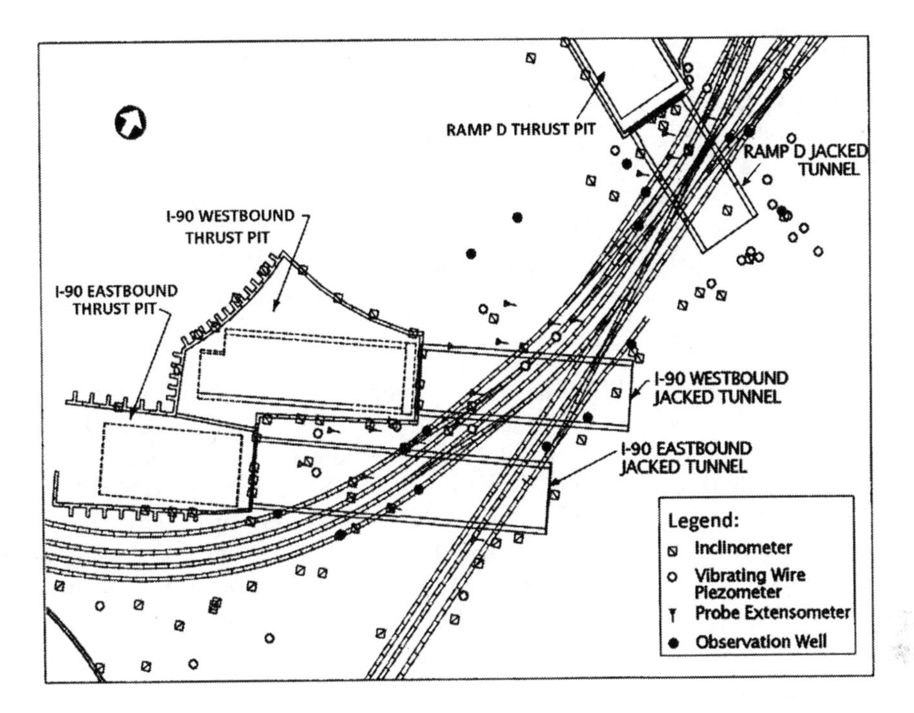

Figure 7.22 Arrangement of geotechnical instrumentation.

surface movements due to ground freezing and tunnel jacking. These were read weekly during ground freezing and two to three times a week during tunnel jacking. Ground temperatures were monitored by 90 sensors located strategically between selected freeze pipes. They were also read once per week during the freezing period. In addition, 25 vibrating wire piezometers and 15 observation wells were installed at various locations to monitor groundwater pressures. They were used to assess both short- and long-term fluctuations in groundwater levels and pressures associated with construction activities. Further details of the monitoring and instrumentation are given by Rodwell (2001) and Daugherty (1998).

7.7 Managing Interfaces

7.7.1 Tunnel Launch Pads

Fundamental to the success of the tunnel jacking was the primary need to avoid any interruption to railway operations and to maintain an acceptable level of safety throughout the construction and post construction period. Secondly, it was important that the jacked tunnels could be installed comfortably within alignment tolerance. A key aspect of this was the planning, design and construction for the thrust pits. They have a close interface with the railway and the installed jack tunnels and need to be able to adequately sustain the jacking forces that were predicted to reach in excess of 24,000 tonnes (Powderham *et al.*, 2001). One key aspect is the interface between the jacked tunnel boxes and the thrust pit base slabs. The top surface of these had to be constructed to exceptionally high tolerances for such massive reinforced concrete elements as they were the launching platforms for the jacked boxes.

7.7.2 Jet Grouting

As noted in Section 7.5.2, the jet grouting caused significant ground movements and pressures on the diaphragm walls. It did, however, perform extremely well in controlling lateral movement at depth during excavation for the thrust pits, while the low cut-off diaphragm walls, although intrinsically more economical, led to greater wall movements because of the practical difficulty in achieving an intimate contact with the thrust pit side walls. (Ground improvement using lime columns was also evaluated in field trials but rejected in favour of the other methods adopted.)

7.7.3 Ground Freezing

The ground freezing, along with design changes for the diaphragm walls, such as pre-stressing, was developed by the main contractor working closely with the specialist design team for the OM as a value engineering (VE) alternative to the extensive programme of ground treatment initially designed. While bringing major benefits overall, the ground freezing imposed the greatest loads and deflections on the thrust pit walls. To avoid these walls and their upper support becoming overloaded, a system of hydraulic jacks was installed, as shown, on the thrust pit headwalls together with pressure relief holes drilled close on the outside of the thrust walls (Figures 7.15 and 7.26). Figure 7.25 shows the cumulative effects on the headwall of ramp D together with the actual and predicted displacements prior to the effects of ground freezing.

7.8 Fostering a Collaborative Approach

7.8.1 Teamwork

There were many aspects of close teamwork that contributed to an ongoing and integrated approach to successfully deliver the alternative of tunnel jacking. Since this was such a radical departure from the base case, it required a complete reassessment of the design and construction issues for

Figure 7.23 Ramp D interface with the railway showing installation of the overhead electrification currently well advanced with the jacked tunnel.

Photo credit www.bigdig.com.

this contract. This presented the opportunity for the design consultant to introduce construction expertise and work closely with the MTA and their management consultant from the start of the concept stage. This in turn helped to progress the scheme development with other key authorities such as the FHWA and the railway operators. The introduction of tunnel jacking also allowed British expertise in the technique to be comprehensively engaged. The consultants' specialist designers for the jacked tunnels remained fully involved right through to completion of construction, working closely, as noted, with the contractors' team on VE alternatives as well as providing construction advice and supervision services. The contractor, supported by his tunnelling specialists, brought further enhancements through the VE process. This included fully developing the option to globally ground freeze beneath the railway that brought a range of important benefits. These and further aspects of teamwork are described below.

Partnering was encouraged on the project (Angelo, 1996) and what developed through the SCIDRAT was effectively a form of this (see Sections 7.3.2

Figure 7.24 Combined thrust pit interface with the railway. The first stage element of I-90 EB tunnel is ready to be jacked and parked.

Photo credit www.bigdig.com.

and 7.4.1.). As early issues were discussed during the pre-construction phase, strong personal relationships were formed within the SCIDRAT team well before construction activities created any significant effects on the railway. The team continued to meet over a period of two years, spanning the 18 months of construction work for the jacked tunnels. This provided the opportunity for the individuals and the organisations they represented to develop a good level of trust and rapport. SCIDRAT meetings were set on a regular weekly basis with occasional extra meetings to address particular incidents. It was very successful in managing the key issue of ground movements and railway operation. Over-reaction was avoided and a proactive and constructive approach maintained. Based on trends, experience and judgement, trigger levels were addressed and changed, if acceptable, on a progressive step-by-step basis – rather than in an inflexible and prescriptive approach. A confrontational environment was avoided.

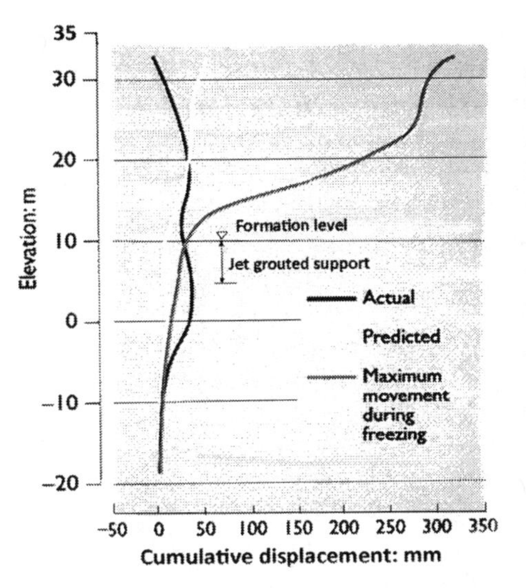

Figure 7.25 Headwall deflections – ground freezing had far more effect than excavation.

Figure 7.26 Headwall pressure relief system.

7.8.2 Extending the Limits of Progressive Modification

The practical maximum limits for adjustment of the rail tracks to maintain them within acceptable tolerances were not precisely known in advance. The general process to maintain alignment was to add ballast around the heave zones, rather than to remove ballast within these zones and so keep the tracks near their original level. So, the progressive increments of movements were additive. One ultimate limit for upward movements of the tracks was the vertical clearance under railway infrastructure such as bridges and the alignment of the tracks with station platforms. In practice, such factors did not prove critical. If the movement trends had indicated that this limit would have been reached, then the contingency measure would have been to correct it by extensive removal of ballast and thus restoring the tracks closer to their original lower level. This would have involved a re-ballasting train operating during night-time possessions and so should not have led to any significant suspension to railway operations. No such contingency proved necessary. However, the magnitude of the heave was such that the initial limits set on ground movement were agreed through the SCIDRAT to be progressively increased. (See discussion on teamwork in Section 7.8.1.) This was an unusual but very effective example of progressive modification being concurrently applied to both the protected facility, namely, the rail tracks and the limits set for control.

The overall success was demonstrated by there being no interruptions to railway operations and no associated delays to tunnelling operations throughout the entire construction and post construction phase. This was achieved even though, as noted above, the accumulative ground movement significantly exceeded the original maximum set by the railway authorities. This was initially set at 100 mm with the amber and red trigger levels for the traffic light system being set at 50 and 75 mm, respectively. The measured ground movements were directly compared with the resulting effects on track tolerances. Both were monitored on a daily basis, and it soon became apparent that the ground movements, both vertically and laterally, could be accommodated while comfortably maintaining rail tolerances within acceptable limits. The key here was that the rate of movement was very slow (see Figure 7.21) and the effects on the railway could thus be effectively addressed by periodic track maintenance. In practice, the original upper limit on ground movement was widely exceeded by a factor of 2 and locally up to 3 times. This would have been the equivalent of dramatically exceeding a red limit but here, with the feedback from monitoring the actual performance, the traffic light criteria were progressively increased, eventually reaching three times the original trigger levels.

The introduction of tunnel jacking with its associated innovations was a very helpful factor as a basis for developing this successful co-ordination team. This was because a key safety concern for the railway was the avoidance of track relocation and all the associated risks with the signalling system. The original cut and cover scheme involved five track relocations, while the tunnel jacking did not require any. The railway authorities advised the SCIDRAT of all their issues and concerns relating to railway operations right from the start. Thus, the basis of a win–win approach to successfully deliver this complex and challenging project was established.

7.9 Results

The alternative of tunnel jacking delivered a range of major achievements to the project. In this success, the OM played a fundamental role. The key benefits were as follows:

- Time and cost: The jacked tunnels were installed within schedule and budget while contributing major cost savings. The construction programme for the original base case design was around six years. It would have required the phased relocation of the railway that involved two additional contracts, which were eliminated by the tunnel jacking alternative. The track relocations would have also prevented any early start of the overhead line electrification (OLE), as shown in Figure 7.23. The tunnel jacking enabled a fixed interface with the railway to be maintained so that the OLE could be progressed in parallel with the tunnel construction. Construction of the thrust pits commenced in mid-1997, with the jacking of the tunnels progressing in phases from early November 1999. This started with Ramp D and was completed in February 2001 with the breakthrough of I-90 EB (Figure 7.27). The tunnel jacking alternative was thus delivered in two years less than the original programme while also allowing the OLE to be completed much earlier than planned.

- Community: The improvements to the I-90/I-93 Intersection have dramatically improved transport mobility in the notoriously congested downtown Boston. It has also helped to reconnect neighbourhoods cut-off by the old elevated highway as well as to improve the quality of life in the city beyond the confines of the new expressway. But with the existing congestion, the CA/T placed an additional burden during construction, increasing dependence on public transportation. The tunnel jacking solution brought real relief to the community in this context by ensuring the continuous safe operation of the railway throughout the construction period. As noted, it also allowed the early upgrading of the railway service with the new OLE system. Figure 7.10 shows the breakthrough of the I-90 WB tunnel with the installation of the OLE well advanced. Awareness and involvement of the local community were achieved through the projects' Outreach Program which included site visits, newspaper articles and a series of documentaries on the local television network. The crowning glory is the Rose Fitzgerald Kennedy Greenway which now breathes new life into Downtown Boston providing a series of parks and public spaces over one and half miles in place of the elevated and aging infrastructure of the interstate highways.
- Safety and environment: The system of tunnel jacking combined with the ground freezing provided a particularly safe working environment. Heavy temporary steelwork multi-strutting was avoided or substantially reduced and tunnel faces were robustly stable. Also, since the ground between the railway and the jacked tunnels was left in place, there was added environmental benefit in reducing bulk excavation by around 100,000 m^3 minimising transportation of heavily contaminated materials through the city.
- Quality: Apart from the overall benefits during construction, the tunnel jacking also led to enhanced quality in the completed works. Pre-casting the tunnel elements in the thrust pits resulted in high-quality construction with a minimum of joints providing a low-maintenance solution.

Figure 7.27 Jacked tunnels completed with existing I-90 extending westwards in the background.

Photo credit www.bigdig.com.

7.10 Conclusions

Apart from the range of additional major benefits described above, this case history provides a prime example of the key role of the OM in assuring safety, protecting infrastructure and successfully addressing uncertainties. It also demonstrates the effectiveness of the OM as an enabler for innovation through the wide range of the unique features developed for this project.

The change from traditional cut and cover construction to tunnel jacking enabled three full-size interstate highway tunnels to be placed under a complex operating railway with no track relocations or interruptions to the service and with an excellent safety record. The jacked tunnels totalled over 240 m in length with the combined elements forming a single tunnel weighing up to 30,000 tonnes. The site conditions presented an unusually onerous range of challenges. The comprehensive success has established a new benchmark for the technique and the widespread recognition significantly raised its profile internationally.

The awards that the project has received have underlined this. These include: the American Consulting Engineers Council Grand Conceptor Award for the excavation support to the thrust pits; the Royal Academy of Engineering MacRobert Award for Innovation – Finalist, and winning the Quality in Construction Award for International Performance, the Building Award for International Achievement, and the ASCE, CERF Charles Pankow Award for Innovation.

References

Angelo, J. A. (1996). Digging it in Boston. *Engineering News Record, Cover Story*, August 19, **1996**, pp. 24–26.

Daugherty, C. W. (1998). Monitoring movements above large shallow jacked tunnels, Proc Geo-Congress 98, Boston, MA, ASCE Jacked Tunnel Design and Construction, pp. 39–60.

Powderham, A. J. (2004). Jacked tunnels – open heart surgery on Boston, Ingenia, Issue 19, Royal Academy of Engineering, UK.

Powderham, A. J. (2009). The Vienna Terzaghi Lecture, the observational method – using safety as a driver for innovation.

Powderham, A. J., Howe, C., Caserta, A., Allenby, D. and Ropkins, J. (2004). Boston's massive jacked tunnels set new benchmark, Proc. Institution of Civil Engineers, Civil Engineering, London, May, 2004.

Powderham, A. J., Taylor, S., Hitchcock, A. and Rice, P. M. (2001). Ground movement control for tunnel jacking under railway, Proc. International Conference 'Response of buildings to excavation-induced ground movements', CIRIA, July 17–18.

Powderham, A. J., Taylor, S. and Winsor, D. S. (2003). Moving structures – tunnel jacking for the Boston Central Artery, Proc. Institution of Structural Engineers, September, 2003, pp. 28–33, UK.

Rodwell, J. H. (2001). Instrumentation monitoring on contract 9A4 of the big dig in Boston, Geotechnical News, December.

Ropkins, J. W. T. (1998). Jacked box tunnel design, Proc Geo-Congress, Boston, MA, Geotechnical Special Publication No. 87, ASCE, pp 21–38.

Soudain, S. (1999). Digging Boston out of a jam. *Ground Engineering*, **32**, No. 7, 22–25.

The Economist. (2003). November: 60.

Wheeler, P. (1997). Boston's heart bypass. *Ground Engineering*, **30**, No. 4, May 1997, pp. 30–31.

Chapter 8

Irlam Railway Embankment (1996–1998)

8.1 Introduction

The UK's railway infrastructure was largely constructed during the 19th and early 20th centuries, and the network's effective operation is critically dependent on many structures which are more than 150 years old. Hence, there is now a considerable part of the UK's civil engineering industry which is dedicated to the assessment and maintenance of this elderly and deteriorating infrastructure. This case history describes an innovative solution for the replacement of an old railway bridge which had become a maintenance liability. Throughout the construction of the replacement structure, the railway had to remain operational (a common requirement across the UK's busy network), except for a single 100-hour track possession. The application of the observational method (OM) was critically important, both to ensure the operational safety of the railway during the construction works and that a robust replacement structure could be built. The replacement structure involved the design and construction of the UK's first expanded polystyrene (EPS) embankment for railway use, it is also one of the world's largest EPS embankments. The river channel, which the bridge had crossed, had been cut-off following construction of the adjacent Manchester Ship Canal. Hence, the bridge became redundant and the river channel had subsequently become infilled with waste from local industry, which had been dumped into the channel. The waste materials were underlain by soft alluvium. The geotechnical behaviour of the dumped waste (which included a soap-works waste) was uncertain, although it was (compared with conventional soil mechanics metrics) likely to be highly compressible. The waste was also chemically contaminated. Structural investigations had identified that the bridge superstructure was in a fragile state and supported on timber piles which were also potentially in a poor condition. Therefore, the project team were faced with a high level of uncertainty, whilst being responsible for maintaining the operational safety of the railway and planning, designing and constructing a cost-effective replacement, whilst minimising any environmental impacts during/after the engineering works. The application of the OM provided a safe and effective way to untie this particular 'Gordian Knot'.

8.2 Key Aspects of Design and Construction

8.2.1 Overview

The site is located in the north-west of England, on the western edge of Manchester close to an old industrial area known as Irlam. Bridge 193 was built in the mid-19th century to cross the River Irwell and was part of the Trans-Pennine route. It remains an important route for local freight and passenger traffic. Figure 8.1 is an archive drawing of the bridge design, which comprised a 3-span steel bridge deck, supported by two masonry piers (either side of the river channel) and two masonry abutments. The masonry piers and abutments were founded on numerous driven timber piles. At the end of the 19th century, the Manchester Ship Canal was built and the canal cut-off the meander of the River Irwell which had flowed beneath Bridge 193. Hence, the bridge effectively became redundant. During the 19th and early 20th centuries, the area to the west of Manchester was subject to considerable industrial activity,

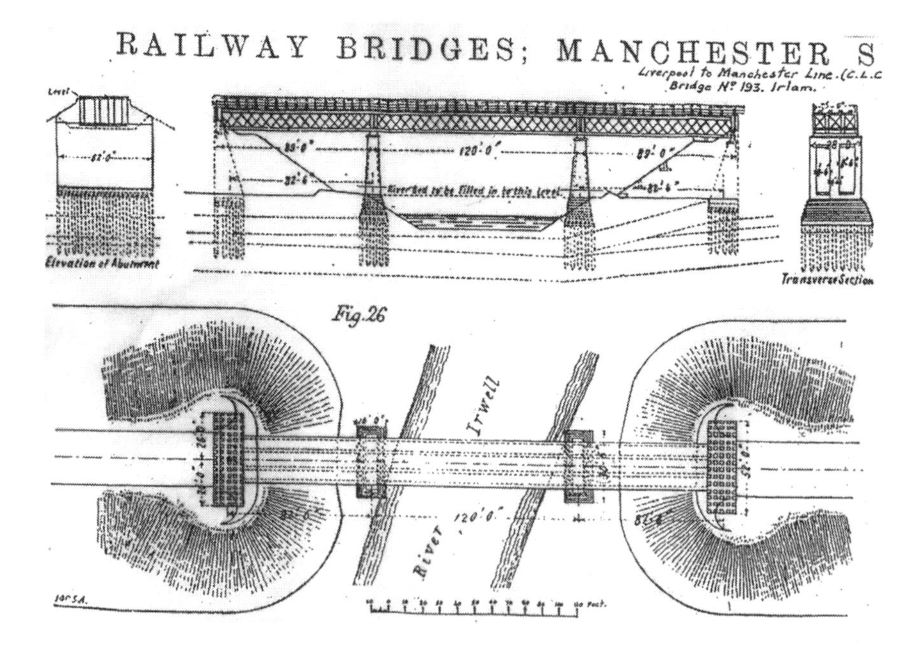

Figure 8.1 Archive drawing of Bridge 193.

and the area adjacent to the bridge site was used by numerous factories. Historical maps indicated that the river channel was infilled during the early part of the 20th century. During this period, there was no regulation of industrial waste disposal and based on local experience, it was suspected that the old river channel was likely to be infilled with a variety of non-engineered and chemically contaminated materials. The bridge had had a long history of maintenance problems and a general 20 mph speed restriction had been put in place due to concerns about its condition (reduced to only 10 mph for freight trains), Figure 8.2 shows a photo of the bridge in 1997.

8.2.2 Geology

Figure 8.3 provides a simplified summary of the ground conditions. Two phases of investigation were implemented: phase 1 comprising static cone and piezocone penetration tests, boreholes and trial pits; phase 2 comprising a footing load test and inclined rotary coring to sample the timber piles. Typical soil descriptions and index parameters are summarised in Table 8.1.

The channel infill deposits were up to 8 m thick with varying thicknesses of loose to very loose sandy silts and soft to very soft clayey silts (understood to be dredged materials from the adjacent canal). Within the channel infill deposits, the likely behaviour of the 'soap-works waste' was a particular concern. The soap-works waste had the appearance of a very soft putty chalk (Figure 8.4). The investigations indicated that the soap-works waste was likely to be of maximum thickness just to the north of

Figure 8.2 Bridge 193 prior to replacement works.

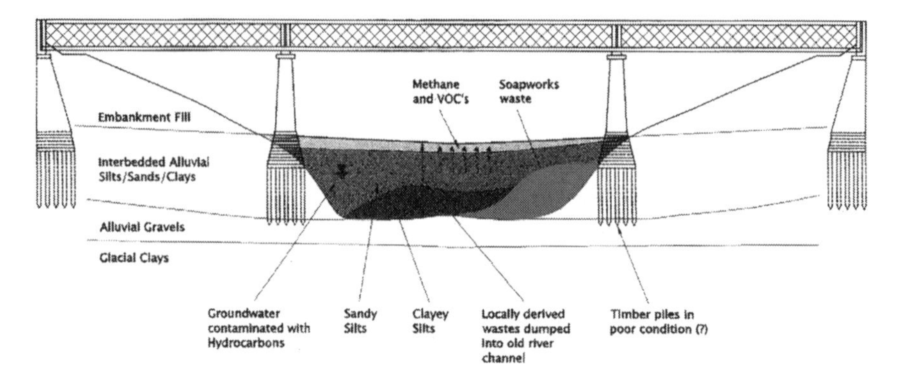

Figure 8.3 Bridge configuration and ground conditions.

Table 8.1 Summary of index properties for channel infill deposits.

Typical description	*moisture content (%)*	*plasticity index (%)*	*liquid limit (%)*	*liquidity index (%)*	*CPT qc (MPa)*
Dredged silts. Loose to very loose black clayey SILT with brick, coal, glass and timber becoming silty Clay at depth.	16–30	9–15 (12)	23–38 (29)	+0.3–+0.7	0.3–0.8
Soap-works waste. Soft to very soft, white clayey SILT with oily odour, becoming silty CLAY at depth.	26–62	12–30 (22)	27–63 (46)	+ 0.4–+1.5	0.3–0.6

Notes: 1. Data are for materials below dessicated crust, greater than 2 m below ground level. 2. Figures in brackets are mean values. 3. Clay fraction varied between 2% and 25%, grading of dredged silts and soap-works waste similar.

the bridge's eastern pier, thinning to the south and west. This was consistent with desk study information that had indicated that a soap-works factory had been active in the early 20th century to the north of the bridge site, close to the eastern bank of the river channel. It seems likely that this material had probably been end-tipped into the channel from the back of the factory site. Mineralogical analysis by scanning electron microscope and x-ray diffraction indicated that the material was predominantly calcite (calcium carbonate). A pair of boreholes was drilled at selected locations (one borehole for continuous high-quality piston sampling, drilled with a mud slurry to minimise disturbance and the adjacent borehole for closely spaced field vanes and in situ permeability tests). Borehole shear vane tests in clayey strata recorded peak undrained shear strengths between 15 and 30 kPa and a strength sensitivity of about 5–10 (i.e. remoulded strengths of only 2–3 kPa). A particular challenge with construction on industrial wastes is the assessment of long-term settlement due to creep (Watts and Charles, 2015). Due to inherent heterogeneity, test data from small scale laboratory tests are often not representative of field behaviour. Empirically established relationships between penetration tests and compressibility (derived for natural soils) are often unreliable, especially for industrial wastes. Therefore, the decision was made to carry out a footing load test (2 m by 1.5 m plan area, at one metre below ground surface). A bearing pressure of 50 kPa was applied for a period of 260 days. The influence zone of the test is limited with only the upper 1–2 m of strata below the test being significantly stressed. The test data indicated a creep coefficient, $C\alpha$, for the final phase of the test, which was relatively high at 0.012. A key objective of the ground investigation was to assess the coefficient of consolidation of the clays and silts, and the time required for primary consolidation to occur. Laboratory oedometer and Rowe cell tests were used to assess the influence of stress level and anisotropy, whilst in situ tests (Piezocone and constant head permeability) were used to assess representative field behaviour at the initial void ratio and stress state (with laboratory data used to assess reductions in the coefficient of consolidation at higher stress levels). LeRoueil *et al.* (1995) have outlined some of the challenges associated with the interpretation of the consolidation characteristics for lightly over-consolidated clays and silts. The derived parameters are summarised in Tables 8.2 and 8.3. Underlying the weak channel infill deposits were more competent layers of medium dense to dense alluvial sands and gravels (SPT '*N*' between 26 and 45), about 4 m thick, underlain by very stiff to hard glacial clays (SPT '*N*' between 36 and 79). The groundwater table level varied between 1.5 and 2 m below ground level during the monitoring period. The ground was contaminated with elevated concentrations of arsenic, groundwater was contaminated with petroleum hydrocarbons.

Methane (from a nearby landfill) and volatile organic compounds (from volatilisation of petroleum compounds) were present in the gas phase.

8.2.3 Bridge Assessment

Structural investigations of the bridge deck highlighted significant areas of corrosion and a limited residual design life, even under the applied speed restrictions. Hence, in order to be able to raise the line speed and reduce maintenance liabilities, it was recognised that bridge replacement or substantial repairs were urgently required. There was concern that the timber pile foundations may have deteriorated and be in a potentially unstable state. Hence, coring of the timber piles was carried out and samples were successfully obtained, which were then sent to the Timber Research and

Figure 8.4 Soap-works waste (white material clearly visible in trial pit).

Table 8.2 Summary of compressibility characteristics, channel infill deposits.

Soil Type	*Compression Index, Cc*	*Recompression Index, Cr*	*Creep Coefficient, Cα*
Dredged Silts	0.06 to 0.2 (0.14) **0.11**	0.004 to 0.03 **0.025**	(0.01) **0.006**
Soap-works waste	0.3 to 0.34 (0.3) **0.21**	0.005 to 0.01 **0.07**	0.012 (0.024) **0.015**

Note: Quoted range based on GI tests, figures in brackets were assumed for design. Figures in **bold,** based on back analysis of pre-load embankment settlement. Over-consolidation ratio, OCR, assumed 1.0; back analysis indicated an OCR of 1.4 at 5m below ground level (b.g.l).

Table 8.3 Summary of consolidation characteristics, channel infill deposits.

Soil Type	*Permeability (m/s)*	*Lab Cv, (nc) m2/yr*	*Lab Cv (oc) m2/yr*	*Lab Ch (nc) m2/yr*	*Insitu Ch (oc) m2/yr*
Dredged Silts	1x10-6 to 1x10-7	1.3 to 9.3	10.7 to 35.8	6.3 to 17.8	180 to 460
Soap-works waste	1 to 2 x 10-8	0.26 to 1.7	14.6 to 28.6	2.7 to 9.3	120 to 320

Note: nc: normally consolidated; oc: over-consolidated. Permeability derived from variable and constant head borehole tests. For design: coefficient of consolidation (normally consolidated state),Cv, assumed 20m2/yr for soap-works waste and 50m2/yr for dredged silts (i.e. between 6 and 10 times lower than insitu Ch measured by piezocone). Back analysis consistent with assumed design values.

Development Association for specialist analysis. The timber was identified as a true Pine from the Taeda species (common name 'pitch pine'). Measured density varied between 685 and 720 kg/m^3 (consistent with denser species, such as Caribbean Pitch Pine). Some of the samples had suffered softening of the outer timber surface. However, overall, the timber was judged to be in a sound condition and was unlikely to have lost much of its original strength. The piles were below the lowest seasonal water table level, and it was considered that the elevated concentrations of arsenic in the channel infill may have assisted in preserving the timber (historically arsenic was used as a timber preservative). The pile groups supporting the central piers comprised 50 piles (5 rows of 10 piles), and the pile cap (comprising a grid of horizontal timbers) was at about 2 m below ground level (i.e. just below the groundwater table level).

8.2.4 Project Constraints

Table 8.4 summarises the major site and operational constraints. An important requirement for Railtrack (the organisation which pre-dated Network Rail and who was responsible for managing the track infrastructure) was to keep the railway operational throughout any repair/replacement works, except for a 100-hour track possession. This meant that only the bridge deck could be removed and the old bridge piers (and foundations) would be left in place within any new structure. Feasibility studies considered the following options:

1. replace with a new bridge;
2. replace with a new embankment;
3. refurbish the existing bridge to take ballasted track.

The outcome of the feasibility study indicated that options 1 and 3 were too expensive, especially when assessed on a whole life cost basis.

Table 8.4 Major project constraints.

Type of constraint	*Description of constraint*	*Implications*
Site	Headroom limited below bridge deck	Limited options for piling.
Structure	Bridge deck in poor condition. Bridge had to remain operational, except for a single 100 hour track possession.	Works carried out before possession had to minimise potential for further deck deformation. Limited time available to carry out works during possession.
Soil	Deep layers (up to 8 m thick) of weak compressible soils. High water table.	Large displacements likely if conventional fill used up to underside of bridge deck.
Safety	Soils and groundwater contaminated. Gas (methane and VOCs).	Health and safety implications if contaminated soils excavated. Disposal costs.
Sustainability	EA wished to avoid any works which penetrated into underlying gravels (risk of contamination spread?)	Impacts on neighbours?

VOC, volatile organic compound.

Construction of a new bridge or refurbishment of the old bridge would also leave Railtrack with a long-term maintenance liability. Therefore, more detailed assessments of option 2 were carried out, taking into account the constraints outlined in Table 8.4.

8.3 Evaluation of Embankment Options

8.3.1 Identifying Viable Options

Railtrack's objectives were to maintain safety, keep the railway operational, eliminate future maintenance, achieve differential settlement criteria for track, and to minimise adverse environmental effects (during/after the works). A wide range of embankment options were assessed against these objectives and a selection of options are summarised in Table 8.5. Only two solutions were considered viable, these were:

1. conventional granular fill supported on a piled raft;
2. ultra-lightweight fill (EPS) and minimise piling/ground improvement.

The requirement to keep the bridge deck in place, until the 100-hour track possession, meant that any piling or ground improvement works had to be carried out beneath the bridge deck with a headroom of less than 10 m. The limited headroom placed severe restrictions on the range

Table 8.5 Options for replacement embankment.

Potential solution	*Main concerns*
Conventional embankment (granular fill) on piled-raft	Inadequate headroom for conventional piling plant, e.g. CFA piling. Potential contaminant migration pathway. Low headroom driven piles, vibration and high risk of damage to existing bridge and foundations. Low headroom bored piles, pile bore stability (and associated risk of local movements adjacent to timber piles) and integrity of pile shaft, handling contaminated arisings/groundwater, gas venting.
Conventional embankment without piling/ground improvement	High risk of instability and excessive deformation. High risk of destabilising existing bridge during embankment construction.
Encapsulation of bridge deck supported by part-height embankment	Long-term maintenance liability. Ground stabilisation still required prior to embankment construction.
Embankment with EPS core and upper embankment/shoulders with conventional fill. No pre-loading.	Excessive differential settlement in long term, due to settlement of soap-works waste and dredged silts in old river channel.
EPS core, as above, with pre-loading prior to permanent works.	Risk of damage to existing bridge and foundations if pre-load embankment construction poorly controlled.

of piling and ground improvement options which could be used, since available CFA piling and commonly used ground improvement (such as vibro based methods) required a head room in excess of 10 m. The installation of driven piles posed an unacceptable risk of damage to the existing bridge and its foundations. Given the ground and groundwater conditions there were significant concerns associated with the use of low headroom bored piles, in particular the Environment Agency raised concerns about the potential risk of the piles forming a pathway for migration of contaminants. Irrespective of this potential risk, during bored pile installation, contaminated arisings and groundwater would have to be handled and disposed of, and gas risks safely managed. The EPS embankment, although highly innovative (and hence Railtrack were initially rather sceptical), did have several compelling advantages:

- minimal disturbance to the existing ground and hence minimal environmental impacts;
- minimal disturbance to the existing piled foundations;

- the EPS blocks could be easily placed in low headroom below the bridge deck (in fact the EPS embankment could be built practically with near zero headroom, by sliding the blocks in sideways, beneath the main bridge deck girders). This minimised the works necessary during the track possession, a major advantage compared with conventional embankment fill requiring compaction;
- the wedge-shaped embankment areas between the central piers and abutments (Figure 8.3) were relatively easy to construct with EPS but highly problematic for low headroom piling and conventional fill;
- cost-estimates for both options were broadly similar, however, commercial reviews indicated that there were more contractual risks if option 1 (conventional fill on piled-raft) was adopted, based on local piling experience in similar ground/groundwater conditions.

Nevertheless, there were concerns with the EPS embankment. Because the bridge piers had to be left in place, these would act as 'hard spots' and the freight train loads (with high axle loads) meant that relatively high stresses would be imposed on the EPS and differential settlement was a concern. Conventional fill would need to be used on the embankment shoulders and above the EPS core (to protect the EPS and minimise the applied stresses imposed by rail traffic), which meant that the replacement EPS embankment would still have sufficient weight to generate significant long-term settlement. This, in turn, necessitated a pre-load embankment to minimise time-dependent settlement beneath the permanent embankment. The design concept is outlined in Figure 8.5, this aimed to minimise:

- works during the 100-hour track possession;
- long-term settlement.

8.3.2 Potential Adverse Effects on Bridge Foundations

A pre-load embankment, between a height of 4 and 4.5 m, was judged to be a highly effective means of reducing long-term settlement. The applied bearing pressure was about 20% higher than the permanent embankment (with an EPS core) and was expected to reduce long-term settlement (during a 120 year design life) from about 300 mm (without any pre-load) to less than 50 mm. There was a substantial period of time available (about 9 months) for the weak ground to consolidate, relative to the anticipated consolidation characteristics of the soft clayey silts. From a practical perspective, the fill used for the pre-load could be re-used to form the shoulders of the final embankment and provide protection to the EPS. However, the pre-load embankment would inevitably impose additional stresses on the existing bridge pier and timber piles. Soil–structure

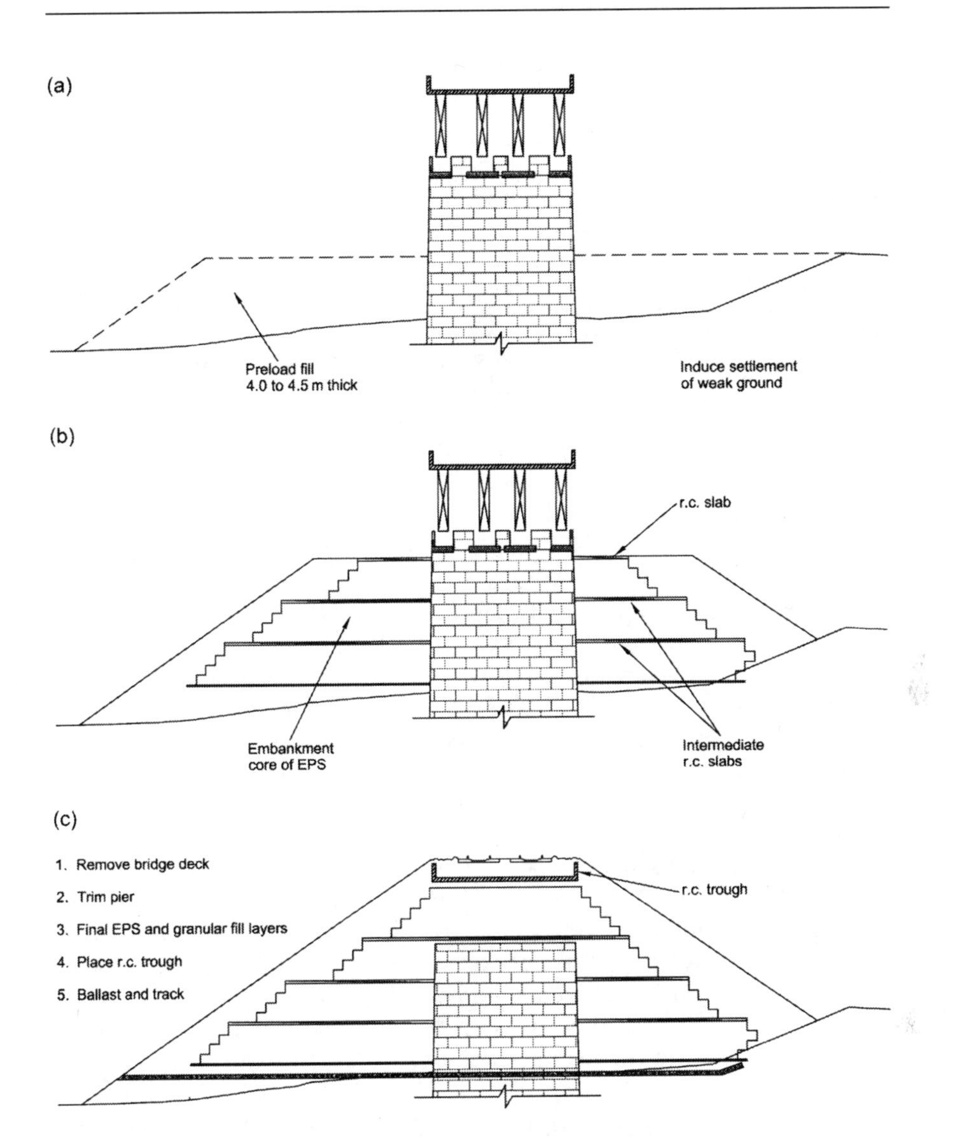

Figure 8.5 Sequence of construction for replacement embankment.

interaction analysis of the existing pile groups (using PC-MPILE) and a structural check (using BS 5268) indicated that the timber piles had a sufficient safety margin (comparing the stresses imposed by the bridge against the permissible stresses allowed by BS 5268). However, the safety margin reduced significantly under the imposed load of the pre-load embankment (Table 8.6).

Table 8.6 Stress analysis of existing timber piles.

Design Scenario	*Average utilisation (%)*	*Maximum utilisation (%)*
Existing condition	52	73
Negative skin friction during pre-load	77	95
Lateral soil movement during pre-load	93	111

Note: utilisation is ratio of imposed stress to allowable stress, based on BS5268; values>100% imply pile is over-stressed. 'Average' is calculated for perimeter piles, 'maximum' is for most highly stressed perimeter pile. Negative skin friction calculated on effective stress basis, assuming negative shear stress on pile shaft equivalent to a quarter of mean effective overburden stress, this method is more consistent with observed values than total stress methods: (Meyerhof, 1976; Fellenius, 2006).

The potential additional imposed loads from the pre-load embankment comprised the following:

1. Down-drag force on piles (negative skin friction) – the intent of the pre-load is to induce consolidation and creep settlement of the soft compressible channel infill deposits. The soil settlement will be resisted by the timber piles, with the down-drag force being taken mainly by the piles on the perimeter of the pile group, (Fellenius, 2006, discusses negative skin friction effects). There was no practical method to mitigate the adverse effect of negative skin friction.
2. Down-drag force on masonry piers – as the compressible soil consolidates the pre-load embankment fill will settle relative to the masonry piers. This could have imposed a significant additional down-drag force (about 20% larger than 1 above). However, this adverse effect was minimised by placing 2 layers of low friction geomembrane (with a specified interface friction of less than 15°) around the piers, together with a ring of 600 mm thick EPS blocks between the geomembrane and the pre-load embankment fill (to reduce the horizontal stress from the fill). Acting in tandem, these measures were estimated to reduce the down-drag force acting directly on the piers to less than 10% of the original estimated values.
3. Out-of-balance earth pressure from pre-load embankment – during construction of the pre-load embankment, some out-of-balance horizontal load could be applied, if the fill level on one side of the pier is allowed to increase relative to the opposite side. This could be minimised through strict on-site control and minimising the thickness of each layer of placed fill.
4. Lateral loads from horizontal movement of soft soils – potentially the most dangerous loads which could be imposed on the piles were those due to lateral soil movements under the pre-load embankment weight (soil–structure interaction mechanisms are discussed by Ellis

and O'Brien, 2012). The bending moments potentially induced in the timber piles by lateral soil movements were calculated by two different methods; an empirical method described by Stewart *et al.* (1994) and a theoretical method outlined by Springman and Bolton (1990). Both methods gave similar results and as indicated in Table 8.6, these lateral loads and the associated combined bending and compressive stresses induced in the timber piles were more significant than the effects of negative skin friction.

There was considerable discussion about the risk associated with the potential over-stress of the timber piles due to the combined effects of 1 and 4 above. A pessimistic interpretation of the analysis results and BS 5268 would indicate a risk of progressive failure of the pile group. A more realistic assessment indicated this was unlikely. CP 112, which pre-dated BS 5268, gave permissible stresses which were 20% and 40% higher for bending and compression, respectively, than those required by BS 5268. Perhaps more importantly, the ultimate strength of Pitch Pine (based on BRE bulletin 50) is about 3 times higher than the permissible stress values given in BS5268. The permissible stress given in BS5268 is intended to avoid creep deformation of the timber; hence, once the timber stress exceeds the permissible stress, say in an over-stressed pile, it would yield and begin to creep (hence its mobilised structural stiffness would reduce) and there would tend to be a redistribution of load into adjacent stiffer piles. Therefore, any plausible failure mechanism was likely to be ductile. The large size of the pile groups also provided significant redundancy. Nevertheless, there was considerable concern amongst the project team given the need to maintain the operational safety of the Railway. The lateral displacement of soft clay during and after embankment construction has been studied by Tavernas and Leroueil (1980), and key observations (based on analysis of several case histories) are summarised in Figure 8.6. Figure 8.6(a) summarises behaviour during embankment

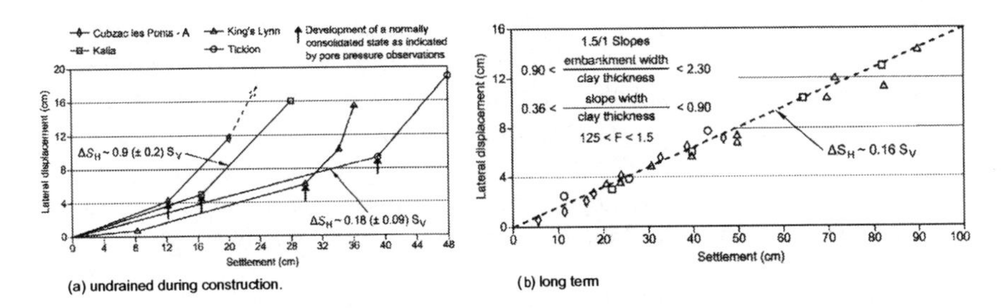

Figure 8.6 Case history observations of lateral displacement and settlement for embankments constructed on soft clay (after Tavernas and Leroueil, 1980).

construction when the clay is predominantly undrained (due to rapid embankment construction) and identifies a threshold (denoted by an arrow) beyond which lateral displacement is practically the same as settlement. Prior to this threshold horizontal displacements are much smaller than settlements. This threshold coincides with loading the soft clay beyond its pre-consolidation pressure. Figure 8.6(b) plots long-term horizontal displacement and settlement as the soft clay consolidates, during this phase, lateral displacement is much smaller than settlement. Hence, for the planned works, it was concluded that it was essential to avoid rapid, undrained, loading of the channel infill deposits in order to minimise the risk of large lateral ground displacements and excessive loading of the timber piles. Therefore, careful control of the rate of pre-load embankment construction would significantly reduce the risk of damage to the timber pile foundations.

8.4 Achieving Agreement to Use the OM

In parallel with the conceptual design of options 1 and 2 (conventional fill and EPS embankments, respectively), detailed risk assessments were carried out, each risk was identified, and mitigation measures were developed. Cost estimates for various mitigation measures were prepared. The Railtrack teams responsible for track safety and operations were briefed and the pros/cons of the options discussed. In general, it was considered that option 2 (with a pre-load embankment) was preferable to option 1. The designer recommended the use of the OM to control risks during construction of the replacement embankment. The OM could be readily applied via progressive modification, simply by modifying the pre-load embankment construction sequence. The pre-load embankment would be constructed in several stages or increments – the amount of fill placed in a single stage could be reduced and/or the 'rest' period between stages of fill placement could be increased, if the OM design team deemed that the risk to the existing bridge, in particular the bridge foundations, was becoming too high. Hence, the primary risk mitigation was to avoid undrained loading of the soft ground, allow it to consolidate and gain strength. Several additional mitigation measures were also developed, and these are summarised in Table 8.7. However, the probability of needing to implement these additional measures was judged to be low. Commercially, Railtrack had framework contracts in operation which meant that demobilising and remobilising an earthworks contractor to site to place fill in a controlled manner over a period of several months was relatively economic. A key aspect of gaining agreement was to demonstrate that risks for all plausible scenarios could be kept as low as reasonably practical. It was also necessary to show that mitigation measures could be practically implemented before track alignment and operational safety were adversely impacted. To facilitate this, a series of

Table 8.7 Potential mitigation measures.

Mitigation measure	*Comment*
Reduce rate of embankment construction	Reduce potential disturbing force due to excessive lateral ground displacement, allow weak ground to consolidate and gain strength
Shore piers	Maintain bridge safety, if excessive pier displacement
Support deck	Maintain bridge safety, if excessive pier displacement and/ or excessive differential settlement of deck
Local strengthening of deck	If local distress observed

bridge and foundation failure modes were evaluated, and a series of flow charts developed. These supported quantitative risk analyses (using event-tree analyses) and detailed planning for implementation of the OM. It was particularly important to brief and gain agreement with the various parties involved so that roles and responsibilities (as outlined in Table 8.8) were clear, especially for implementation of any potential mitigation measures. Railtrack gave their approval for implementation of option 2, primarily because operational safety could be demonstrated in an objective and transparent manner to all stakeholders through the application of the OM.

Table 8.8 Project team, roles and responsibilities during the OM implementation.

Team Member	*Role*	*Responsibility*
Designer	Engineering consultant (OM designer)	Inspect bridge and monitor condition. Check and interpret instrumentation. Provide technical advice to Railtrack during the OM implementation.
Railtrack	Project manager and track infrastructure owner	Safe operation of railway. Manage works and instruct contractor. Communications with P-Way contractor.
Preload contractor	Build pre-load embankment (OM contractor)	Safe implementation of works. Carry out mitigation measures as instructed.
Instrumentation contractor	Install instruments, provide factual data	Check instruments functioning correctly. Replace faulty instruments, report factual data to OM designer, contractor and Railtrack.
Permanent way contractor	Track maintenance	Maintain track alignment within Railtrack prescribed tolerances for twist.

Note: designer, contractor and Railtrack nominated a senior engineer to the OM team who had authority to agree and authorise the pre-planned mitigation actions if trigger levels breached.

8.5 Implementation of the OM

8.5.1 Trigger Levels and Contingency Measures

Simple analyses were adopted to establish initial values for trigger limits. Because of the heterogeneous ground conditions and the deteriorated and uncertain state of the structure, there was at best limited value in carrying out highly sophisticated analyses and arguably could have been misleading. To assess ground movements, the classical one-dimensional method was used to calculate the time-dependent settlement of the channel infill together with empirical correlations by Tavernas and LeRoueil (described above) to assess horizontal displacements due to settlement. A structural frame analysis was used to assess the stress changes across the bridge truss steelwork associated with imposed pier displacements (settlement and/or horizontal displacement).

Figure 8.7 shows three flow charts for: pier tilt, track alignment and sub-surface ground movement. These flow charts were used together with Table 8.9 (which provides trigger levels for specific bridge and ground displacements) to operate an amber/red traffic light system. Pier tilt was identified by the OM design team as the critical observation for management of the OM. Sub-surface movement (ratio of maximum horizontal displacement to settlement) was a secondary observation and was mainly used to enable more informed judgements of ground/bridge interaction. It was deemed to be necessary given the uncertain behaviour of the soapworks waste. Pier tilt was judged to be the most relevant and critical observation for maintaining bridge safety during the pre-load embankment construction, because the pier tilt could be directly affected by embankment construction if poorly controlled. However, for Railtrack track operations, the monitoring of track alignment was their normal means of confirming track safety. Hence, for Railtrack, this was non-negotiable as being a critical observation, as well as pier tilt. The project team had some reservations about the usefulness of monitoring track alignment. The track alignment was likely to be influenced by several factors, independent of embankment construction, such as the poor condition of the timber sleepers. Implementing the OM with two sets of critical (or primary) observations is not ideal because of the additional complexity. In this case, it was feasible because the initial mitigation measures were the same, for example, if a red trigger was breached on either pier tilt or track alignment then embankment construction would be stopped, and line speed would be reduced to 5 mph (by the P-Way contractor). The subsequent actions would vary depending on the outcome of an inspection of the bridge deck and piers (refer to Table 8.7 and Figure 8.7).

Exceeding an amber trigger level on primary readings (pier tilt or track alignment) or a red trigger level on a secondary reading (i.e. sub-surface movement) required the rate of raising the embankment height to be

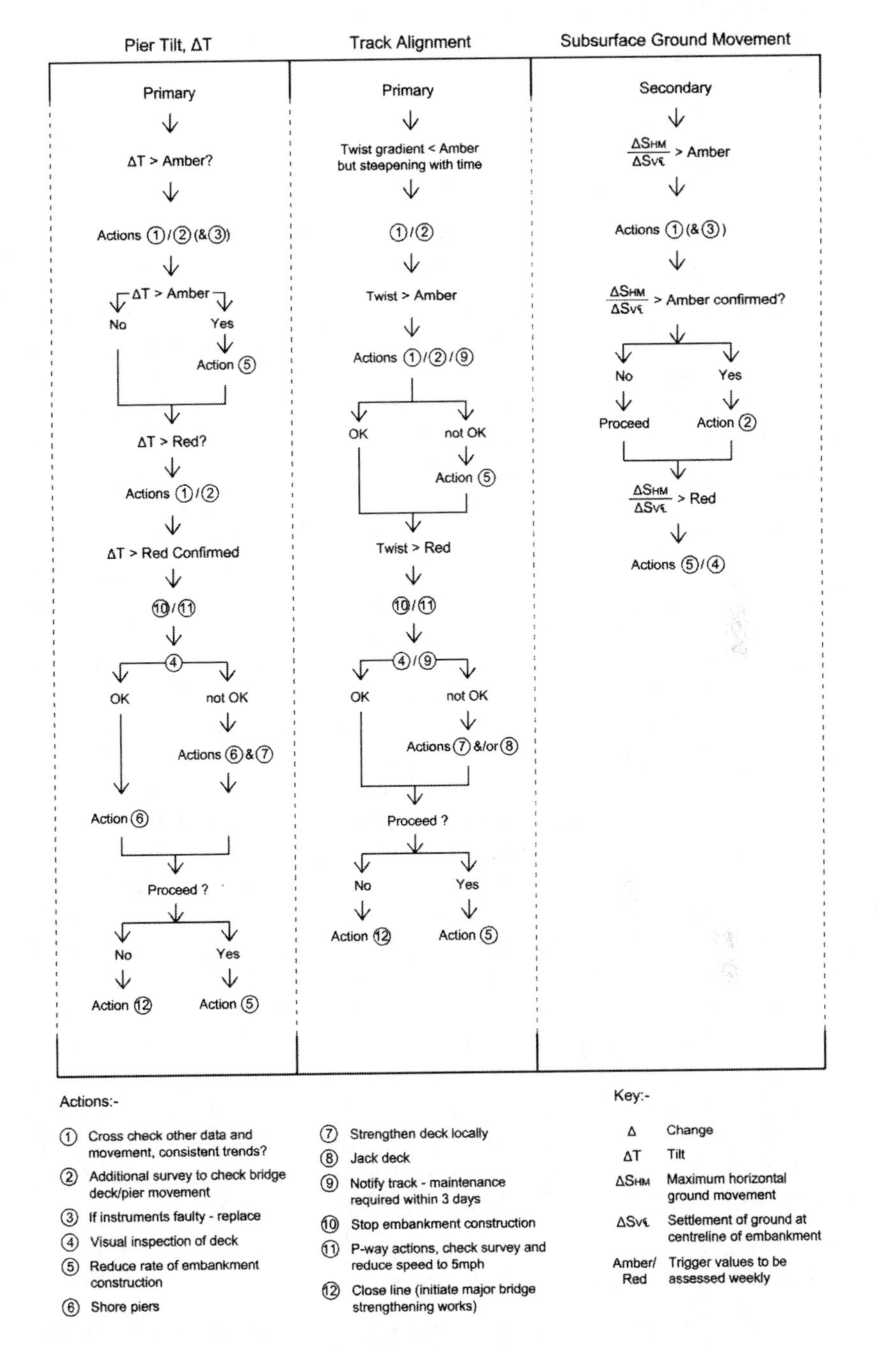

Figure 8.7 The observational method flow chart.

Table 8.9 Trigger levels for the OM implementation.

Parameter	*Amber*	*Red*
Pier tilt (degrees)	0.05	0.14
Track deflection ratio	1:240	1:120
Maximum horizontal ground movement/settlement	0.2	0.4

Note: red limits were 40% higher than the red trigger level

reduced. Slowing embankment construction was a simple action and was expected to be effective in maintaining safety (in terms of the embankment stability and hence impacts on the bridge piers).

As noted above, because the contractor building the pre-load embankment was working through a framework contract, de-mobilising from site (and subsequently re-mobilising when appropriate) was straight-forward and inexpensive to manage through the framework. Similarly, actions by the P-Way contractor were easy to manage through existing contract arrangements (which required the P-Way contractor to visit various sites in the neighbourhood for track maintenance). The implementation of the OM through three different sets of traffic lights would not have been practically manageable, if the mitigation measures had been different for each traffic light and if the works on site had not been managed through a multi-project framework agreement and been intrinsically simple (i.e. placing fill).

There was a dedicated OM team on site throughout the works. They were in full control of taking measurements of ground and bridge movements, interpretation and communicating collated observations (within a standard format) to senior staff in the locally based offices of the designer, contractor and Railtrack. The P-Way contractor monitored the track deflection. The site OM team were authorised to implement amber actions. A senior staff member from each of the designer, contractor and Railtrack led the OM team and regularly reviewed observations from the site team. If a red trigger level was breached, then they would be immediately notified to authorise the previously agreed actions (i.e. stop embankment construction and inform the P-Way contractor). Communications were effective throughout the implementation of the OM. This was facilitated by frequent face to face meetings, which were quick to arrange because all the OM senior staff were locally based, both to the site and their offices.

8.5.2 Instrumentation and Monitoring

Careful thought was given to the instrumentation, including the types of instruments, their location and purpose. All members of the OM site team were briefed so they were aware of the specific reasons for each instrument installation. The sub-surface instrument layout and the specific objectives of each instrument are summarised in Figure 8.8 and Table 8.10, respectively. Advantages and disadvantages were carefully considered, for example, the ground response could be assessed by monitoring ground deformation (horizontal displacement and settlement) or pore water pressure. From a theoretical perspective, monitoring pore water pressure could be considered to be a more fundamental parameter (e.g. to differentiate between undrained and drained behaviour). However, given the nature of the ground conditions, severe problems were anticipated in obtaining reliable and representative measurements of pore water pressure (due to the presence of ground gas and the variability between naturally derived clays/silts and industrial wastes). In contrast, horizontal ground displacement could be linked to increased pile bending and the ratio of horizontal displacement to settlement could be linked to the onset of undrained behaviour and the risk of ground instability (as discussed by Tavernas and Leroueil, 1980). Importantly, ground deformation was likely to give a more representative measure of global ground response (since deformation reflects cumulative strain in each sub-layer), whereas pore water pressure measurements tend to reflect the localised response of particular sub-layers.

Prior to embankment construction, instruments were installed and a series of readings were taken during a period of about 8–10 weeks to verify that they were functioning correctly and identify any environmental effects. These baseline readings proved to be extremely important and enabled an important facet of the bridge behaviour to be determined. Figure 8.9 shows a plot of the bridge pier displacement with time (based on inclinometer measurements), also shown is the recorded

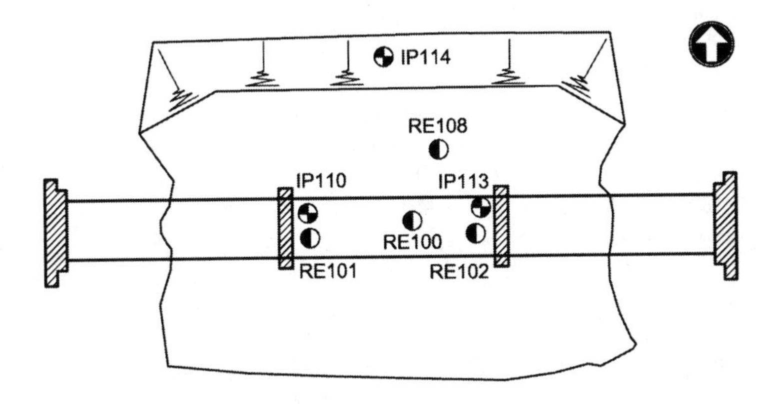

Figure 8.8 Sub-surface instrumentation layout.
Notes: IP= inclinometer; RE= rod extensometer.

Table 8.10 Instrumentation design.

Instrument type	*Purpose*	*Primary or Secondary*	*Comments*
Pier inclinometer	Tilt of pier	Primary	Electro-levels, continuous monitoring of piers, identify movement trends. Critical observations for controlling pre-load works
Precise survey	Tilt of pier and settlement	Primary	Intermittent data, so mainly used to cross-check pier inclinometers
Track recording vehicle	Track alignment	Primary	Railtrack required to monitor track twist, however excessive twist could be due to several causes, including poor condition of timber sleepers supporting track.
Inclinometer	Subsurface lateral displacement	Secondary	Excessive lateral ground displacement could lead to risk of damage to timber piles but, depending on timber pile capacity and load redistribution across pile group, ground movement may not lead to critical condition for bridge.
Extensometer	Subsurface settlement	Secondary	Ground settlement important for monitoring ground consolidation induced by pre-load, ratio of lateral displacement to settlement an important indicator of overall ground behaviour (i.e. undrained or drained).
Piezometers	Groundwater pressure	–	Considered to be non-representative, and likely to be unreliable, due to presence of gas and local variability.
Strain gauges on bridge deck	Changes in stress in deck	–	Considered to be too difficult to interpret, due to poor bridge deck condition and highly complex load paths.

Note – temperature recorded continuously, this proved to be invaluable for interpretation of pier displacement.

air temperature. Precise survey measurements confirmed the overall magnitude of displacements indicated by the inclinometers. The structural survey of the bridge identified that the bridge bearings were in poor condition and therefore there was some uncertainty about their effectiveness. The observations, prior to pre-load embankment construction, provided strong evidence that the east pier bearing had effectively locked-up, whereas the west pier bearing remained functional and was able to move freely. The east pier displacement closely reflected fluctuations in air temperature (due to thermal bridge deck displacements exerting push/pull on the east pier, via the locked bearing, as temperatures increased or decreased). In contrast, the west pier exhibited negligible displacement.

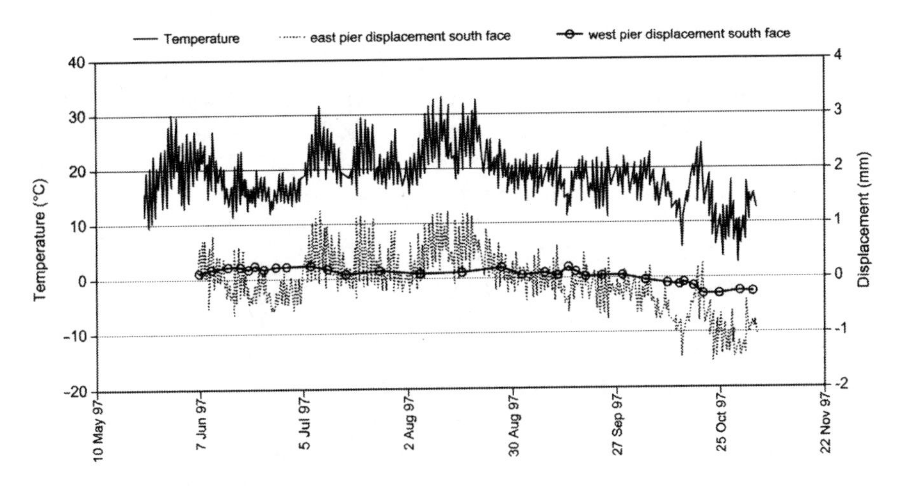

Figure 8.9 Displacement of bridge pier, before and during pre-load embankment construction.

8.5.3 Pre-load Embankment Construction

During the first stage of pre-load embankment construction, an amber trigger was exceeded for sub-surface deformation, which necessitated additional surveys of pier and deck deformation to be carried out, Figure 8.7. An adverse trend was then identified for exceedance of a red trigger. Although it was recognised that the reason for the adverse trend was relatively small settlement (since the trigger level was a ratio of horizontal displacement and settlement) rather than excessive horizontal deformation (which was the primary concern regarding damage to the timber piles). A prudent decision was made to implement the red actions before the red trigger was exceeded (conventional stability calculations indicated a factor of safety in excess of 1.3). However, because of the unusual nature of the soap-works waste, a cautious approach was judged to be appropriate. Pre-load embankment construction was extended over an additional 2 weeks from that originally planned, with three additional smaller embankment lifts of 0.8 m (rather than two lifts of 1.2 m). During subsequent embankment construction stages, the horizontal ground displacement did not increase significantly, whereas settlements did increase (hence the ratio reduced). The additional structural surveys and bridge inspection did not identify any adverse changes of the piers or bridge deck. Overall, the instrumentation and monitoring functioned effectively, red trigger levels were not exceeded for the primary instruments (for piers and track). Occasional high readings were observed, exceeding amber trigger levels, but re-checking indicated that these were due to survey error, and reliable data for the primary instruments remained in the green zone.

The timing of key construction activities is given in Table 8.11. Pre-load embankment construction comprised 4 stages (an initial fill height of about 1.8m, followed by three 0.8m high stages), with intermediate rest periods, over a 9½ week period. The pre-load was then left to consolidate for a further 9 months. Settlement during and after pre-load embankment construction is shown in Figure 8.10. At the extensometer location, there was about 2 m of silt underlain by 4 m of soap-works waste, underlain by 2 m of soft clay. A noticeable feature is the stiff initial response during the first stage of embankment construction, then (after the embankment height exceeded about 1.2 m) settlement started to increase significantly (probably after the pre-consolidation pressure was exceeded, at about 20–25 kPa in excess of the initial vertical effective stress). Although somewhat surprising for a 100-year-old dumped fill, a pre-consolidation effect had possibly been induced by water table level fluctuations and/or chemical bonding effects within the industrial waste. At the end of embankment construction, settlement of the channel infill deposits had nearly reached 90 mm. However, careful construction control ensured horizontal displacements were less than 20% of settlement (consistent with Figure 8.6(b) for drained embankment construction). Table 8.12 summarises the observed ground displacements. Back-analysed compression indices can be compared with those determined from the ground investigations in Table 8.2 (a fairly close match is apparent, perhaps this is fortuitous to some extent, but also suggests the efforts to obtain high quality GI data were worthwhile). The main difference was that it was assumed that the channel infill was normally consolidated, whereas the observations indicated it was lightly over-consolidated (with an over-consolidation ratio of about 1.4 at 5m depth). It was the lightly over-consolidated state which led to relatively small settlement during the early stage of embankment construction.

Table 8.11 Timing of key construction activities.

Activity	*Date*
Pre-load embankment construction	18/8/97 to 21/10/97
Consolidation of pre-load	21/10/97 to 2/7/98
Pre-load removal	2/7/98 to 16/7/98
Construction of permanent embankment to underside of bridge deck	31/7/98 to 28/9/98
Possession works, Bridge deck demolition and completion of embankment	2/10/98 to 6/10/98
Track ballasting and raising line speed	6/10/98 to 6/10/99

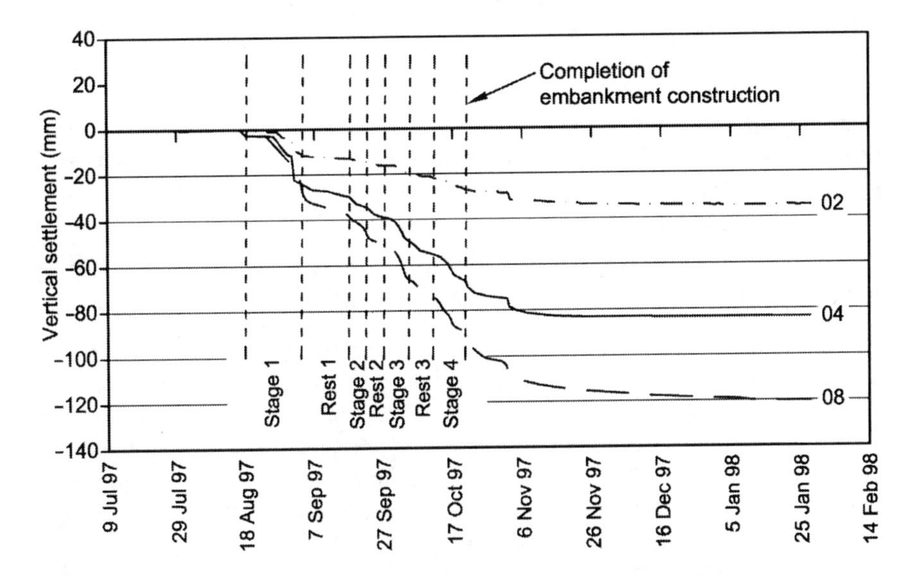

Figure 8.10 Channel infill deposits – settlement versus time, under pre-load embankment. Note: numbers 02, 04, 08 refer to settlement recorded at 2m, 4m, 8m bgl respectively.

Table 8.12 Summary of observed ground displacement.

Displacement	*End of construction (EOC)*	*EOC + 3 months*	*EOC + 7.5 months*
Settlement of Channel Infill deposits	88 mm	115 mm	128 mm
Total settlement	95 mm	127 mm	140 mm
Horizontal displacement at ground surface	8 mm	17 mm	18 mm
Maximum horizontal displacement	16 mm	17 mm	18 mm

Note: maximum observed displacement. By end of June 1998 (EOC+9 months) total settlement = 145 mm.

The pier settlements at the end of the pre-load period were 12 and 18 mm for the east and west pier, respectively, these settlements are consistent with settlement estimates for the alluvial sands/gravels and glacial deposits underlying the channel infill deposits together with a small additional settlement of the pile foundations due to the down-drag force acting on the perimeter of the pile group.

Small track displacements were measured throughout the works, and these did not exceed background values prior to embankment construction. At the end of the consolidation period, settlement had increased to about 145 mm, primary consolidation was practically complete about 3 months after pre-load construction and secondary consolidation (creep) was developing. Back-analysis of the settlement data confirmed that the soap-works waste was subject to significant creep (about double the value expected for a comparable natural inorganic silt).

8.5.4 Removal of Pre-Load and EPS Embankment Construction

At the end of the pre-load consolidation period, the pre-load embankment was carefully removed to avoid out-of-balance loading of the bridge piers. Final embankment construction then commenced with construction of a geogrid reinforced granular mattress and gas venting blanket. The EPS core comprised 13,000 m^3 of EPS, in 18 layers, arranged in a staggered pattern to avoid continuous vertical joints. Five grades of EPS were used with nominal densities varying between 20 and 55 kg/m^3. At that time, only conventional EPS grades of up to 30 kg/m^3 had been used for highway applications. Rail loading is much higher than highway traffic loading, and the strength and stiffness of these conventional EPS grades were likely to be inadequate to resist the stresses imposed by train loads. However, discussions with an EPS manufacturer indicated that new EPS grades of 40 and 55 kg/m^3 could be manufactured, which would be potentially much stiffer and stronger than conventional EPS grades. During an initial review of published data, it also became apparent that there were deficiencies in conventional EPS testing methods which would lead to EPS stiffness being under-estimated. An extensive laboratory testing programme was designed (using local strain gauges to more accurately measure EPS stiffness) which led to an improved understanding of the static and dynamic properties of EPS, the results are described in detail by O'Brien (2001a). The new 55 kg/m^3 EPS grade was about 5 times stiffer than the conventional 20 kg/m^3 EPS (Figure 8.11). The availability of more reliable data for the strength and stiffness characteristics of the EPS also enabled a more rational basis for EPS design and quality control to be developed, as described by O'Brien (2001a). This was similar to Norwegian design practice for highway embankment design (as outlined by Frydenlund and Aaboe, 1996), but the new design method extended across a wider range of EPS grades and to the higher cyclic stresses imposed by rail loading.

Numerical modelling (FLAC) was used to assess the stresses imposed on the EPS during train loading and to optimise the layout of the different EPS grades (O'Brien, 2001b). A bespoke non-linear stress-strain model (based on Puzrin and Burland, 1996) was developed to characterise the EPS

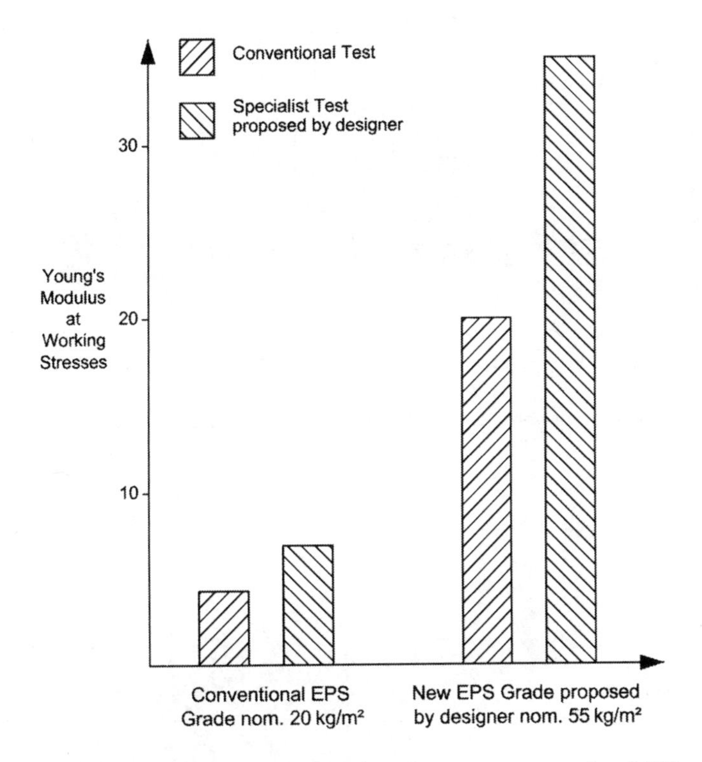

Figure 8.11 Young's Modulus of conventional grade versus new grade of EPS and influence of test method.

behaviour. The data used to derive the input parameters for the modelling were obtained from the laboratory testing programme outlined above. The main focus for the numerical modelling was minimising differential settlement adjacent to the buried bridge piers and ensuring that the EPS in the vicinity of the piers was not over-stressed. The main features of the final design are summarised in Figures 8.12 and 8.13 and Table 8.13.

Although beyond the scope of this case history, considerable thought was given to minimising potential long-term maintenance liabilities for the new embankment. An important practical aspect of the design was to protect the EPS from oil spills, fire and vandalism or burrowing animals. The level of protection exceeds that normally provided for highway embankments with a core of EPS. This was achieved by using a reinforced concrete trough lined with an HDPE liner, with MDPE protecting the sides of the EPS core. Intermediate reinforced concrete slabs provided additional protection. The relatively thick shoulders of granular fill provided an initial protective barrier to EPS damage from

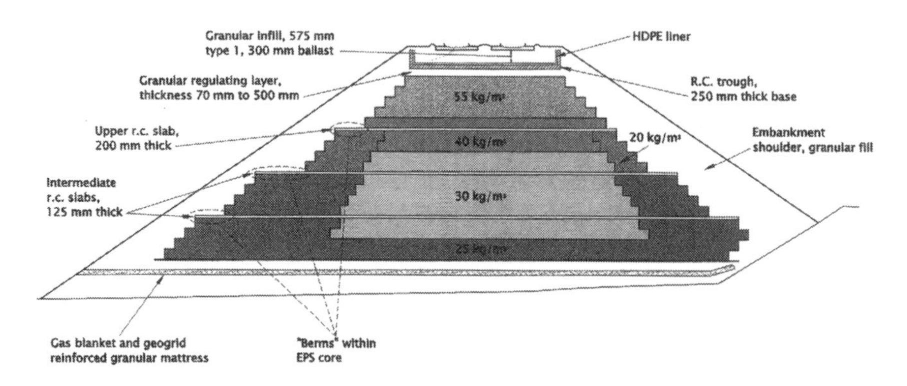

Figure 8.12 Main components of embankment design, typical cross-section.

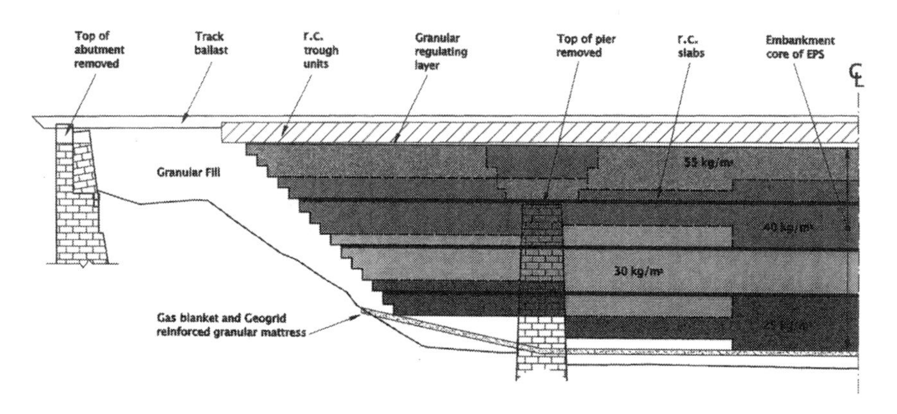

Figure 8.13 Main components of embankment design, long-section.

a range of potential causes. All outer EPS blocks were a fire-retardant grade. These protective features are also summarised in Figures 8.12 and 8.13 and Table 8.13.

Train operations continued unhindered during the embankment construction (Figure 8.14). Construction of the EPS core and embankment shoulders was completed up to the underside of the bridge deck, and the upper reinforced concrete slab was poured prior to the track possession (Figure 8.15). Then, a major 100-hour track possession was used to remove the bridge, finish the embankment, place the pre-cast concrete troughs and reinstate the track. A 1000 tonne crane was used to carry out the lifting operations during this major logistical exercise. Trains began running across the embankment immediately after the possession, with

Table 8.13 Main features of replacement embankment.

Component	*Comment*
Granular fill and ballast within upper trough	Provide conventional track bed for rails.
Pre-cast reinforced concrete troughs	Provide 'container' for HDPE liner and lateral restraint for track bed. Efficient means of reducing concentrated stresses from train loads. Quick to place during possession.
Structural pin connections between troughs across piers	Minimise differential settlement.
HDPE and MDPE liners	Protect underlying EPS from oil spills.
Intermediate reinforced concrete slabs	Levelling screed for buildability; stress redistribution layer and horizontal restraint through EPS blockwork. Also provides protective barrier to underlying EPS.
EPS core geometry, use of berms	Ensure yield stress of underlying soils not exceeded.
EPS core of different grades, nominal density from 20 kg/m^3 to 55 kg/m^3	Minimise costs; minimise differential settlement; avoid over-stressing low-grade EPS in highly stressed zones. Fire retardant grades of EPS used for all outer blocks of EPS core.
Geogrid reinforced granular layer	Form stable base for overlying embankment, reduce tensile forces across interface between conventional fill/EPS core.
Gas venting blanket	Controlled venting of methane/VOCs migrating through channel infill deposits.
Granular fill for embankment shoulders	Provides protective layer to EPS core from accidental causes and vandalism.

the line speed progressively raised to 85 mph. Settlement monitoring during a 17-month period after embankment completion indicated the ground settlement was about 5 mm, i.e. only about 3% of that measured during consolidation of the pre-load embankment. This excellent performance highlights the effectiveness of pre-loading to reduce the long-term settlement of compressible soils.

8.6 Conclusions

The replacement of deteriorating 19th century railway infrastructure poses significant challenges, not least because rail operations normally have to be maintained except for short track possessions. Maintaining safety is essential, despite numerous uncertainties, both in the ground and

Figure 8.14 Embankment construction.

Figure 8.15 Completion of EPS and upper concrete slab prior to Possession Works.

specific structures involved (including the residual strength of their deteriorated condition). This case history outlines how the OM was used to safely control construction works which were an essential pre-cursor to replacement of an old bridge. The construction of a pre-load embankment was required to consolidate and strengthen weak ground adjacent to the bridge foundations, prior to construction of the replacement embankment. The OM was used to control these works and minimise the risk of damage to the bridge and its timber pile foundations. The OM was successfully implemented, and there was negligible movement of the bridge during the pre-load construction. The observations did identify displacement of one of the piers, but this was due to a malfunctioning bridge bearing and thermal expansion/contraction of the bridge deck. The replacement embankment was noteworthy for several reasons:

- Originality: it was the UK's first EPS embankment for railway use.
- Elegance: the solution overcame a host of conflicting design, construction and programme constraints.
- Innovation: new stronger and stiffer grades of EPS were used for the first time. New quality control procedures were developed, and the composite EPS/reinforced concrete trough design had not been previously used.
- Contribution to advanced analysis: new EPS test procedures and a new stress–strain model was developed for characterising the EPS behaviour.
- Simplicity: 7,000 blocks were placed by hand, with the final 1,600 placed in a 50-hour period during the track possession works.

The overall result was the removal of a bridge that had become a significant maintenance liability and the removal of a speed restriction on the main Trans-Pennine route.

References

Ellis, E. A. and O'Brien, A. S. (2012). Global ground movements and their effects on piles, ICE Manual of Geotechnical Engineering, Chapter 57. Published by I.C. E, pp. 887–898.

Fellenius, B. H. (2006). Results from long term measurement in piles of drag load and downdrag. *Canadian Geotechnical Journal*, **43**, 409–430.

Frydenlund, T. and Aaboe, R. (1996) Expanded polystyrene – the light solution, Proceedings of EPS Tokyo Conference, Japan, October 1996, pp. 32–46.

LeRoueil, S. (1995). Practical applications of the piezocone in Champlain Sea clays, Proc Int Symp on CPT. CPT-95, pp 512–522.

Meyerhof, G. G. (1976). Bearing capacity and settlement of pile foundations. *ASCE Journal of Geo Eng*, **102**, No. GT3, 197–208.

O'Brien, A. S. (2001a). EPS behaviour during static and cyclic loading from 0.05% strain to failure, Proc of EPS Geofoam 2001, 3rd International conference, Salt Lake City, USA.

O'Brien, A. S. (2001b). Design and construction of the UK's first polystyrene embankment for railway use, Proc of EPS Geofoam 2001, 3rd International conference, Salt Lake City, USA.

Puzrin, A. M. and Burland, J. B. (1996). A logarithmic stress-strain function for rocks and soils. *Geotechnique*, **46**, 157–164.

Springman, S. M. and Bolton, M. D. (1990). The effect of surcharge loading adjacent to piled foundations, UK TRRL Contractor report 196.

Stewart, D. P., Jewell, R. J. and Randolph, M. F. (1994). Design of piled bridge abutments on soft clay for loading from lateral soil movements. *Geotechnique*, **44**, No. 2, 277–296.

Tavernas, D. P. and Leroueil, S. (1980). The behaviour of embankments on clay foundations. *Canadian Geotechnical Journal*, **17**, No. 2, 236–260.

Watts, K. and Charles, A. (2015). *Building on Fill, Geotechnical Aspects*, 3rd edition, CRC Press, UK.

Chapter 9

Heathrow Airport Airside Road Tunnel

9.1 Introduction

The Airside Road Tunnel (ART), completed in 2005, formed an important strategic element of the airport's development and addressed two main functions. In the short term, prior to the opening in 2008 of the new Terminal 5 (T5), it provided unrestricted access to remote aircraft stands, thus reducing surface traffic. In the long term, the ART forms an integrated transport route between the Heathrow Central Terminal Area (CTA) and T5 (Figure 9.1).

9.2 Key Aspects of Design and Construction

9.2.1 Portal Structures and Tunnels

The ART comprises 1.2 km of twin bored tunnels running east-west between two portal structures (Figure 9.2). The project, costing £140 million, was managed by the owner BAA using an integrated team approach (Powderham and Rust D'Eye, 2003).

The portal structures were designed and constructed to meet the requirements of the bored tunnelling programme. A 9 m diameter earth pressure balanced tunnel boring machine (TBM) was used to construct both tunnels (Darby, 2003). To accommodate the installation and removal of this large machine involved excavations up to 17 m deep between large diameter contiguous piled walls. Bored tunnelling was undertaken in two phases between June 2002 and June 2003, with both drives starting from the west and ending at the east portal. The west portal site is constrained by aircraft stands and the interface with the existing Piccadilly Line loop tunnel. This was constructed between 1984 and 1986 and is operated by London Underground (LU). It passes beneath the west portal at two locations (Figures 9.2 and 9.3) and is a 4.1 m diameter precast concrete segmentally lined tunnel. At the junction area, the excavation was less than

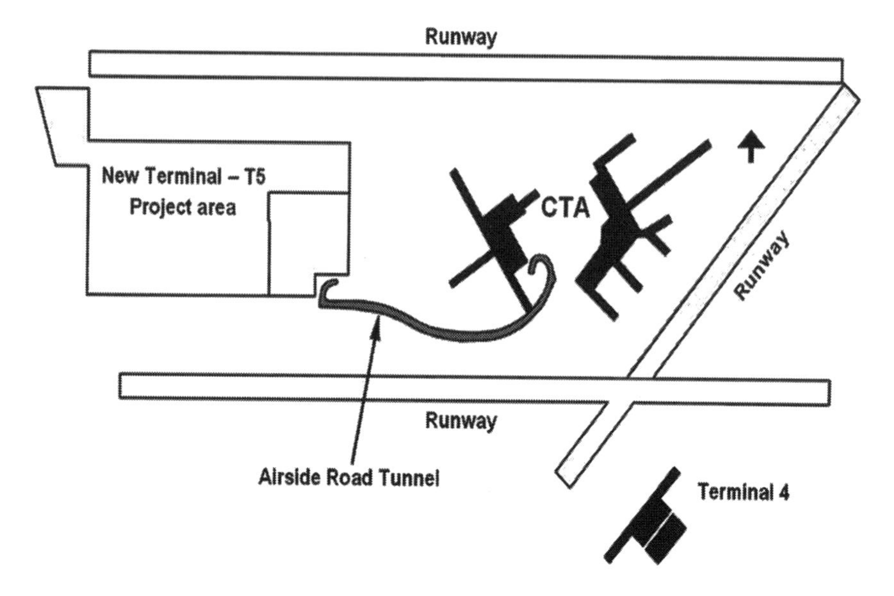

Figure 9.1 Plan of airport.

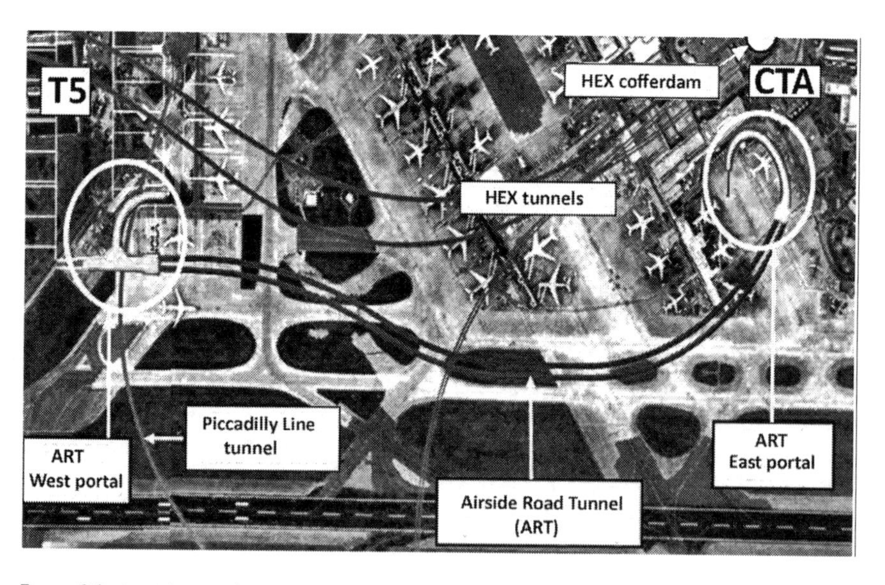

Figure 9.2 Aerial view showing ART portals in relation to T5 and CTA.

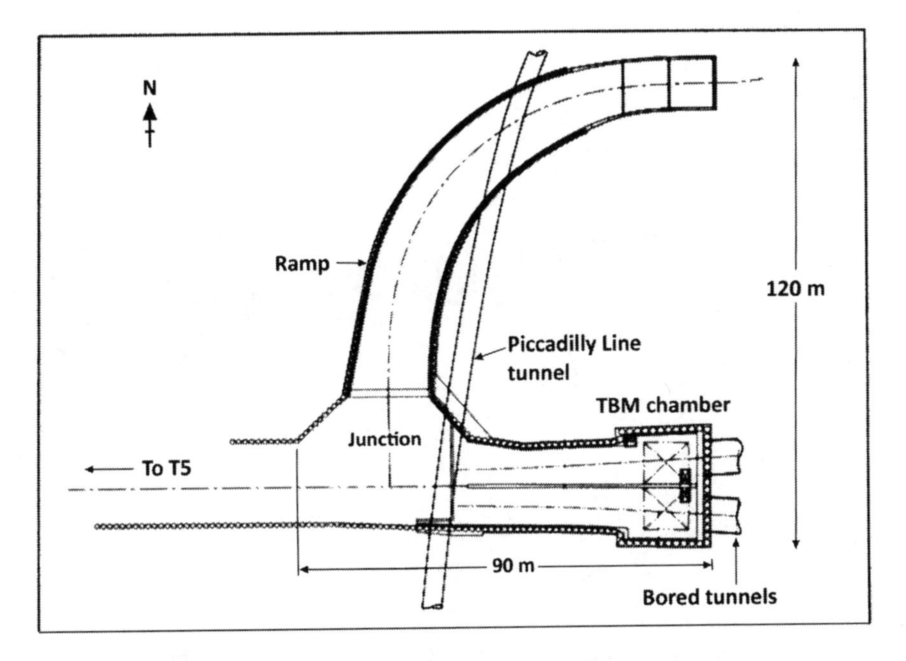

Figure 9.3 ART plan of west portal. The TBM chamber is 17 m deep and measures 20 × 30 m in plan.

5 m above the tunnel crown, with retaining wall pile lengths curtailed so as not to enter the exclusion zone of 3 m around the tunnel set by LU (Figure 9.4). All airport facilities and the Piccadilly Line tunnel had to be kept fully operational throughout the ART construction.

The portal structure comprises a cut and cover section which includes the TBM launch chamber, a junction and a ramp (Figures 9.3 and 9.24). These elements were addressed sequentially for each application of the observational method (OM).

9.2.2 Geology

Ground conditions at Heathrow Airport were well documented following four phases of ground investigation carried out between 1995 and 2000 for the T5 development. A geotechnical database was created to collate all available information. This combined earlier site data with that from the geotechnical investigations following the Heathrow Express (HEX) tunnel collapse in October 1994.

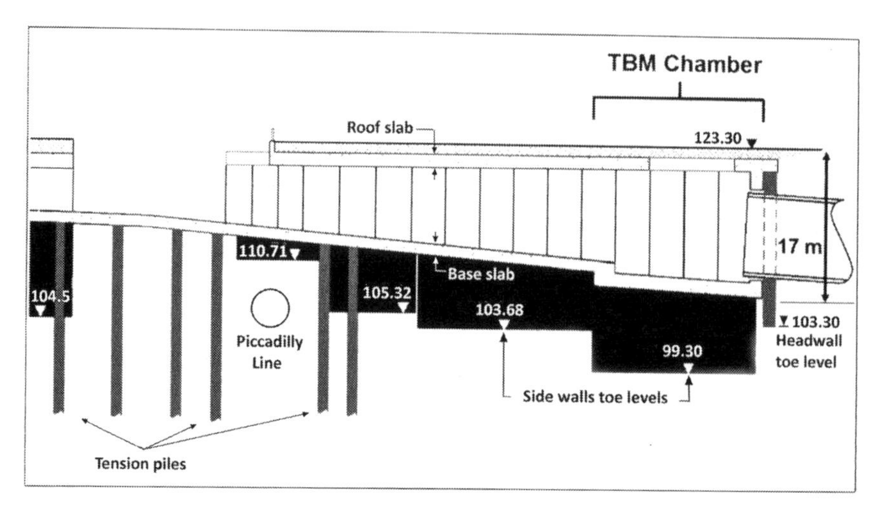

Figure 9.4 ART longitudinal section at west portal.

The ground conditions are typically made ground overlying Terrace Gravel for the first five metres above London Clay. This, extending to a depth of around 60 m, overlies the Lambeth Group and the Upper Chalk. The water table is generally around 2 m below ground level within the Terrace Gravel.

9.2.3 Base Case Design

The base case design for the ART portal structures was undertaken during 2000–2001, with the west portal design being followed by the east – the same order in which they would be constructed. Based on BAA's key requirements, the aim was to achieve a technical solution that could be constructed to meet the bored tunnelling programme and that was demonstrably safe and economic. It also needed to have a minimal impact on both airfield and LU operations during its construction. Simplicity and robustness were used to enhance safety and facilitate quality and ease of construction. Both portals were designed to utilise the same construction methods, plant and equipment. Sizes of the structural elements were standardised and overall cross-sectional dimensions kept constant where practicable. Summaries of the geotechnical design parameters used for the ART base case are provided in Tables 9.1 and 9.2, where z is depth below top of London Clay surface.

9.2.4 Construction Sequence

For the west portal TBM launch chamber, top-down construction provided the most economical and practical solution. By incorporating as much of the temporary works as possible into the permanent design, costs could be reduced and the construction programme shortened, allowing for the earliest re-instatement of the adjacent aircraft stands. The base case design for the cut and cover sections required one level of temporary propping at the mid-height of the piled walls. This was designed in structural steelwork and, for the TBM chamber, had a total weight of 60 tonnes (Figures 9.5–9.7).

9.2.5 Retaining Wall Design

Design of the retaining walls for the ART portal structures was carried out in accordance with the recommendations of CIRIA Report 104 (Padfield and Mair, 1984), adopting 'moderately conservative' parameters as listed in Tables 9.1 and 9.2. While respecting the need for robustness, various economies were used in the design.

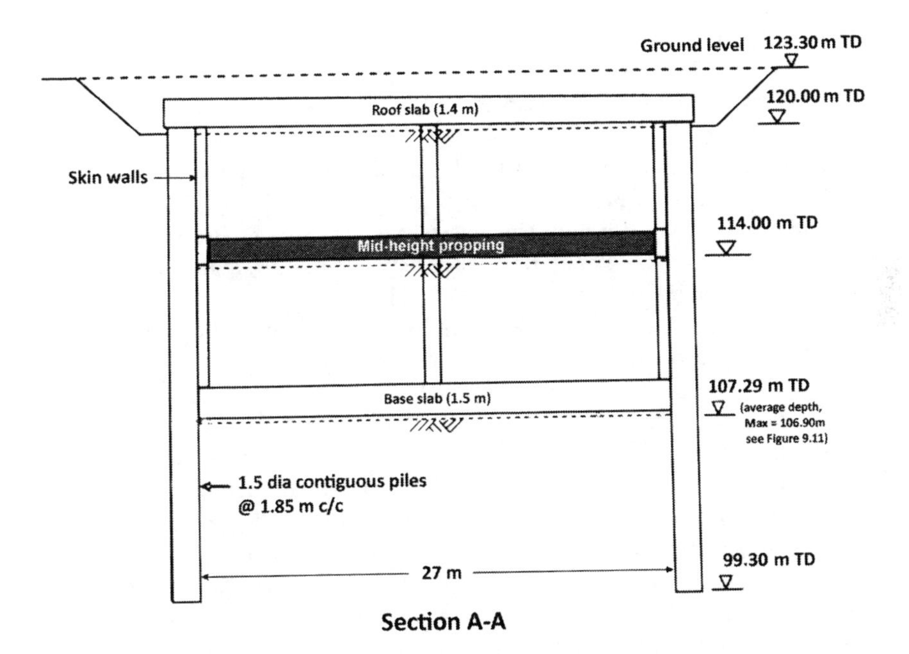

Figure 9.5 Base case design for west TBM launch chamber.

Table 9.1 Summary of geotechnical design parameters.

Stratum	γ_b *(kN/m³)*	ϕ'_{peak} *(deg)*	ϕ'_{crit} *(deg)*	c' *(kN/m²)*	E' *(MN/m²)*	K *(m/s)*	K_0
Terrace Gravel	19.5	38	35	0	30	1×10^{-4}	0.4
London Clay	20.0	23	20	8	$0.77E_u$	5×10^{-10} (h) $0.5\ k_h$ (v)	1.5

Table 9.2 ART base case design parameters for London Clay.

Parameter	*Base case design value*
Undrained shear strength, s_u	67.5 + 6z (kN/m²)
Undrained Young's modulus, E_u	27 + 2.4z (MN/m²)
Drained Young's modulus, E'	21 + 1.8z (MN/m²)
In situ earth pressure coefficient, K_0	1.5

CIRIA Report 104 recommends:

- Applying a minimum construction surcharge of 10 kN/m^2 to the retained soil.
- Allowing for an over-excavation by the lesser of either 0.5 m or 10% of the retained height.
- Reducing the undrained shear strength, s_u, to zero over the top 1 m to allow for excavation disturbance and dissipation of excess pore water pressures at excavation level.
- Reducing the s_u profile by between 20 and 30% to account for potential softening of the soil in front of the wall.

The first three factors were applied, but the s_u profile was not reduced on the passive side. This economy was based on previous experience of similar excavations in London Clay. (Though not in current guidance at the time, this was subsequently recommended in the design guide, CIRIA Report C580 (Gaba *et al.*, 2003)). To take further advantage of the temporary conditions during construction, two further design factors were included that reduced predicted loading and wall movement and increased support. A mixed analysis was selected using effective stresses on the retained side and total stress conditions for the passive support. Also, since the gaps between the contiguous piles would allow seepage, a 50% reduction in pore water pressure on the retained side was assumed before internal faces of the piles were sealed with skin walls.

The reinforced concrete frame at the top of the TBM launch chamber is shown in Figure 9.8. It illustrates the significant constraints to installing contingency temporary propping in an acceptably short period. In the event of adverse movement trends, the three piled walls of the chamber would have to be expeditiously supported. The simplicity of utilising individual single props as applied at Limehouse Link was not applicable here. To satisfy timely contingencies using steelwork propping, the only effective option would have been to install the steel framework (Figures 9.6 and 9.7) fully fabricated and held ready just below the reinforced concrete frame at roof level. While this would have enabled its prompt deployment, it would have defeated the prime objective of the OM to eliminate this temporary propping. How this key issue was resolved is discussed in Section 9.3.

9.3 Achieving Agreement to Use the OM

9.3.1 *Promoting Innovation*

BAA brought an exceptional focus to safety and risk management, but, while respecting these essential requirements, the team was also encouraged to seek improvements through value engineering and innovation. Such encouragement by a client as an integral part of design development is unusual and BAA was rightly commended for such an enlightened

Figure 9.6 Tubular steel props ready for fabrication.

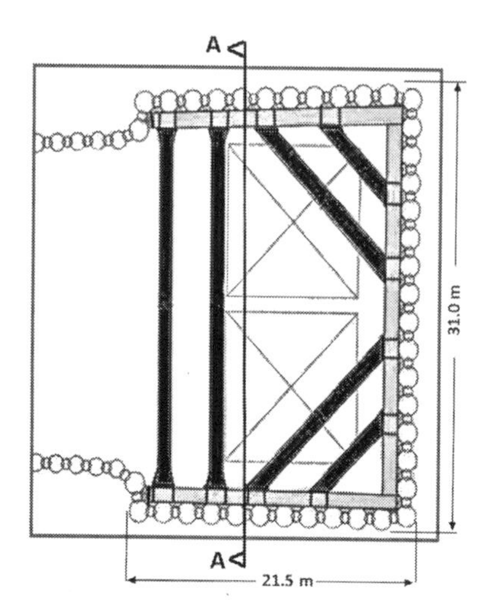

Figure 9.7 Layout of mid-height propping.

Figure 9.8 Access spaces above launch chamber.

approach. Within such a conducive design and construction environment, achieving agreement to adopt a technique as compatible with such objectives as the OM would have appeared straightforward. The potential benefits of the OM were already well recognised and appreciated by the client – not at least from its successful application to the Heathrow cofferdam (Chapter 5). In fact, such was the interest that the client and its local design team had carefully appraised recent case histories in advance.

9.3.2 Addressing the Issue of Precedent

Of the recent case histories, the application of the OM at Limehouse Link was viewed as most relevant. However, that apparent similarity presented an immediate problem. It was perceptively observed that, while the success at Limehouse Link set an impressive precedent, a similar application of the OM through progressive modification for the ART was not feasible. This was because the much smaller scale of the individual elements comprising the portal structures did not allow enough room for initial trial sections. At Limehouse Link, such sections were used to commence the OM for each working front. As noted in Section 9.2.5, even the installation of the structural steel framework at roof level in the launch chamber for contingency deployment would have eliminated most of any potential benefits. Thus the immediate challenge was to identify an acceptable way forward without the need for a trial section or having heavy steelwork frames ready as contingency propping. This required thinking outside the box or rather, in this case, within it. For the ART, it led to a novel application of the OM to minimise the amount of temporary support in very limited and discrete spaces (Powderham, 2009). The success also demonstrated the potential for applying the OM on small projects, or at least within a small footprint, where a linear progressive approach commencing with trial sections is not feasible. The implementation and the innovations developed from it are described later.

9.4 Implementation of the OM

9.4.1 A Big Challenge at a Small Scale

The base case design for the ART was robust and satisfied all of the project criteria of cost and programme. Consequently, any application of the OM would not be a 'best way out' category (Peck, 1969). Moreover, the use of the OM had not been originally planned as part of the design and so would be neither strictly be '*ab initio*' in Peck's terminology – although in practice, the applications for the ART were closer in spirit to this category. The integrated team maintained a keen interest to consider value adding changes throughout the project. Team members were well aware of the potential of the OM to retaining walls in similar ground conditions in the United Kingdom. Their initial expectation, however, was that the ART portals were too restricted and variable to allow any significant benefits from implementing the OM. They would simply be negated by the costs of resources arising from additional design, instrumentation, supervision, and monitoring. As noted above, there was also no obvious way to effectively provide the usual contingency of temporary steelwork propping. The challenge was therefore to develop an adaptation of the OM that would overcome these spatial and logistical constraints.

Common to previous applications of the OM to retaining walls, the critical observations would be the wall deflections. The typical measure adopted to address adverse trends during excavation is by means of emergency lateral support through contingency propping. However, it must be demonstrated in advance that such propping can be installed in a safe and timely fashion. Alternative flow charts comparing the base case design with the OM design are shown in Figure 9.9.

For the ART, the western TBM launch chamber was the first element to be constructed. It measured around 21.5 × 30 m in plan with a maximum

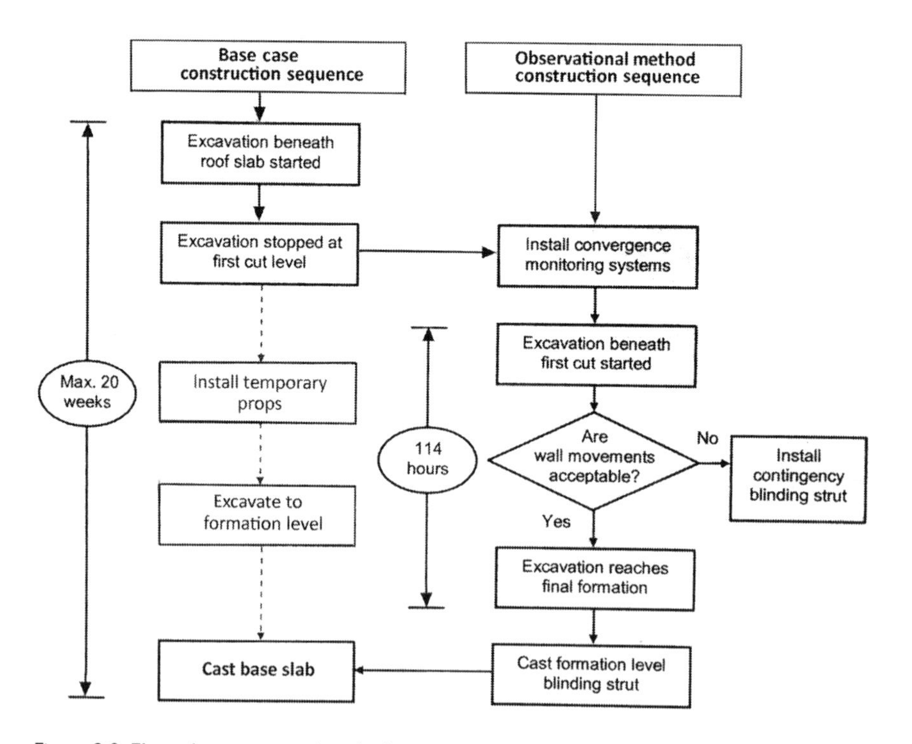

Figure 9.9 Flow charts comparing the base case with the OM design.

depth of excavation of 17 m (Figures 9.10 and 9.11). While it was technically feasible to install contingency steelwork propping at the roof level within the chamber, such an approach would have offered no significant advantage over the base case design – even if the wall movements proved acceptable and contingent measures were unnecessary. The presence of the heavy steel framework would also complicate construction activities and inevitably slow the excavation, again obviating one of the key benefits of the OM. There was a further concern that the deployment of the 60 tonnes steelwork as contingency propping within the confined space of the TBM chamber risked taking too long, even with it preassembled and held in readiness just below roof level. Thus, proposing what was seen as a previously established and successful approach for the OM was a non-starter.

9.4.2 The Breakthrough Concept

For the ART the key to unlocking the restrictive spatial constraints was the development of using intermediate blinding struts for contingency lateral support. For the project team, this simple but very effective innovation transformed the OM from being unacceptable into a very attractive opportunity. The usual key requirement for the prompt installation of the contingency support, should adverse trends in wall movement develop, still applied. This involved developing a suitable concrete mix design for early strength gain and the use of strictly

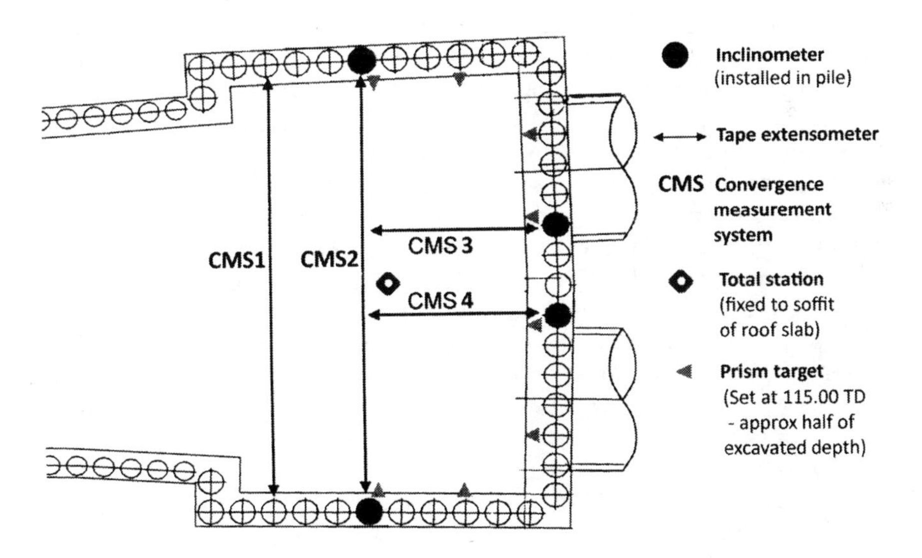

Figure 9.10 TBM chamber instrumentation locations.

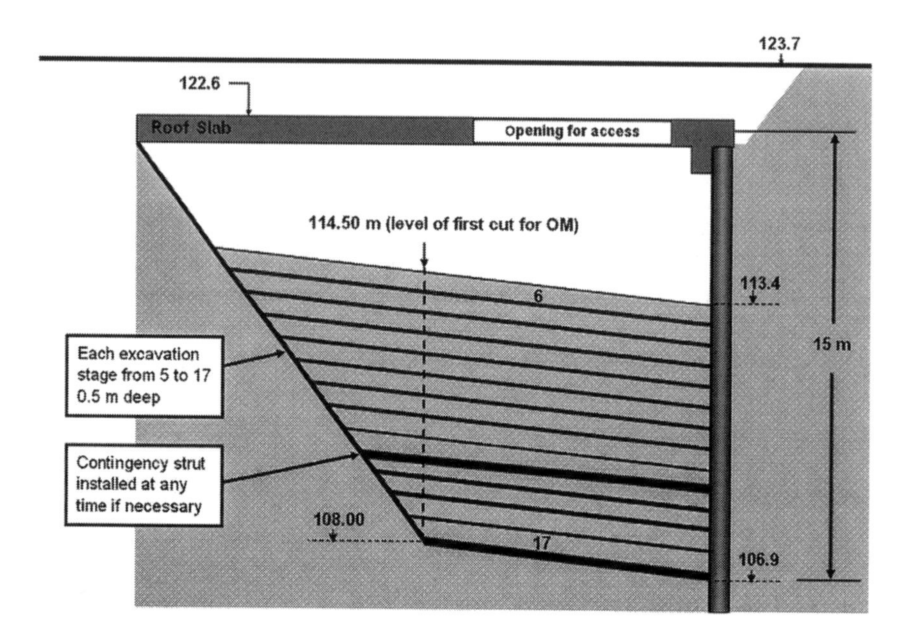

Figure 9.11 Laser controlled excavation sequence initiated at ART west TBM chamber.

controlled excavation sequences using lasers (Figure 9.11). The innovation was a literal adaptation of the step by step approach (Powderham, 1998). Simply put, the breakthrough concept basically involved a change in the direction of travel by 90° where, instead of a horizontally orientated sequence, progressive modification was applied vertically downwards.

The application of the OM to the TBM chamber, initiated in 2001, was fully successful. It was then applied in sequence to all of the other elements of both portals. The widely varying geometry and structural arrangements presented different cases of soil–structure interaction in each element. These included both propped and cantilevered retaining walls (Figure 9.24). The adoption of blinding struts as the contingency measure was the common enabling factor to each application of the OM. This development clearly established the potential for applying the method to small projects, where spatial constraints do not accommodate trial sections or allow the practical provision and installation of steelwork for contingency propping. For the ART it also created a further benefit, since it provided the basis to develop an innovative solution at the complex interface with the operating Piccadilly Line tunnel.

9.4.3 Instrumentation, Trigger Levels, Contingencies and Measured Performance

For the OM, the instrumentation must reliably capture the critical information. For efficiency and clarity, it should be as simple as practicable and not generate an undue amount of data. This was achieved on the ART project by specifying three independent systems: primary, secondary and primary back-up (Table 9.3). The control chart for implementing the OM during construction is shown in Figure 9.12 with assigned contingency actions for the traffic light control zones.

The maximum allowable deflection of the walls of the launch chamber, at full excavation depth, was 45 mm, giving a total convergence of around 90 mm. This was based on the structural capacity of the walls, but it was desirable to achieve lower values to reduce the potential ground movements outside the chamber. Accordingly, the red trigger level for wall convergence at completion of excavation was set at 55 mm and the amber trigger level for this stage was set at half this value. These maximum levels and the overall boundaries marking the three traffic light zones, along with the measured performance of the walls, are shown in Figure 9.14. The two plots of convergence are for the side walls and show a variation between locations of up to 3 mm. Some of this difference would have derived from CMS2 being in the centre where there was the minimum benefit from soil arching (see Figure 9.10). In any case, all measured wall convergence remained well within the green zone. Similar results were obtained for the headwall where the maximum recorded deflection was around 5 mm. This is directly comparable with the convergence of the side walls of around 10 mm at the end of excavation – that is, around 5 mm deflection for each wall. This indicates that the 4 m shallower embedment of the headwall did not adversely affect its performance. The measured movement trends were so encouraging that, after 48 hours from the commencement of the OM, the rate of excavation was increased from one to three 0.5 m thick layers over the 24-hour cycle. This significantly accelerated the overall programme for the construction of the TBM chamber and, despite the threefold rate, no

Table. 9.3 OM monitoring systems used on the ART project.

OM monitoring system	*Instrumentation type*	*Monitoring output*
Primary	Tape extensometers (at CMS 1 to CMS 4)	Side wall convergence and headwall deflection
Secondary	Inclinometers	Wall displacement profile
Primary back-up	Survey of prism targets (by robotic total station)	Wall displacements

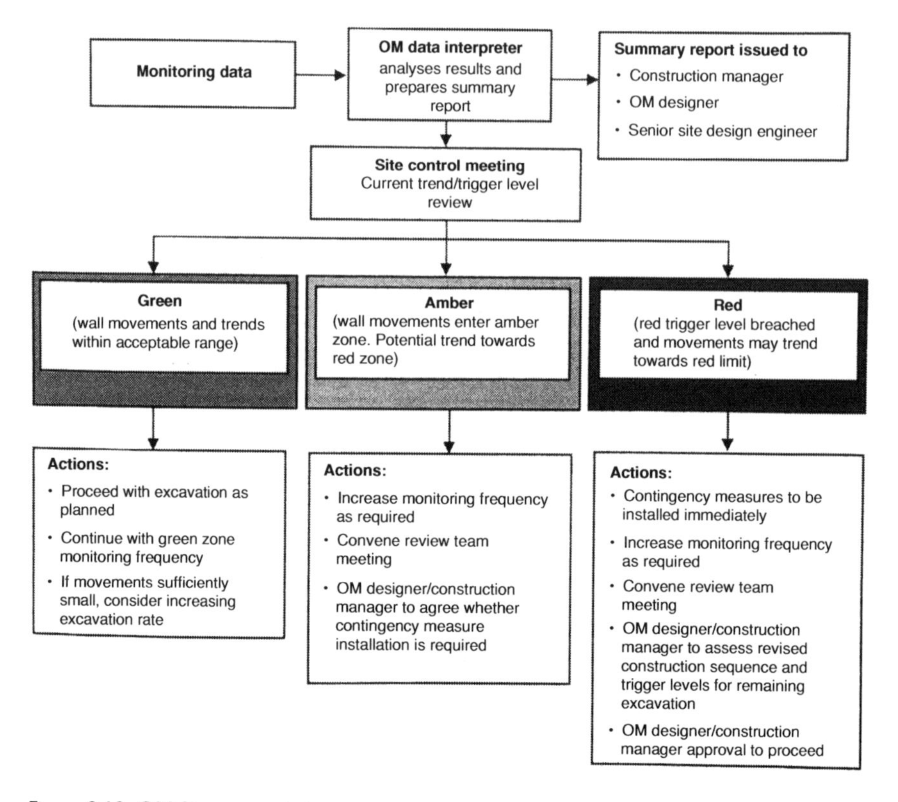

Figure 9.12 OM Site control chart.

significant increase in the rate of wall movements was detected. The completed excavation is shown in Figure 9.13

It is interesting to note that the maximum wall deflections recorded here were very similar to those recorded for the HEX cofferdam when the excavation there had reached the same depth of 15 m (see Figure 5.17 in Chapter 5). For the ART TBM chamber, it can be seen in Figure 9.14 that the wall convergence increased by around 2–3 mm during the 36 hours after the blinding strut at formation level was cast. A further 5 mm of convergence was recorded during the period of one month before the base slab was completed. While many factors may have influenced these minor trends, the dominant ones were attributed to the blinding strut. There would have been some bedding down in the contact between the wall piles and the strut to which the thermal contraction of the blinding strut would have added further potential for movement. For example, a temperature drop of 20°C from the maximum heat of hydration would have caused around a 5 mm contraction over the 27 m internal width of the chamber.

Figure 9.13 TBM launch chamber excavation completed without mid-height propping.

Supplementary checks were also made by monitoring the level in the centre of the blinding strut to assess whether there was any upward movement from heave of the clay below the excavation. This monitoring was also continued on the base slab for 6 months after its construction. No movements were detected. (Further details of the application of the OM for key elements of this case history, namely the TBM chamber of the ART West Portal and the East Portal Ramp, were reported by Hitchcock, 2005).

9.5 Innovation in Blinding Struts

9.5.1 Interface with the Piccadilly Line Tunnel

The success of the OM at the western TBM chamber encouraged the team to extend its application to the other elements of the ART portals. The positive and open culture towards innovation also led to

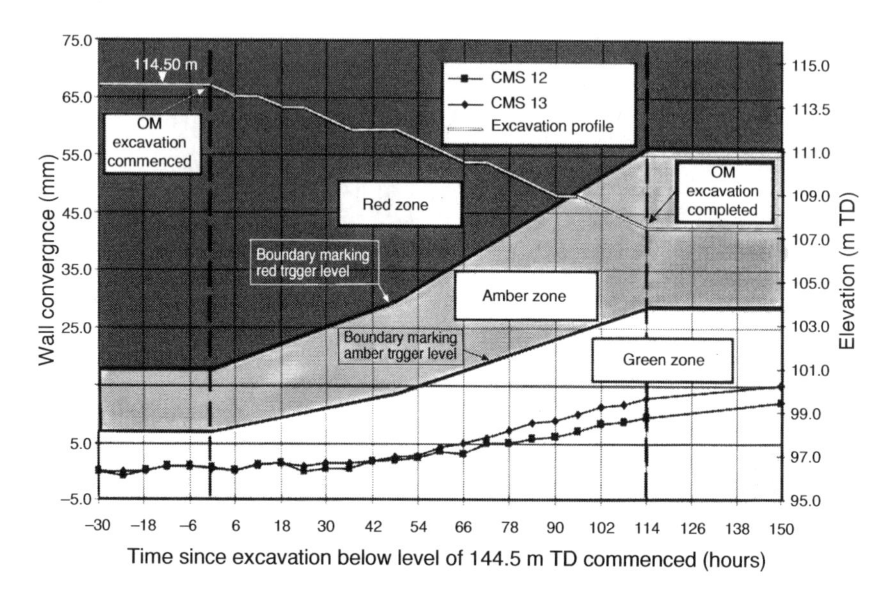

Figure 9.14 OM traffic light control zones.

the development of a new approach to addressing the challenge presented by the interface with the Piccadilly Line. This involved the innovative use of blinding struts that completely replaced the need for ground anchors or the heavy steel props and associated walings. It was, essentially, an extension of the contingency measure developed for the application of the OM in the adjacent sections of the ART. As noted, the application of the OM for the ART was implemented on the basis of adding value and definitely not through the need to address a crisis. However, the complex geometry with the Piccadilly Line did present a substantial technical challenge. The exclusion zone of 3 m specified by LU (Figure 9.15) placed stringent limits on the allowable depth of embedment of the contiguous piles above the tunnel. These sections of wall therefore needed much greater temporary support than those with the much deeper embedment outside the 3 m zone.

The base case design used mid-height support through a combination of ground anchors and steel props. Both of these conventional methods of temporary support had benefits but also presented serious disadvantages. The ground anchors were more adaptable to the complex geometry of the junction, but limitations to their proximity to the Piccadilly Line tunnel were imposed by LU. The steel propping avoided this constraint but was unattractive due to the heavy loads and large spans involved and the

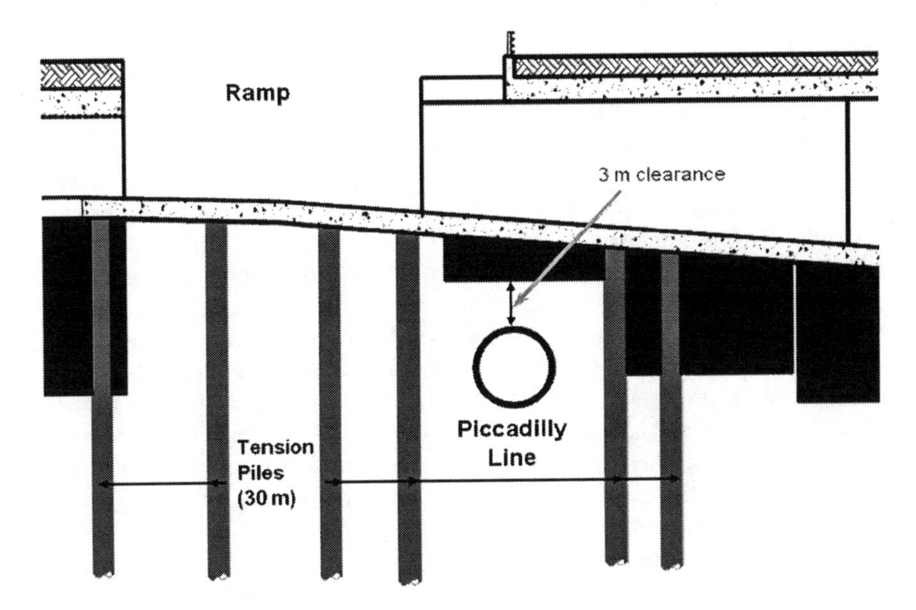

Figure 9.15 Exclusion zone around Piccadilly Line tunnel.

safety risks associated with its installation and subsequent removal. As part of such a temporary support system, massive and heavily reinforced concrete temporary waling beams would also be needed that would be time consuming and onerous to construct (Figures 9.16 and 9.17).

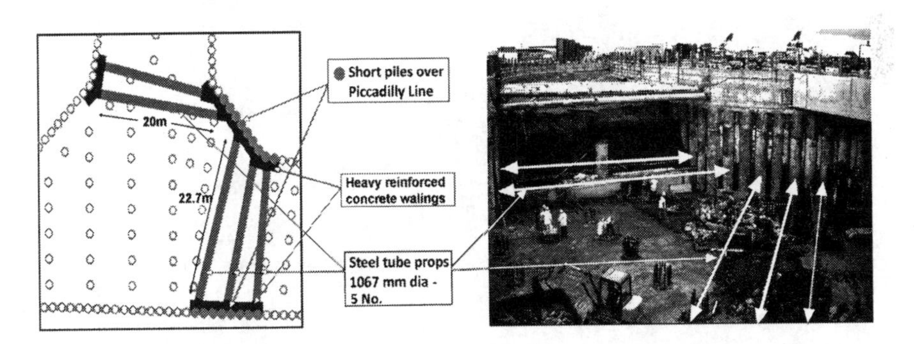

Figure 9.16 Original proposed alternative with heavy mid-level tubular props instead of ground anchors.

Figure 9.17 The low-level blinding strut cast around tension piles.

9.5.2 Implementing Innovation

The solution developed used a sequence of intermediate blinding struts at mid-height and lower ones at formation level as shown in Figures 9.18 and 9.19. Laser control for level and thickness was used for the blinding struts at both levels (Figures 9.20 and 9.21). The intermediate blinding strut was cast with construction joints and removed in strips as the blinding struts below the base slab level were cast in the 17 elements shown. These lower blinding struts made use of the tension piles where available. Figure 9.17 shows the lower blinding strut cast around the heads of the tension piles in the junction area. The use of the tension piles to provide temporary propping support was essential for the lower blinding strips numbered 1–8 since, until the element 8A and the strip 9 were cast, there was no line of thrust across the excavation between the walls. The beauty of this solution was that the complex geometry did not affect the inherent simplicity of using the two levels of blinding struts. It also made construction safer and easier since there were no heavy temporary works to install, work under and around and then finally remove. This innovation was developed from a novel approach to contingency support in the OM. It is important to appreciate that the performance of the blinding struts could not be effectively monitored by the OM. Their failure in buckling would be sudden and give little or no warning. No simple and safe contingency would therefore be available to address this in a timely way. Thus, this use of twin layers of blinding struts, although innovative, was not itself

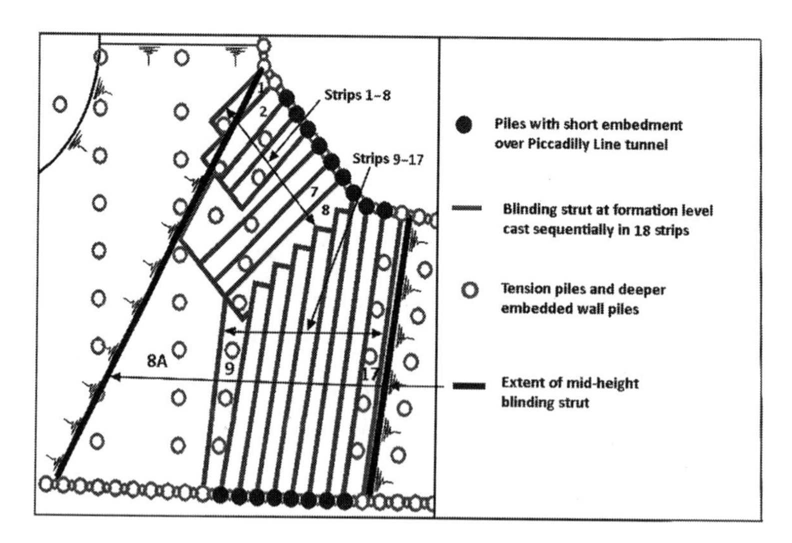

Figure 9.18 Plan showing blinding strut arrangement and sequence.

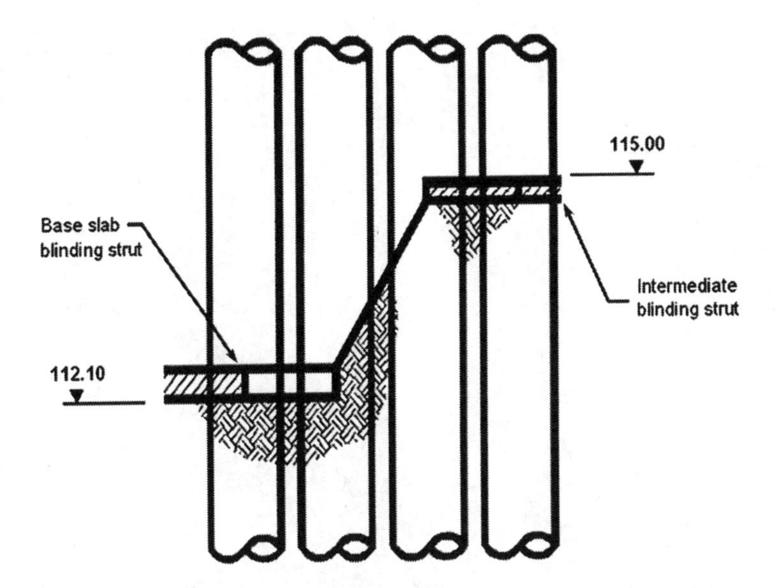

Figure 9.19 Cross section through blinding struts.

Figure 9.20 Installation of the intermediate blinding strut using laser control.

Figure 9.21 Sequential casting of lower blinding struts in strips.

an application of OM and the design of the blinding struts had to be appropriately conservative. Consequently, it had to be demonstrated in advance that this solution was safe and robust and acceptable to all stakeholders. Fortunately, this acceptance was readily achieved, and the proposal implemented with no delays to the project incurred. The ramp was constructed as the last stage of the OM applications resulting in no need for the top props (Figures 9.22 and 9.24).

It was essential to ensure the integrity of the blinding struts. Rigorous checks were made to ensure compliance with the specified tolerances and to avoid any potentially critical defects such as clay inclusions (Figure 9.23).

9.6 Conclusions

The open approach to innovation and close teamwork fostered by the client was fundamental to the acceptance and ongoing support maintained throughout the applications of the OM. Implementation through progressive modification using laser-controlled excavation significantly enhanced managing the risk in restricted spaces with sensitive interfaces.

Figure 9.22 Ramp construction with no upper props.

Figure 9.23 Critical checks on the blinding strut.

Figure 9.24 Ramp excavation completed between cantilevered piled walls without temporary top props.

This led to a comprehensive success, eliminating all of the temporary propping originally required by the base case design. Total savings amounted to 275 tonnes of structural steelwork and 31 weeks in construction time. The novel approach of providing contingency support through blinding struts enabled the OM to be applied successively to the wide range of individual elements of the ART portals. The innovative use of blinding struts was extended to provide a very effective solution to a technically challenging interface with an operating underground railway tunnel.

The success of the application of the OM on the ART was recognised by the Fleming Award of the British Geotechnical Association in 2002. This case history also provided the basis for two substantial research programmes. One was for an engineering doctorate on improving delivery of transportation projects through the application of the OM (Hitchcock, 2005). The researcher was a member of the OM team and was thus afforded detailed involvement with both the design development and the actual construction on site. Further research was undertaken in a PhD programme to assess the nature of the performance of blinding struts, their capacities and failure modes (Abela *et al.*, 2008).

References

Abela, J. M., Vollum, R. L., Izzuddin, B. A. and Potts, D. M. (2008). Short- and long-term effects on upheaval buckling of blinding struts, A-1780, Creating and Renewing Urban Structures, 17th Congress of IABSE, Chicago, IL, September 2008.

Darby, A. W. (2003). The airside road tunnel, Heathrow airport, Proc. Rapid Excavation and Tunnelling Conference, New Orleans, LA.

Gaba, A. R., Simpson, B., Powrie, W. and Beadman, D. R. (2003), Embedded retaining walls – best practice guidance, Report C580, CIRIA, London.

Hitchcock, A. R. (2005). Improving delivery of underground transportation infrastructure – an observational method case history, Thesis for Engineering Doctorate, University of Southampton, School of Civil Engineering and the Environment, Supervisors: Powrie, W. and Powderham, A. J.

Padfield, C. J. and Mair, R. J. (1984). *Design of Retaining Walls Embedded in Stiff Clay*, Report 104, CIRIA, London.

Peck, R. B. (1969). Advantages and limitations of the observational method in applied soil mechanics. *Geotechnique*, **19**, No. 2, 171–187.

Powderham, A. J. (1998). The observational method – application through progressive modification. *Proceedings of the Journal ASCE/BSCE*, **13**, No. 2, 87–110.

Powderham, A. J. (2009). The Vienna Terzaghi lecture, The Observational Method – using safety as a driver for innovation.

Powderham, A. J. and Rust D'Eye, C. (2003). Heathrow express cofferdam: innovation & delivery through the single-team approach – part 1: design and construction part 2: management. *Proceedings of the Journal of Boston Society of Civil Engineers/American Society of Civil Engineers*, **18**, No. 1, 25–50.

Chapter 10

Raising the 133 m High Triumphal Arch at the New Wembley Stadium (2002–2004)

10.1 Introduction

The new Wembley Stadium's main architectural feature is an iconic 133 m high triumphal arch. The arch is 7.4 m diameter and has a total weight of 1,750 tonnes. With a span of 315 m, it is one of the largest and most slender structures of its type in the world. Because the arch is so large, it was fabricated at ground level and then raised into its final position. Raising the arch was an extremely high-profile event (being covered extensively by mass media at the time). It was also a complex and challenging engineering operation. In addition to its striking architectural impact, the arch performs an important structural function. It supports the north roof directly and most of the south roof indirectly. The construction plan required the arch to be built and raised into its final position in order to free the area beneath and allow the stadium construction to be completed. Hence, raising the arch was on the critical path for completion of the project. The foundations for arch raising comprised pile groups which were designed and constructed during the early phase of the project. Detailed design of the temporary works, required for the arch raising, was carried out at a later stage of the project. It became clear that the loads applied onto the pile groups, under some of the design scenarios, were higher than those used for foundation design. This raised concerns about the adequacy of the arch foundations. This chapter describes the application of the observational method (OM) during arch raising, with a focus on the arch foundations.

10.2 Key Aspects of Design and Construction

10.2.1 Overview

The arch comprises thirteen 20.5 m long structural steel sections, together with two 'pencil ends' which connected the arch into permanent pile group foundations (the 'western' and 'eastern' arch base foundations). The structural steel arch is shown in Figure 10.1, with a pencil end being lifted into position by a crane. A turning strut is adjacent to the crane (lying horizontal).

Figure 10.1 Turning strut and arch.

Each end of the arch comprised a hinge joint which allowed free rotation of the arch during its raising. The hinge joint at the eastern arch base foundation is shown in Figure 10.2.

Five sets of hydraulic jacks were attached to the arch via a system of cables to pull it into its final position. Each set of hydraulic jacks is founded on a 'jacking base' and could apply a total force of 14 MN (Figure 10.3).

Each set of jacks comprised four individual units which could be operated independently, if necessary, in order to control the alignment of the arch during raising. Varying the force across the jacks at each base would apply a torsional force on the pile group. Five intermediate struts (founded on 'turning strut bases') were used to give the pulling force a vertical component. As each set of jacks was activated, the turning struts rotate and exert a lifting force on the arch. The temporary turning struts are substantial structures, with the largest strut being about 100 m long (Figure 10.1). The lifting mechanism is illustrated schematically in Figures 10.4 and 10.5.

Once the arch was at 60°, the arch was connected to five further bases (restraining bases) which would prevent the arch falling forward in an uncontrolled manner (Figure 10.4(b)). Figure 10.6 shows the arch raising with the arch close to vertical, and both the jacking and restraining lines connected to the arch. Once the arch went past the vertical position (at 100°) the jacking points and turning struts became redundant and were

Figure 10.2 Pin connection between arch and eastern arch base pile cap.

Figure 10.3 Jacking base used for arch raising.

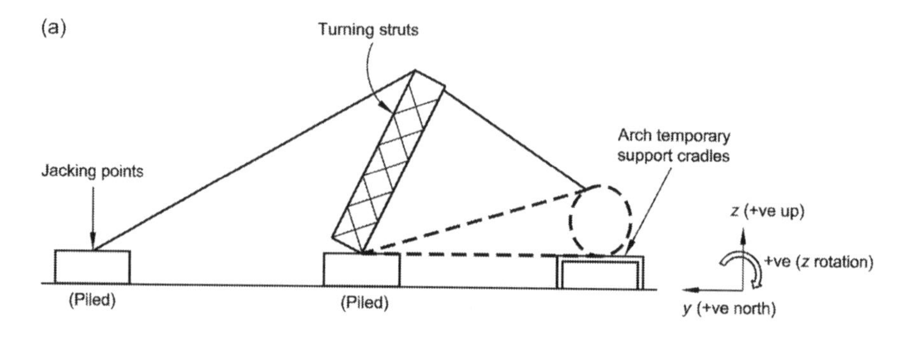

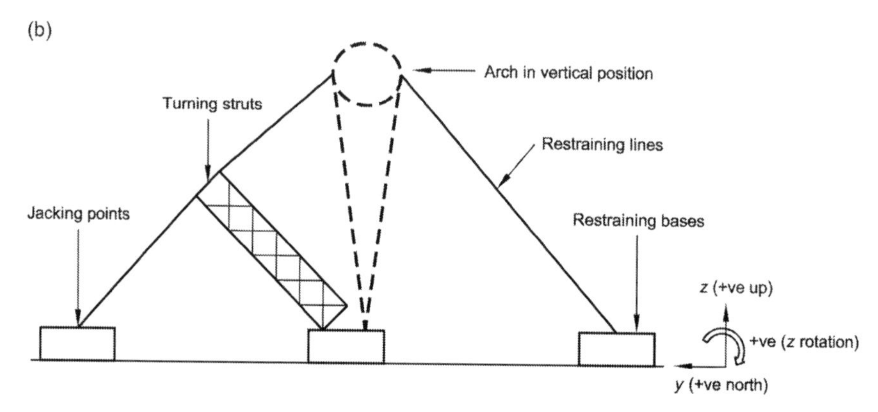

Figure 10.4 Schematic cross section showing arch lifting mechanism: (a) initial arch lifting position and (b) activation of restraining lines.

dismantled. From this position to the final position (112°), the arch was held by the restraining lines. These lines held the arch in place until the roof was constructed and the whole structure became self-supporting.

10.2.2 Foundations and Geotechnics

The piled foundations for the fifteen temporary arch bases (5 jacking bases, 5 turning strut bases and 5 restraining bases) are summarized in Table 10.1. The permanent arch bases (eastern and western) are also constructed on piled foundations. The eastern arch base has nineteen 33 m long piles and the western arch base (which also acted as a 'shear core' foundation for the stadium) has sixty 29 m long piles. All piles were 1.5 m diameter. Part of the reason for the variations in pile length for the temporary and permanent arch bases (between 15 and 42 m) is the changes in founding surface level arising from the bulk earthworks which

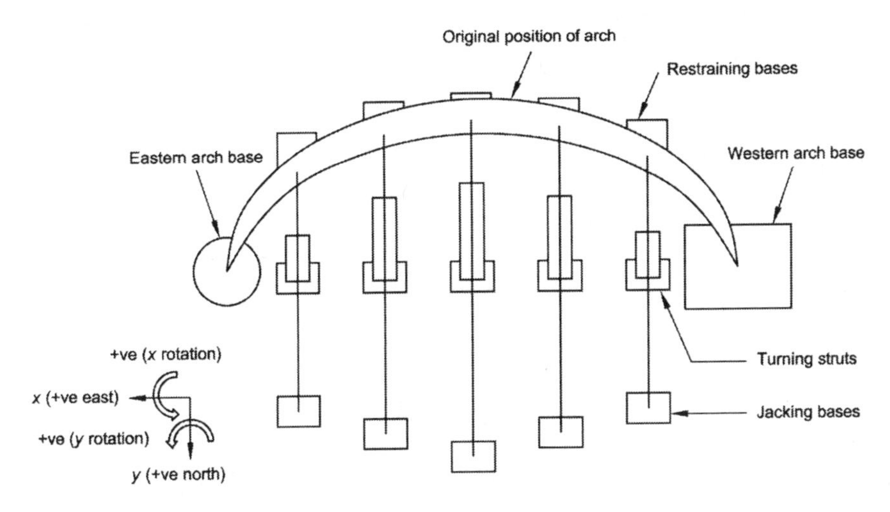

Figure 10.5 Schematic plan view showing arch lifting mechanism.

Figure 10.6 Arch being raised into position.
(Credit: ImageSelect/Alamy Stock Photo.)

Table 10.1 Geometry of pile groups for temporary arch bases.

	Jacking bases				
	J1	*J2*	*J3*	*J4*	*J5*
Number	7	9	12	9	7
Length (m)	32	20	15	17	42
	Turning strut bases				
	T1	*T2*	*T3*	*T4*	*T5*
Number	9	10	10	10	6
Length (m)	24	25	26.5	25	37
	Restraint bases				
	R1	*R2*	*R3*	*R4*	*R5*
Number	6	8	6	8	6
Length (m)	21.5	10	23	11.5	22

were carried out to level the hill which the old stadium was built upon (noting the new stadium footprint is far larger than the old stadium). The bulk earthworks included fill of up to 10 m thickness and excavation up to 7 m deep. These earthworks induced, as expected, large global ground movements across the stadium footprint (O'Brien *et al.*, 2005).

The site geology comprises London Clay over Lambeth Group and the uppermost 8–10 m of London Clay is weathered. The depth of weathering is unusual (e.g. Chandler, 2000) and far deeper than typically encountered, for example, across central London or at Heathrow. The undrained shear strength at the top of the weathered clay is only about half of that typically measured at the surface of the unweathered London Clay in central London. The ground investigations included routine and advanced testing (summarized by O'Brien *et al.*, 2005) and the characterization of the stiffness of the weathered London Clay was an important consideration in the assessment of pile group deformation during arch raising. Figure 10.7 provides a plot of stiffness at very small strain, as discussed by Mandolini and Viggiani (1997) and O'Brien (2012); this is a key parameter for reliable assessments of pile group deformation.

10.2.3 Arch Raising

The arch was constructed at an angle of 1.63° to the horizontal. This was chosen as a compromise between minimising the force initially imposed on the arch and the practical difficulties of fabricating the arch above the

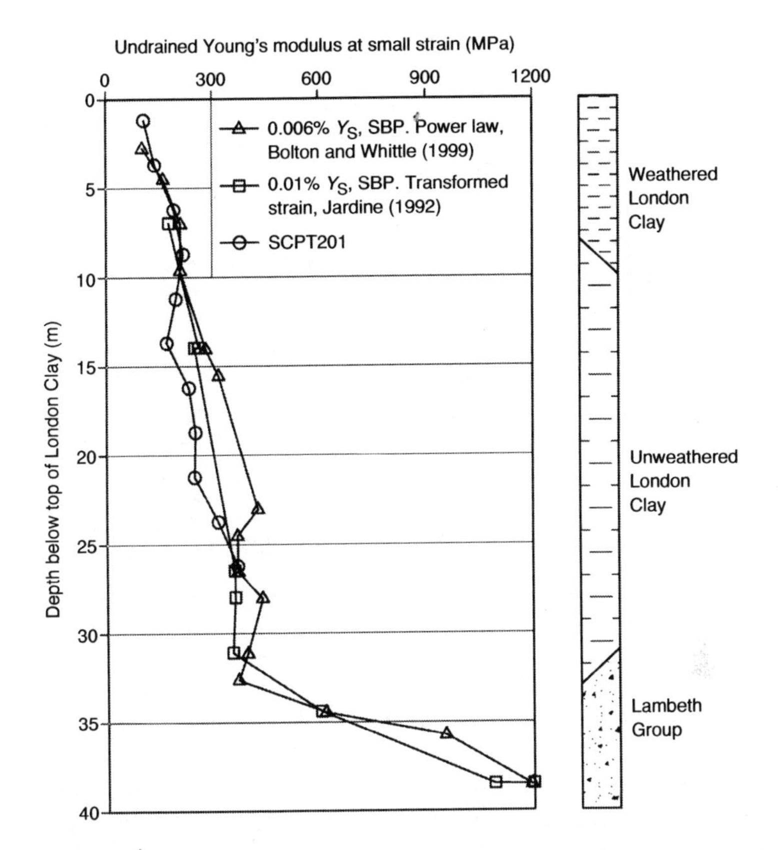

Figure 10.7 Undrained Young's modulus at small strain versus depth below London Clay (γ_s = shear strain).

Note: SCPT = seismic cone; SBP = self-boring pressuremeter; Eu = 3G assumed.

site ground surface. The maximum jacking force was applied to initially lift the arch; however, the maximum force acting on piles within the pile groups typically occurred when the arch reached an angle of between 30° and 40° (depending on the pile group location). The applied loads comprised the jacking force, plus wind and temperature induced changes in loads, which generated up to nine different load cases at each arch angle. Hence, the applied loads on the pile groups were complex and included horizontal and vertical loads, together with moments and torsion. Figure 10.8, for the eastern arch pile group, illustrates the complex variation of loads and moments during arch raising. Horizontal load remained practically constant; however, both applied moment and torsion increased to a maximum when the arch reached an angle of 40°. The horizontal load

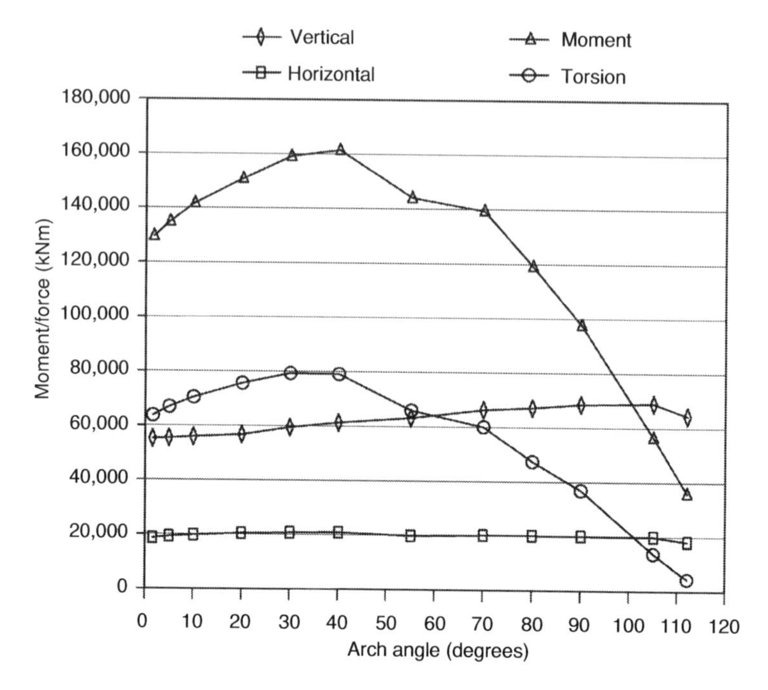

Figure 10.8 Eastern arch base, variation in loading during arch raising.

and moment shown are the resultants of the two horizontal and moment components, and the ratios of these components also varied during arch raising. For example, the northerly horizontal load was about 70% of the easterly component up to an arch angle of 30°; then the northerly load decreased rapidly, becoming a very small southerly load at the final arch angle (less than 1% of the easterly component). Hence the orientation of the resultant horizontal force changed from 55° (east north-east) to 90.5° (practically due east) as the arch was raised.

The arch raising was to be carried out in 5° increments, which meant that there were between 6 and 8 well-defined stages during arch raising before the critical load combinations were applied to the arch base foundations. Hence, this provided an opportunity to apply OM via progressive modification (Powderham, 1998).

10.3 Achieving Agreement to Implement the OM

10.3.1 Creating Opportunity

The early phases of the project, during 2002/2003, which included the pile installation, earthworks and retaining wall construction, progressed relatively smoothly. However, by 2004 (and before arch raising), there were several well-publicized delays and disputes between the main contractor and various sub-contractors for the subsequent construction activities. The steelwork sub-contractor had overall responsibility for the arch raising (they left the project team after arch raising and were replaced by another sub-contractor who completed the steelwork construction). During the detailed design development of the temporary works required for arch raising, it became apparent that the applied loads on the foundations, in particular moments and torsional loads applied to the centroid of the pile groups, could be higher than those originally used for foundation design (during some load scenarios which may occur during arch raising). The pile groups had been designed in accordance with EC7 and BS8110 for geotechnical and structural design, respectively. Therefore, the load increases would ordinarily have led to the requirement for strengthening of the pile groups (e.g. by installing additional piles and connecting into enlarged pile caps). However, the site was extremely congested and some of the superstructure had been built in the vicinity of the arch bases. At this stage, strengthening of the constructed foundations would have led to substantial project delays. A specialist design team from the main design consultant, who were part of the consortium responsible for the stadium's permanent works design, proposed the use of the OM as a 'best way out' application. The focus was on control of safety and minimizing the risk of damage to the arch while raising it to its final position. All parties recognized that maintaining safety during arch raising was paramount. Applying the OM was seen as a potential solution to minimizing risks, while avoiding additional project delays. Prior to agreement being reached, the practical application of the OM was fully evaluated through the following steps:

1. The load–deformation behaviour of the pile groups was re-analysed, using the latest set of load cases and assuming different ground stiffness profiles (i.e. worst credible, moderately conservative and most probable conditions).
2. The potential distortion of the arch was assessed under a wide range of conditions.
3. Options for implementation of contingency measures were considered, both technically and from a practical construction perspective.
4. The existing pile groups and adjacent structures were checked and assessed, in the context of being able to carry out contingency measures and assessing any potential local constraints.
5. A series of workshops were held to discuss the risk management process and for all parties to understand their roles and responsibilities if the OM was implemented.

Central to gaining agreement for the use of the OM was the application of progressive modification. At each stage safety could be demonstrated prior to commencing the next stage of arch raising (i.e. before each five-degree rotation of the arch). If adverse trends were identified, then planned contingency measures could be implemented to reduce risks during the next stage. The use of the OM through progressive modification was fundamental to gaining the confidence of all parties that risks could be minimised as far as reasonably practical. The maximum compressive force was put into the arch at initial 'lift-off' of the arch, but during the prolonged period (anticipated to be between 4 and 6 weeks) required for the arch raising a wide range of different loading scenarios had to be considered. Some of these could have been more onerous in terms of differential loading of the arch (e.g. high winds or potential malfunctioning of a jack or set of jacks). These sorts of scenarios could lead to changes in load across the arch, and indirectly onto the foundations. Hence, it was necessary to assess a wide range of conditions and assure all parties that the arch would be robust under a wide range of circumstances.

10.3.2 Assessment of Foundation Behaviour during Arch Raising

The two permanent arch base foundations needed to be functional and code compliant both in the short term, during arch raising, and then throughout the design life of the stadium. The fifteen temporary arch bases just needed to be functional during the arch raising. The following were important considerations in assessing the potential risk of adverse foundation behaviour during arch raising:

1. Nature of potential structural and geotechnical failure mechanisms: ductile or brittle?
2. A full evaluation of all potential load combinations.
3. For the load combinations, the available code compliant capacity (geotechnical and structural).
4. Redundancy within pile groups?
5. Relationship between pile group deformation and structural forces induced in piles.
6. Axial force distribution across pile groups, in particular, the potential for tensile axial forces to be induced in piles at the 'rear' of pile groups.

The arch is a flexible structure. The jacking cables were also quite flexible with a permissible axial strain of about 0.2%. In contrast, the pile groups are intrinsically stiff, with anticipated deformation (at pile group centroid) of less than 20 mm. The force applied to the arch was partially dependent on the stiffness of the jacking lines (which in turn were dependent on the foundation stiffness). To lift the arch and maintain its alignment within acceptable tolerances, the jacks were operated sequentially, and the amount of jacking force adjusted, depending on the arch deflection. The allowable differential movement of the arch, across each pair of temporary support cables, was far larger than that for the arch base foundations. Hence, moderate variations in pile group deformation (e.g. due to local variations in mobilised ground stiffness) was considered to be acceptable, provided sudden (or brittle) failure and progressive failure modes were avoided. The increases in applied load (between preliminary and detailed temporary works design) were mainly due to increases in moment and torsion. In general, vertical and horizontal loads are shared amongst all piles in a pile group, although soil–structure interaction leads to perimeter piles taking a higher proportion of the load than inner piles (O'Brien, 2012). However, applied moments and torsion tend to concentrate relatively large increases in axial force/moment on the perimeter piles (refer to Figure 10.9). Load cases with high moment and torsion loads typically imposed the most onerous structural forces in the piles. Under these loading conditions, the most heavily loaded piles were located in the corners of the pile groups. Hence, the primary concern was the implications of moment/torsion loads on the piles located around the perimeter of the pile groups and if this would lead to structural over-stress or sudden accelerations in pile group deformation.

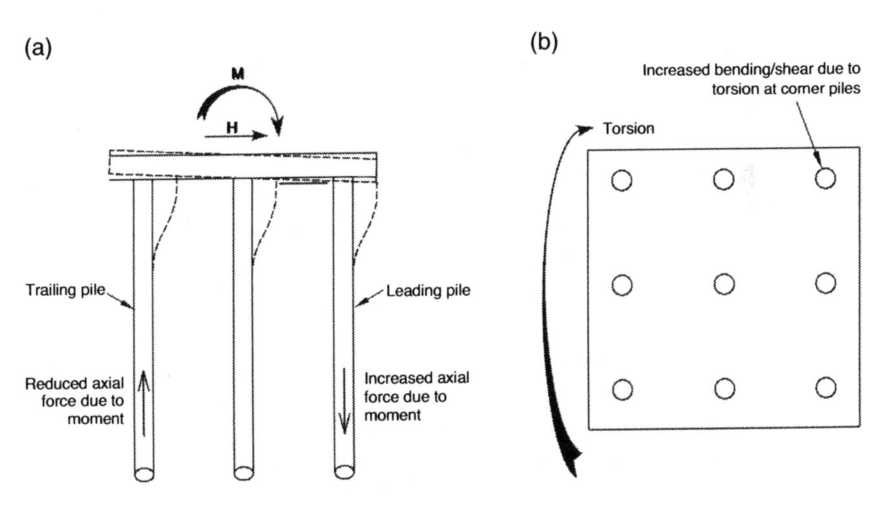

Figure 10.9 Pile group subject to applied moment and torsion: (a) applied moment and (b) applied torsion.

10.3.3 Preliminary Pile Load Tests

At the start of the project a comprehensive load testing programme was undertaken on sacrificial piles. Conventional compression tests, plus tension tests and lateral load tests, were carried out, with mobilised displacements equivalent to about 10% of the pile diameter (for the purposes of the load tests, assumed to be 'pile failure'). All these tests indicated a ductile failure mode. As discussed by Randolph (2003) and Burland and Kalra (1986) most pile foundations, with a few notable exceptions, would be expected to fail geotechnically in a ductile manner. Figure 10.10 provides the load–displacement data for one of the lateral load tests, up to a horizontal deflection of 10% of pile diameter. From a relatively early stage of lateral loading the ground in front of the pile yields and this wedge of failing ground becomes progressively larger (and deeper) as lateral load increases. Towards the end of the pile test, the pile is structurally over-stressed and the reinforcing steel in the pile is yielding. Nevertheless, during the test there is no indication of the onset of sudden (or brittle) failure or a rapid acceleration in pile deformation. This was a key consideration for the safe use of the OM. Even under worst credible conditions,

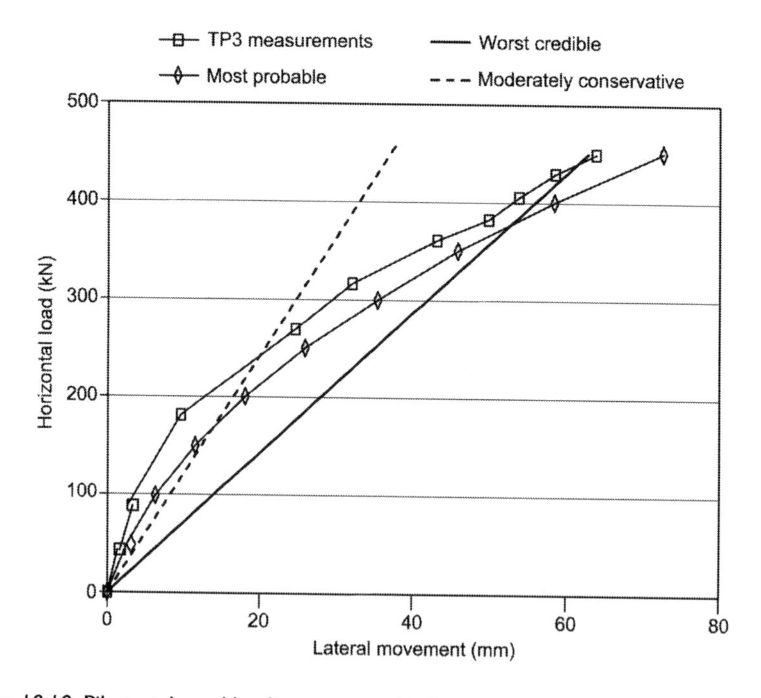

Figure 10.10 Pile test, lateral loading, measured behaviour versus analyses.

the deformation of the pile groups was expected to be far less than the pile deformation experienced during the pile tests.

Following the aforementioned review, it was agreed by all parties that it was possible to implement the OM from both a technical and practical construction perspective. It was also agreed that the OM was the best way forward in terms of minimising project risks.

10.4 Establishing Trigger Levels and Contingency Measures

10.4.1 Foundation Analysis

For each pile group, 121 load combinations were considered covering each stage of the arch raising (at 5° intervals), with up to nine different load combinations at each stage of arch raising. By utilising the symmetry in the pile group layout (Figure 10.5) and identifying the most critical pile groups, the number of temporary arch bases that were analysed in detail was reduced to eight, plus the two permanent arch bases. The critical load cases were selected from the 121 provided for each pile group and pile group analyses run for worst credible, moderately conservative and most probable sets of ground stiffness parameters. Critical load cases were selected, and analyses were carried out to obtain an envelope of pile group displacements and imposed forces (bending moments, shear and axial force) on the piles.

The pile groups could have been analysed by full 3D finite element analysis, taking account of the beneficial resistance provided by the buried pile caps (hence, acting as piled-rafts). However, the pile group analyses were kept as simple as practical, and the resistance provided by the pile caps was ignored (because of their relatively small plan area, the mobilized resistance of the pile caps at working loads was likely to be small). This simplification was essential given the volume of analysis required (as outlined below) and the need for practically real-time analysis during the arch raising process. Pile group analyses relied upon commercially available software (Geocentrix, 2002) and bespoke in-house software. These could simulate non-linear ground stiffness and changes in structural stiffness, both of which were considered essential to capture key features of pile group behaviour (Poulos, 1989). Pile analysis models were carefully calibrated against the suite of pile test data, following the guidance given in Fleming (1992) and England (1999). Further details of the methods of analysis and input parameters are described by O'Brien and Hardy (2006).

To ensure that an appropriate envelope of acceptable pile group movement was identified, two different loading scenarios were considered: firstly, the critical load combinations were applied and each load component was multiplied by a constant factor until one of the piles in the group reached its ultimate structural capacity; and secondly, the dominant load within the load combination (say overturning moment for a turning strut base; torsion for a jacking base) was increased until a pile reached its ultimate structural capacity. Figure 10.12 summarises a sensitivity study for the eastern arch base. The calculated rotation and horizontal movement are given in Figure 10.11(a) and (b), respectively. Figure 10.12 (a) and (b) shows the induced bending moment and shear force, respectively. The induced shear force was not a controlling factor (being well below the code compliant structural capacity for the range of scenarios and arch angles considered) and was relatively insensitive to variations in assumed ground stiffness (worst credible to most probable). In contrast, induced bending moment was a controlling factor, with the code compliant bending capacity being reached under worst credible conditions (assuming zero axial load in the pile). For most probable conditions, the maximum induced bending moment was much lower (about half of the available capacity). It would have been apparent during the first stages of arch raising if behaviour was trending towards adverse conditions. In which case, contingency measures could be implemented before local structural over-stress developed in a pile within one of the groups. Pile group rotation was considered the most relevant deformation mode for defining amber/red trigger levels, and these are shown in Figure 10.11(a). It is important to note that structural yield in bending would be a ductile mechanism, so the arch would not experience a sudden loss of support, and the overall pile group stiffness provided for the arch and supporting cables could be maintained within safe bounds.

The overall geotechnical capacity for the pile groups was well above the applied loading. However, it should be noted that locally some perimeter piles within the pile groups would fully mobilize their

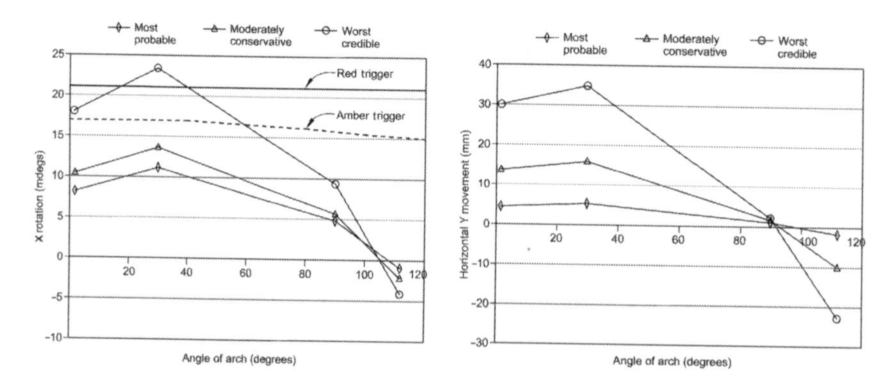

Figure 10.11 Sensitivity studies of pile group deformation, eastern arch base: (a) pile group rotation and (b) horizontal movement of pile group.

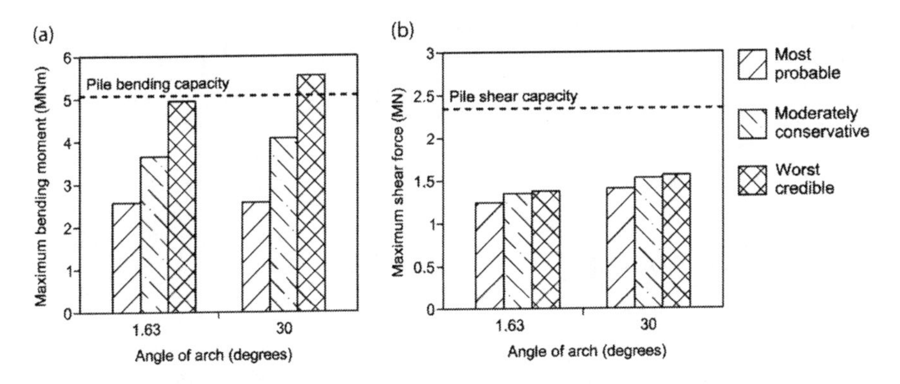

Figure 10.12 Sensitivity studies of structural forces in piles, eastern arch base: (a) maximum bending moment and (b) maximum shear force.

geotechnical capacity. However, this was considered acceptable because the geotechnical failure mode response was ductile, and the pile caps were sufficiently stiff and strong (structurally) to be able to safely redistribute loads across to other piles within the group (under these conditions this is consistent with EC7 requirements). The minimum number of piles within a pile group was six, and therefore they had adequate redundancy (NCHRP, 2004).

Large applied moments could lead to the rear piles in a pile group experiencing tension. The tension pile tests indicated that the mobilized stiffness was considerably lower than under compression, partly due to reduced structural stiffness. In addition, the bending and shear capacity of the piles would be reduced. Based on these considerations, it was decided that tensile axial forces should be avoided. Hence, the distribution of predicted axial loads was carefully reviewed. One of the jacking bases was considered vulnerable, and relatively large tension forces (together with shear and bending) were likely to be induced at the rear piles of the group under some load combinations. A simple remedial measure was used to eliminate this risk prior to arch raising commencing; a large counter-balancing weight was built at the rear of the pile group (simply a 'box' of mass concrete, cast on top of the rear of the pile cap) to ensure the rear piles maintained a small compressive axial force under worst credible conditions.

The arch is a flexible structure, as noted in Section 10.3. The amber and red trigger levels were set by the steelwork sub-contractor to ensure critical differential movement was avoided. The allowable differential movement at the centre of the arch span is about 400 mm or about 1 in 320. Relative to the allowable movement of the pile groups, the allowable differential movement of the arch was not onerous. Differential movement along the arch span could be minimized by varying the jacking force at each jacking base. Further control of the arch differential movement could be exerted through application of torsion at each jacking base (by varying the load at each of the four jacking units on a base, Figure 10.3). Hence, for implementation of the OM, the primary requirement was to minimize the risk of overstress within the pile groups and the associated risk of increased pile group deformation.

10.4.2 Trigger Levels and Contingency Plans

The application of the OM requires the careful measurement of changes in a key parameter (usually deformation) to provide feedback on the performance of the structure and facilitate the timely implementation of contingency measures well before a critical limit is reached. The arch was to be monitored via a precise surveying system. Therefore, this system was extended to include the pile groups which supported the arch raising. A system of precision surveying was devised to monitor the horizontal and vertical movements and associated rotations (in section and plan, i.e. torsional rotation) of each pile group. The critical observations were pile cap rotation and horizontal deformation. Precise levelling was capable of measuring to ±0.1 mm, and total station surveys were capable of measuring to ±1.5 mm. Precise surveys were the primary monitoring system. Electro-levels were used as a secondary monitoring system and were installed on selected pile caps (to check movement trends indicated by the precise surveys). A concern with electro-levels were their potential vulnerability to damage (the site was congested, and a wide range of activities were taking place prior to the arch raising commencing).

Trigger levels were specified for the arch and all the pile group foundations, which, if breached, required the implementation of contingency measures to maintain an acceptable level of safety. A simple traffic light system was used to communicate the risk level to all stakeholders. The arch was far more flexible than the foundations, as noted earlier. Hence, the implementation of the OM was focussed on the pile groups. For the foundations, breaching the amber trigger level required an increased frequency of surveying, together with checks on survey reliability (e.g. damage to survey datum) and the actual applied load combination (including weather conditions and jacking loads/functionality). If there was an adverse trend which was likely to breach the red trigger, then the next arch raising stage would be halted and remedial action would be carried out. For the foundations, the simple and robust remedial measure was to apply a kentledge (as implemented for one of the arch bases prior to arch raising, noted earlier). This remedial measure was also quick to implement (within an 8-hour period, using readily available formwork and mass concrete which could be quickly brought to site). An alternative remedial measure (in the unlikely event of shear translation, manifest as excessive horizontal deformation with negligible pile cap rotation) was to install and pre-stress tie-backs into adjacent structures (previously surveyed and designed, with necessary equipment available on-site). In the event of adverse differential movement of the arch moving towards a red trigger level, then the jacking forces would be adjusted, and torsional loads applied at the jacking bases (with a check on the likely performance of the foundations under the revised loads). A flow chart was prepared which provided a summary of the trigger levels and relevant contingency actions (Figure 10.13).

For the arch foundations, the red trigger level was cautiously defined since the structural capacity was set at the code compliant capacity (i.e. the actual ultimate capacity would be close to double this limit). A cautious value had to be set for the red trigger since time would be required to implement contingency measures. The amber trigger level was more challenging to define; it was assessed from a consideration of four criteria:

i) The deformation monitoring accuracy which could be achieved. This was particularly important for the pile groups, because they are intrinsically stiff structures.
ii) The predicted pile group deformation under the anticipated most likely load combination together with a plausible variation in ground stiffness parameters (typically 75% of the most probable parameters, or the moderately conservative parameters, whichever produced the largest movement).
iii) Ensuring there was sufficient distance between amber and red trigger levels to facilitate timely implementation of the contingency measures.
iv) Ideally the amber trigger level should be sufficiently beyond the anticipated pile group deformation (based on most probable parameters) to avoid regular breaches of the amber trigger. This is a subtle, but important practical consideration, since regular breaches of trigger levels would likely lead to the OM becoming impractical to manage on site.

10.5 Implementation of the OM

10.5.1 Initial Surveys

Precise surveys of the pile caps commenced about three months before commencement of arch raising. Precise surveys were initially very difficult because of the plethora of construction activities being carried out across the site. Following discussions on site, construction activities were rescheduled to create a better environment for precise surveying to be implemented. Survey methods were also progressively improved. During a six-week period, significant improvements in surveying accuracy were

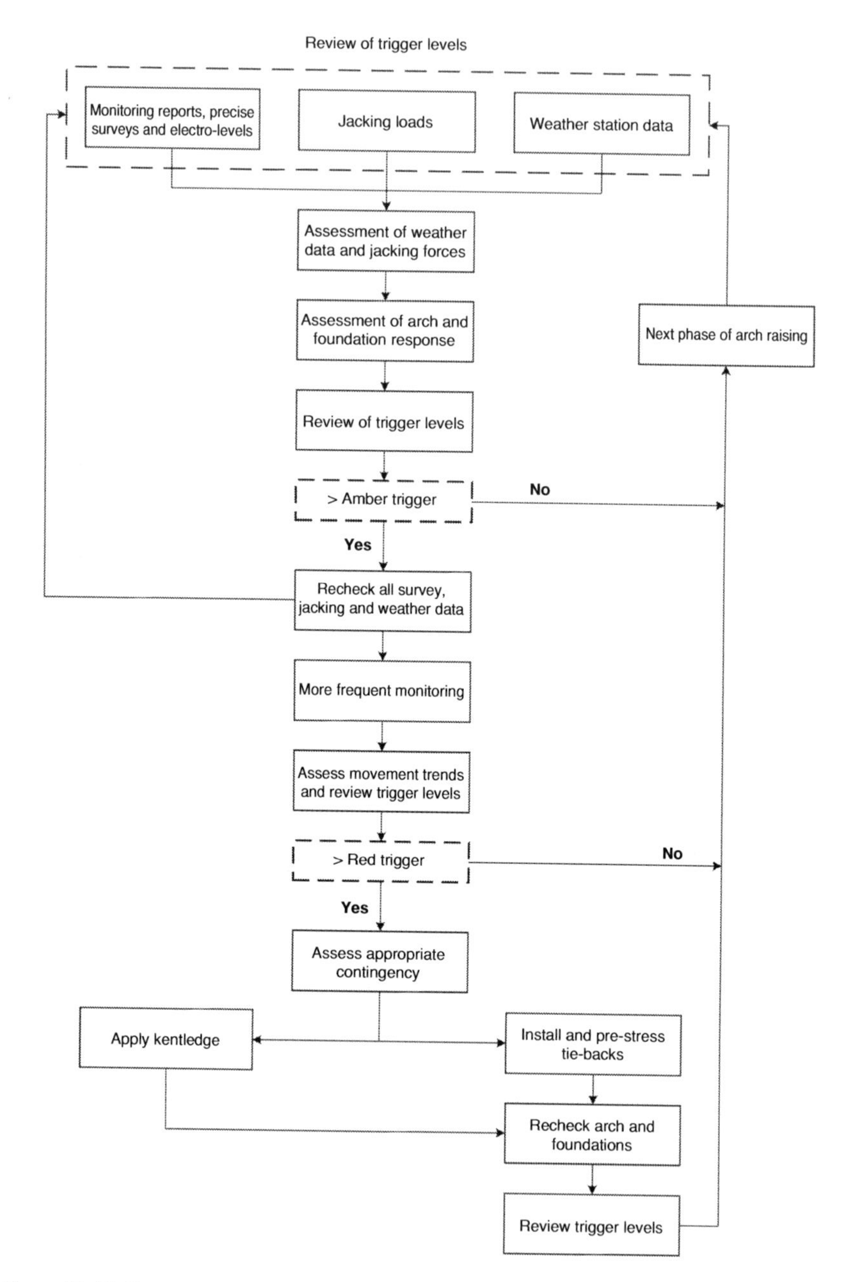

Figure 10.13 Flow chart for implementation of the OM.

achieved, for example for total station surveying from about ±5 mm initially (which was inadequate) to ±1.5 mm. Hence, the prolonged time period available for improving the survey accuracy prior to arch raising was invaluable. It proved to be a crucial element in the success of the OM at this site.

10.5.2 Observed Pile Group Deformation

The critical observations for the pile groups were the rotation and horizontal deformation of the pile caps. Before and after each incremental lift, the six components of movement (three displacements and three rotations) were calculated from the surveying data and compared to the amber and red trigger values. For the temporary jacking and turning strut bases, the most critical load combinations were applied at arch angles of close to 30°, and for the main arch bases at angles of nearly 40°. During this phase, seventy-two sets of data had to be reviewed (six sets of data for each of the ten temporary bases and the two main arch bases) immediately after each stage, to determine if adverse trends were developing. The measured horizontal movement and rotation for the eastern arch base are compared with the calculations from the pile group analyses (Figures 10.14 and 10.15, respectively). Observations during the first couple of arch raising stages indicated that conditions were considerably better than worst credible. The vast majority of the observations were close to the most probable calculated values and occasionally were between most probable and moderately conservative calculated values, as indicated in Figures 10.14 and 10.15 (note that one reading is on the amber trigger, a subsequent check survey indicated readings below the amber trigger). It should be noted that worst credible, moderately conservative and most probable conditions were based on back-analysis of the preliminary pile tests, as mentioned in Section 10.4; the analysis strategy has been discussed in detail by O'Brien and Hardy (2006) and O'Brien (2007). Throughout arch raising, weather conditions were good and wind speeds were low, hence the most onerous load combinations (associated with high winds) were not imposed on the arch. The jacks performed consistently well throughout arch raising and the arch distortion was maintained well within acceptable limits. As a result, it was not necessary to apply significant torsional loads at the jacking bases, which in turn reduced risks to the arch foundations.

During arch raising, amber trigger levels for the foundations were occasionally breached. Resurveys indicated that the high measurements were erroneous due to survey errors or problems with datum points. The amber and red trigger levels for a pile group were dependent on the load combination considered (e.g. combined horizontal and moment, or combined horizontal, moment and torsion). The amber/red trigger levels used during arch raising were based on the most critical load combination (i.e. the load combination which generated the lowest set of deformations to

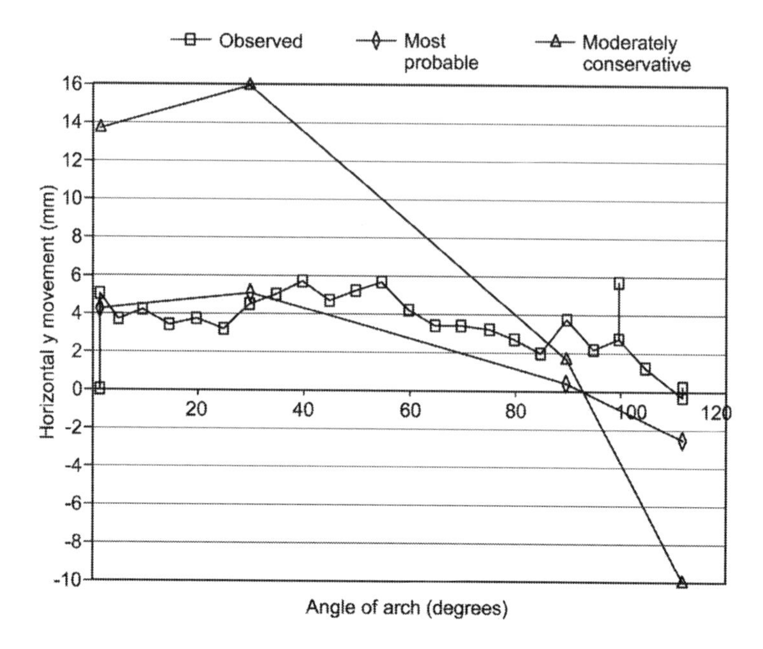

Figure 10.14 Calculated and observed horizontal movement in the *y* direction, eastern Arch base.

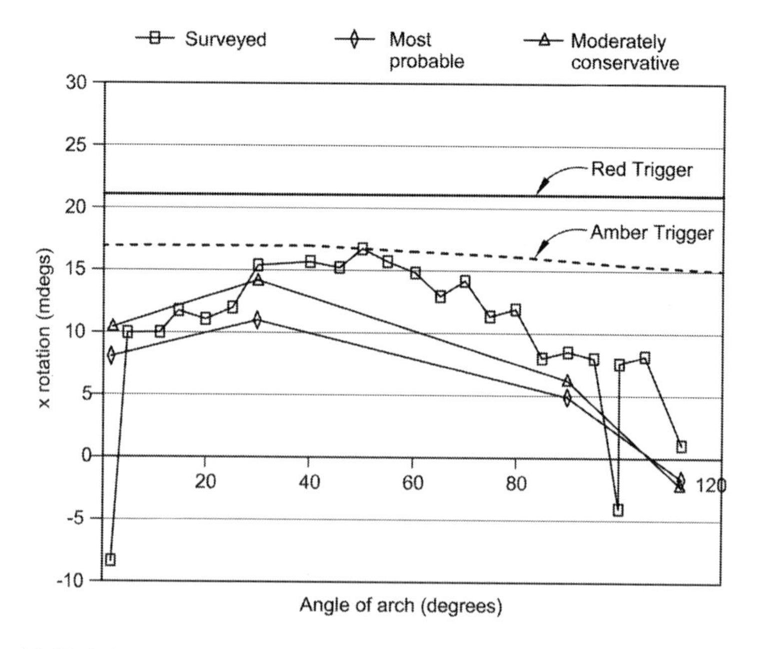

Figure 10.15 Calculated and observed rotation, *x* axis, eastern Arch base.

cause structural over-stress in a pile). It was not necessary to modify these during arch raising. If there had been an adverse trend, it may have been appropriate (depending upon the actual conditions at that stage of arch raising) to have re-assessed the trigger levels based on the relevant load combination (e.g. based on known weather conditions). Using a single set of charts kept the process as simple as possible and provided an extra safety margin throughout the OM application.

Once the arch reached 40°, the applied loads on the arch foundations reduced, although recorded movements did not reduce until an arch angle of about 60° was passed. This apparent anomaly was considered to be due to high ground stiffness being mobilised as loads reduced (i.e. similar to observations during pile tests when a high stiffness is mobilised at the beginning of unloading, following initial loading). Once observed movements decreased, the OM phase was reduced in intensity, although the OM was not completed until the restraining lines were operational (when the arch was at an angle of 100°). Figure 10.16 shows the arch in its final position being held in position by the restraining lines. Completion of the arch raising marked the successful conclusion to a remarkable feat of civil engineering. The iconic arch now forms one of the United Kingdom's most spectacular landmarks.

10.6 Conclusions

The application of the OM provided an effective means of minimizing risk during this unique and challenging civil engineering operation. The arch raising was completed as originally planned. The application of the OM through progressive modification enabled risks to be minimized during arch raising. It was possible to demonstrate that the risk of damage to the arch and the arch foundations was kept as low as reasonably practical. The monitoring and analysis effort, although considerable, was kept as simple as possible. This was important to ensure effective practical implementation of the OM and to maintain effective communication across the project team (which is a key aspect of the implementation of the OM). The application of the OM for pile group foundations is unusual, and it requires a consideration of the following:

a) The trigger levels (for implementation of contingency measures) are dependent on the applied load combinations; hence, a wide range of potential scenarios need to be considered. Nevertheless, once the full range of conditions have been considered, this potential complexity needs to be simplified as far as possible (when e.g. defining trigger levels, movement monitoring and appropriate contingency measures) to facilitate a simple and practical risk management process.
b) Pile groups are intrinsically stiff; hence, the monitoring system has to be capable of reliably measuring small deformations.

Figure 10.16 The arch in its final position.
(Credit: DBURKE/Alamy Stock Photo.)

In common with other applications of the OM, it is essential that potential sudden and progressive failure mechanisms are avoided. For this case history, a programme of pile tests (subject to compression, tension and horizontal loading) taken to failure demonstrated ductile behaviour at deformations well in excess of those which would be mobilized during arch raising. The pile test data also facilitated the calibration of pile group analysis models and the assessment of moderately conservative, most probable and worst credible conditions. Contingency measures need to be quick and practical to implement. Crucially for this case history, a simple contingency (application of a kentledge at the rear of a pile group) was identified which would mitigate the most likely critical conditions.

References

Bolton, M. D. and Whittle, R. W. (1999). A non-linear elastic/perfectly plastic analysis for plane strain undrained expansion tests. *Geotechnique*, **49**, No.1, 133–141.

Burland, J. B. and Kalra, J. C. (1986). Queen Elizabeth second conference centre, geotechnical aspects. *Proceedings of ICE, Part 1, Design and Construction*, **80**, 1479–1503.

Chandler, R. J. (2000). The 3rd Glossop lecture, clay sediments in depositional basins: the geotechnical cycle. *QJEGH*, **33**, 7–39.

England, M. G. (1999). A pile behaviour model, PhD Thesis, Imperial College, University of London.

Fleming, W. G. K. (1992). A new method for single pile settlement prediction and analysis. *Geotechnique*, **42**, No. 3, 411–425.

Geocentrix. (2002). *Repute User's Manual.*

Jardine, R. J. (1992). Non-linear stiffness parameters from undrained pressuremeter tests. *Canadian Geotechnical Journal*, **29**, 436–447.

Mandolini, A. and Viggiani, C. (1997). Settlement of piled foundations. *Geotechnique*, **47**, No. 3, 791–816.

NCHRP Report 507. (2004). *Load and Resistance Factor Design for Deep Foundations*, Transportation Research Board, Washington, DC.

O'Brien, A. S. (2007). Raising the 133 m high triumphal arch at the New Wembley Stadium, risk management via the observational method, Proc 14th European Conf on SMGE, vol. 2, Madrid, Spain, pp. 365–370.

O'Brien, A. S. (2012). Pile-group design. *ICE Manual of Geotechnical Engineering*, **2**, chapter 55, 823–851.

O'Brien, A. S. and Hardy, S. (2006). Non-linear analysis of large pile groups for the new Wembley stadium, Proc 10th Int conf on Piling and Deep Foundations, Amsterdam, pp. 303–310.

O'Brien, A. S., Hardy, S., Farooq, I. and Ellis, E. A. (2005). Foundation engineering for the UK's new national stadium at Wembley, Proc 16th Int Conf SMGE, vol. 2, Osaka, Japan, pp. 1533–1536.

Poulos, H. G. (1989). Pile behaviour – theory and application. *Geotechnique*, **39**, No. 3, 365–415.

Powderham, A. J. (1998). The observational method – application through progressive modification. *Proceedings Journal of ASCE/BSCE*, **13**, No. 2, 87–110.

Randolph, M. F. (2003). Science and empiricism in pile foundation design. *Geotechnique*, **53**, No. 10, 847–875.

Chapter 11

Crossrail Blomfield Box (2012–2015)

11.1 Introduction

Construction dewatering is an essential temporary works requirement for many civil engineering schemes. Terzaghi (1929, 1955) recognised the substantial difficulties of predicting the behaviour of groundwater and wrote about the causes of major accidents and delays in foundation and underground construction:

> *"It is more than mere coincidence that most failures have been due to the unanticipated action of water, because the behaviour of water depends, more than on anything else, on minor geological details that are unknown".*

This theme has been taken up by several eminent geotechnics practitioners, notably Rowe (1968, 1972) who describes a range of challenges arising from groundwater flow and changes in excess groundwater pressure. The problem has also been succinctly summarised by the observation that groundwater can seep through any theory. Therefore, there is always a risk that any conventional design of a dewatering system may not be adequate. Hence, the use of the observational method (OM) should be seriously considered whenever construction dewatering is necessary, especially when the consequences of dewatering failure are unacceptable. This case history describes the application of the OM for one of the deepest retaining wall excavations on Crossrail; the 42 m deep Bloomfield box is located at Crossrail's Liverpool St station. The OM, using progressive modification, enabled the dewatering to be successfully implemented, despite significant geological variability, and provided assurance of site safety and also facilitated time and cost savings.

11.2 Key aspects of Design and Construction

11.2.1 Overview

The box is located in a congested urban area (Figure 11.1). To the north the Bloomfield box is bounded by the Circle, Hammersmith and City and

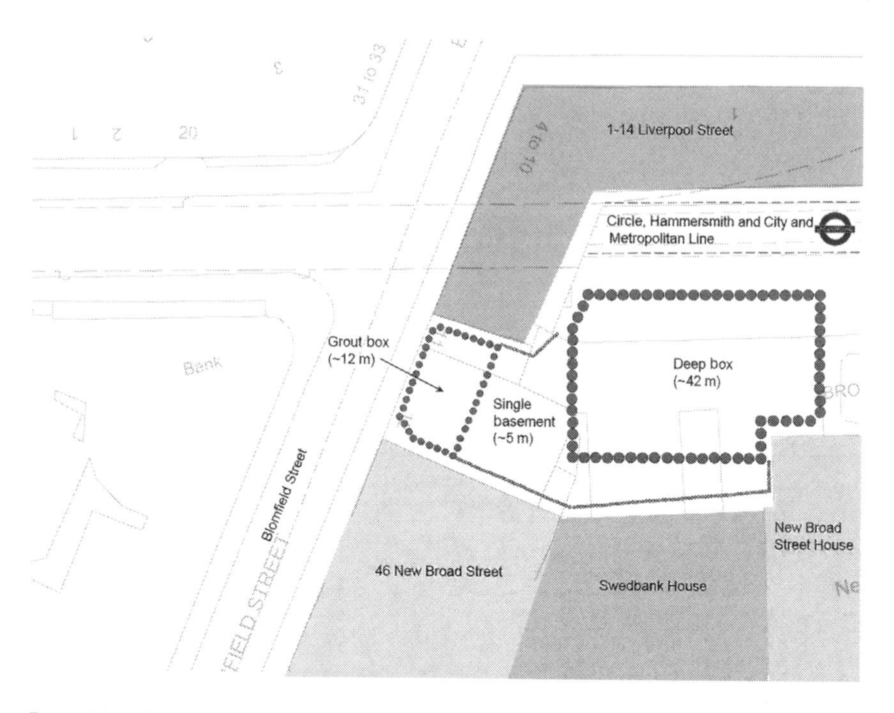

Figure 11.1 Plan of Blomfield Box and adjacent area.

Metropolitan line; to the south it is bounded by 46 New Broad Street House (Figure 11.1). The box comprises three retained excavations, two of which are relatively shallow (a single-level basement, about 5 m deep, and a box to facilitate compensation grouting for adjacent tunnelling works, about 12 m deep) and the deep box. It is only the deep box which will be discussed in this chapter. The overall plan dimensions of the deep box are approximately 30 m long and 20 m wide. The box is initially constructed through the London Clay, but the final 5–6 m of the box needed to be constructed through the Lambeth Group sequence (which comprised silt and sand layers under high groundwater pressures). The retaining walls for the box are formed of 1.2 m diameter contiguous bored piles spaced at 1.4 m centres. The pile toe level is at 61 m ATD, that is, about 52 m below ground surface level. Contiguous bored piles had to be used, rather than diaphragm wall panels due to the severe space restrictions at the site.

Given the excavation depth it was considered that attempting to form secant piles would be unsuccessful. The limiting depth for secant pile walls is typically about 25 m, due to verticality tolerances and the piling rigs' ability to rotate a casing to cut into the installed primary piles.

Adjacent to and penetrating into the box were a complex arrangement of sprayed concrete lined (SCL) tunnels, for the main tunnel, pedestrian access and escalator tunnels (Figures 11.2 and 11.3). Figure 11.4 shows a cross section through the box, together with an outline of the geology. The main SCL tunnel shown is about 5 m from the box and its invert is located within the Lambeth Group. The SCL works for this tunnel also required dewatering through the Lambeth Group; however, the SCL tunnel excavation would take place well before the excavation for the box, as outlined later. The box was to be excavated top-down and the retaining walls were supported by six levels of permanent reinforced concrete beams and walers, as indicated in Figure 11.4.

11.2.2 Geology

The ground surface level is at about 113 m ATD. The ground conditions comprise 2.5 m of made ground and 4 m of Terrace Deposits, overlying a 30-m-thick layer of London Clay. The Harwich Formation (about one metre thick), not shown in Figure 11.4, underlies the London Clay and the Lambeth Group (up to about 18 m thick) underlies the Harwich Formation. Both the Harwich Formation and Lambeth Group are highly

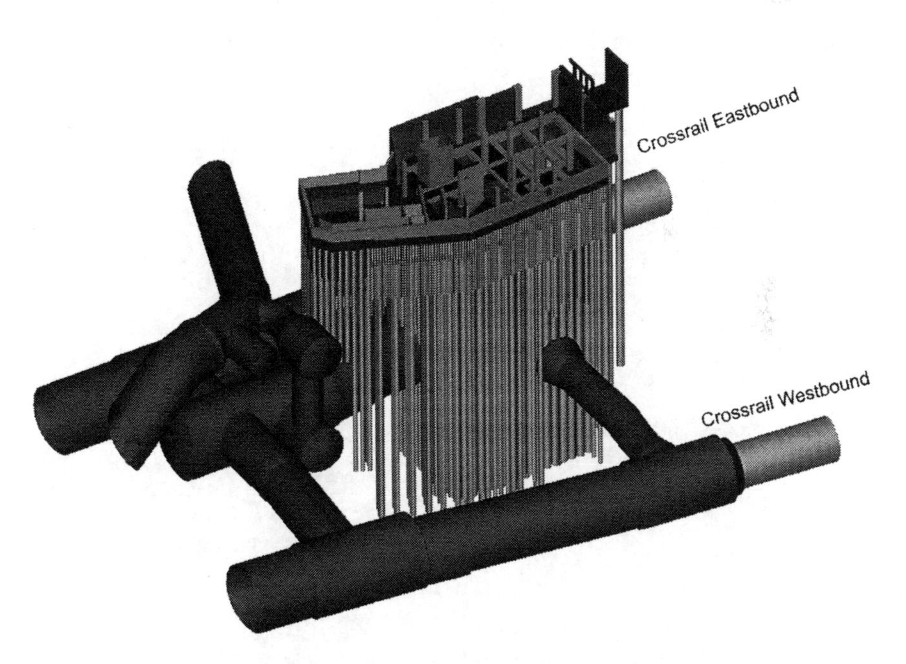

Figure 11.2 Three-dimensional view of Blomfield Box and SCL tunnels.

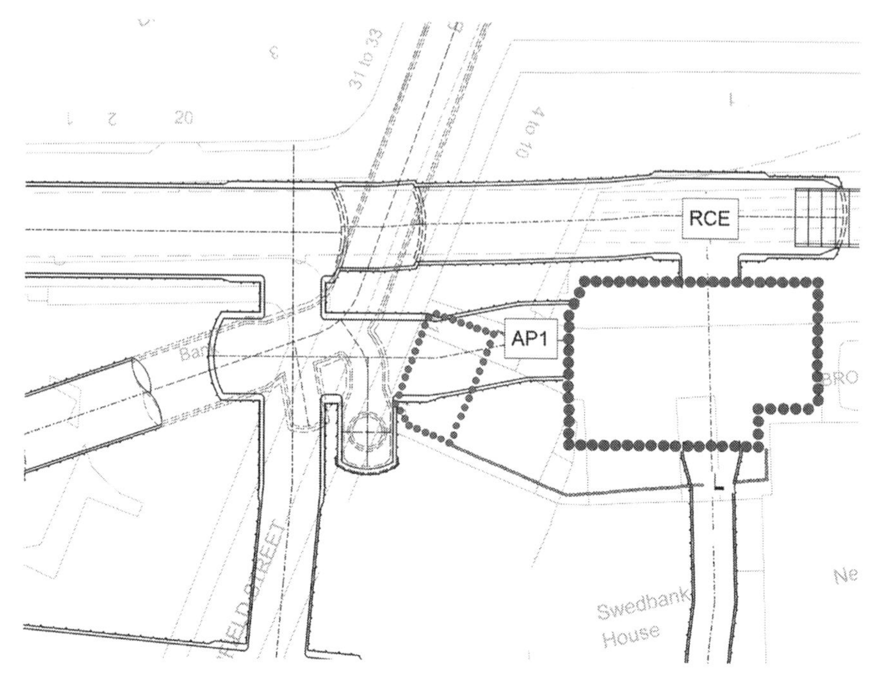

Figure 11.3 Plan of Blomfield box and adjacent SCL tunnels.

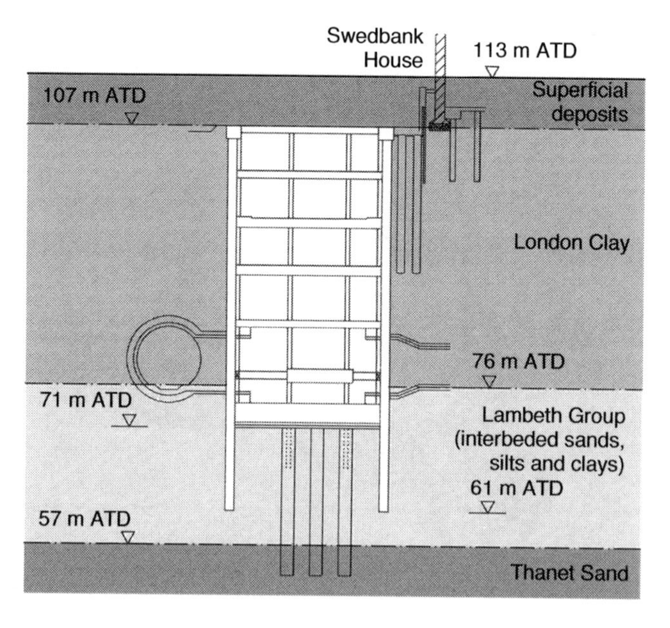

Figure 11.4 Simplified geological profile at deep box.

variable, varying from very stiff clays to very dense sands and gravels. Groundwater level is approximately at 106 m ATD (the River Terrace deposits, predominantly sandy gravels at this site, are practically dry). At depth the groundwater pressure profile is non-hydrostatic (Figure 11.5) due to historical pumping from the Chalk aquifer, which underlies the Thanet Sand. The lower unit of the Lambeth Group (the Upnor Formation, typically silty sands), is in hydraulic continuity with the Thanet sands and Chalk aquifer. However, the groundwater pressure in the upper units of the Lambeth Group is relatively high, measuring up to 250 kPa in the sand channels and the sand/silt horizons in the Laminated Beds. Hence, there was a risk of excavation base instability once the excavation reached an elevation of about 89 m ATD (i.e. as the overburden pressure reduced towards the groundwater pressure in the sand/silt layers, located a few metres below the Lambeth Group surface). The significant hazards associated with excavations in the Lambeth Group and the complexity of the geometrical configuration of higher permeability layers of

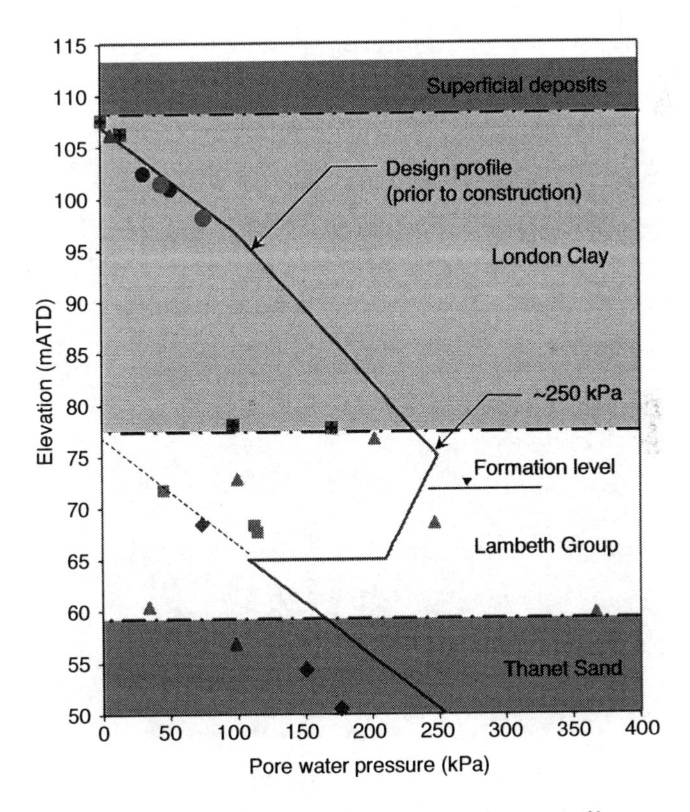

Figure 11.5 Measured groundwater pressures and assumed design profile.

sands and silts interbedded with clay layers have been discussed in detail in CIRIA report C583. The Lambeth Group was formed in an estuarine environment; hence, the nature of the ground and groundwater conditions can be highly variable (Figure 11.6). Figure 11.7 shows a photo of an exposure through the upper Lambeth Group, near the invert of one of the

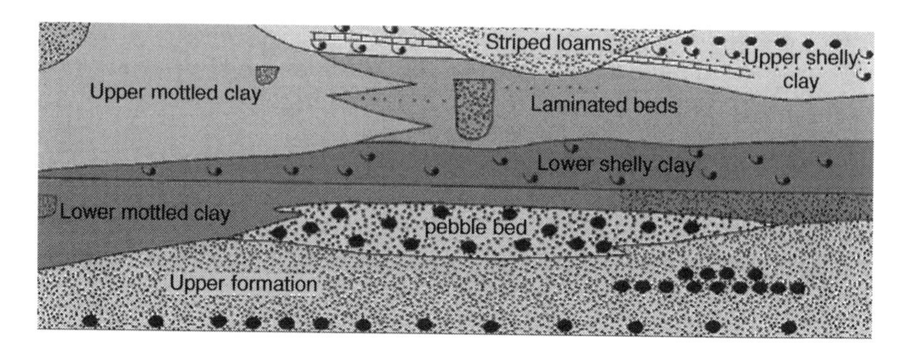

Figure 11.6 Potential variability of Lambeth Group (after CIRIA C583).

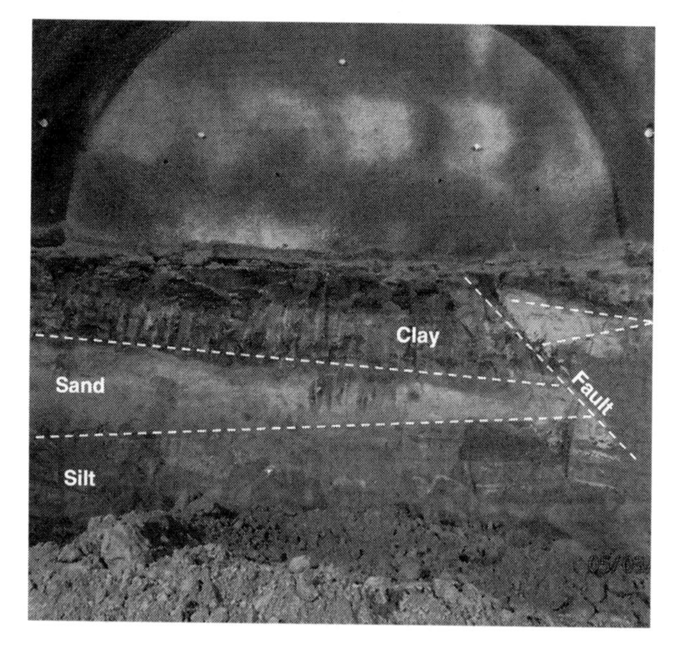

Figure 11.7 Exposure of Lambeth Group at invert of SCL tunnel.
(Note : Fault at right-hand side and interbedded sand/clay/silt layers.)

SCL tunnels, with interbedded clay, sand and silt together with a fault (at right-hand side of photo). It was recognised that rapid changes in soil type could occur across horizontal and vertical distances of a few metres. Therefore, excavations though the Lambeth Group pose a very different challenge to the overlying more homogeneous and low-permeability London Clay. Base instability risks would be increased during the final 5 m of excavation (through the Lambeth Group, refer to Figure 11.4) due to the potential risk of ground unravelling or flowing through the gaps between the contiguous bored pile walls of the box (due to the high groundwater pressures in sand/silt layers). Hence, the key design challenge was how the groundwater pressures could be reduced in the sand and silt layers within the Lambeth Group.

11.3 Achieving Agreement to Use the OM

11.3.1 Potential Hazards

The site, as noted earlier, was located within a heavily congested urban environment. This severely constrained ground investigations, due to the presence of existing buildings, underground utilities and structures. It was recognised at an early stage that this would exacerbate uncertainties in the ground and groundwater conditions. However, given the nature of the geology, even if it had been possible to carry out extensive investigations, there would still have been considerable residual risks. These included base instability, excessive loads on the retaining walls and unacceptable wall deformation. The excavation construction sequence is outlined in Figure 11.8(a)–(e). Dewatering was not necessary during the initial excavation stages (Figure 11.8(a)). However, below an elevation of about 89 m ATD (about 24 m below ground level), the groundwater pressures would need to be reduced in sand layers in the Lambeth Group, in order to maintain an adequate factor of safety against base instability (Figure 11.9(b)). Once the excavation exposed the sand/silt layers, the gaps between the contiguous piles would need to be filled to prevent the exposed sand/silt unravelling through the gaps (Figure 11.9(c)). Preliminary ground–structure interaction analysis (using PLAXIS) also indicated that there would be considerable benefit in reducing groundwater pressures in the active zone behind the piled retaining walls, to make the prop forces more manageable. The use of closely spaced props towards the base of the excavation was anticipated to be counterproductive. Installing heavy props at depths of more than 30 m raised safety concerns, given the limited space at the base of the box excavation. Also, installation of props and waling beams would be time consuming and during this time period the Lambeth Group clays would be vulnerable to softening, owing to the short drainage path lengths with interbedded sand and silt layers. This would risk increased wall deformation and structural loads. It was

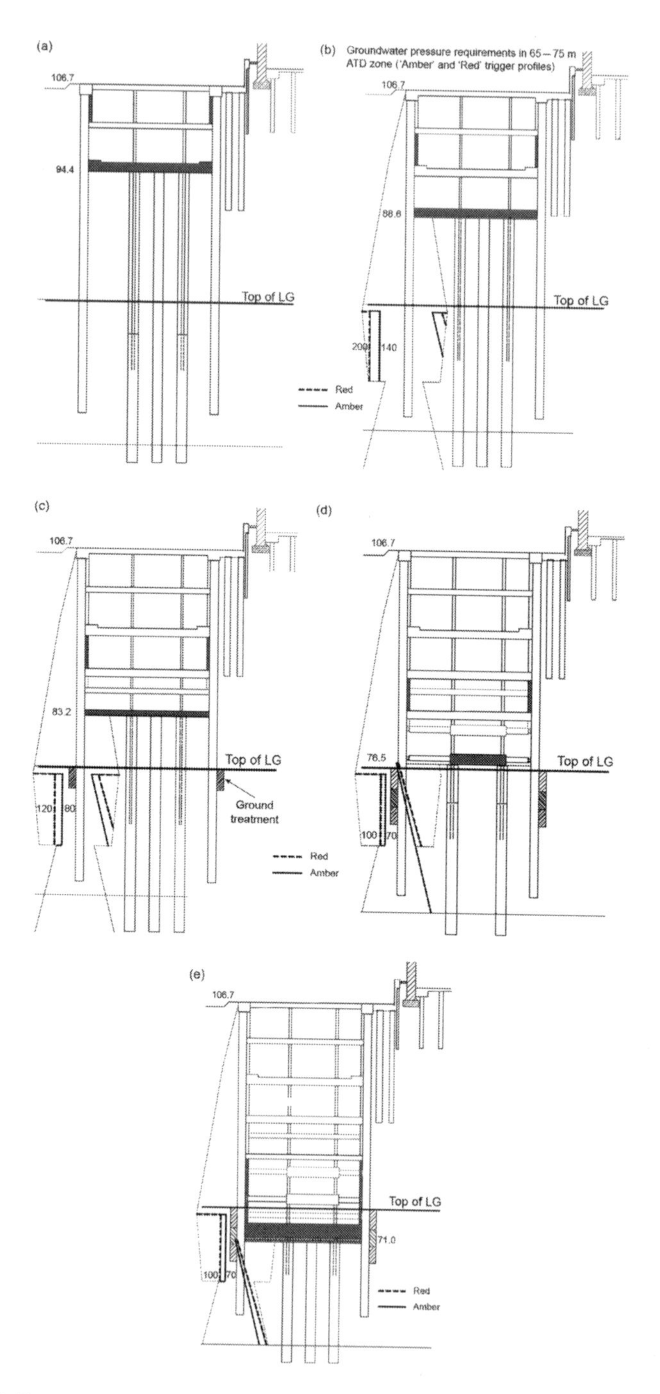

Figure 11.8 Excavation construction sequence and the OM amber/red trigger levels for groundwater pressure: (a) excavation at 94 m ATD, (b) excavation at 88 m ATD, (c) excavation at 83 m ATD, (d) excavation at 77 m ATD, (e) excavation at final formation level, 71 m ATD.

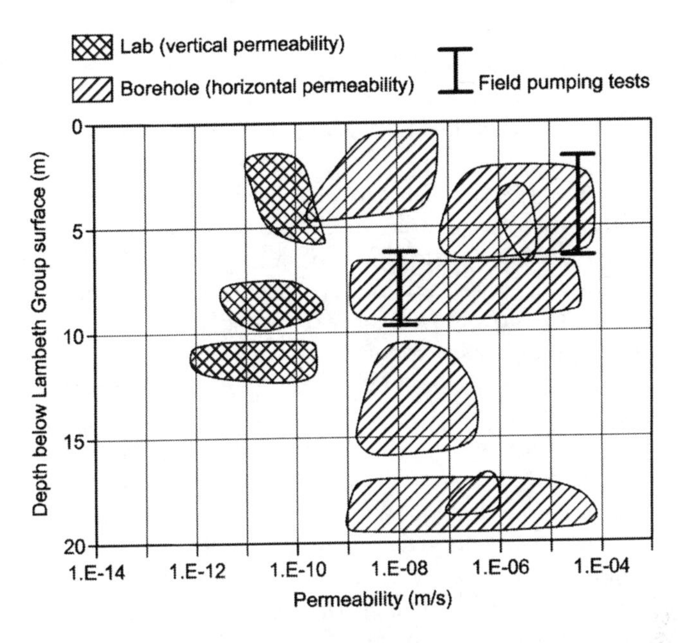

Figure 11.9 Permeability test values for Lambeth Group (lab, borehole and pumping tests).

therefore decided that a dewatering system should be installed on both sides of the retaining wall, to lower groundwater pressures both inside and outside the box to acceptable values.

11.3.2 Design Conditions

A detailed design of the dewatering system could have been carried out and specified. However, this was considered to be unwise for several reasons. Firstly, the limitations associated with the available ground investigation data together with the potential variability of the Lambeth Group, as outlined earlier, meant that there was a considerable difference between pessimistic (or worst credible) and optimistic (or most probable) ground/groundwater conditions which could be assumed for dewatering design. This meant that a conventional design could have been unsafe, if based on an optimistic interpretation of ground conditions, or difficult to build (and unnecessarily costly/time-consuming to implement) if based on a worst credible scenario. Secondly, dewatering design must consider several factors which are intrinsically difficult to quantify, as discussed in Section 11.4. For example, the lateral extent and connectivity of a layer to other permeable layers affects the effective zone of influence of a well and the time period a layer may keep producing water and how quickly groundwater pressure may be reduced and kept low with minimal pumping. Terzaghi's statement (refer to introduction) on the importance of 'minor geological details' is particularly apt for the design of dewatering systems in the Lambeth Group. Worst credible assumptions included both high permeability and intermediate permeability layers, together with some isolated pockets and some permeable layers connected across a wide area (beyond the box footprint). The worst credible scenario required different types of wells and a relatively large number of wells to ensure the excavation would be safe throughout construction. Installation of all the wells anticipated to be necessary in a single phase could have been counter-productive as they would obstruct and slow down the excavation process. More optimistic sets of ground/groundwater conditions were considered as part of the evaluation of the benefits of implementing the OM. However, they could not be justified as a safe starting point for commencing construction. This case history is an example of the huge difficulty in identifying and justifying 'most probable' conditions as a basis for commencing construction, as discussed elsewhere in this book. At this site it would have been impractical to have attempted to identify the 'most probable' conditions and, in the authors' opinion, unsafe to have commenced construction on this basis. The way forward was to implement the OM by progressive modification, based on a cautious estimate of ground conditions, and adjust appropriately on the basis of the observations.

Throughout the Crossrail project, temporary works were the contractor's responsibility. These included any dewatering. However, at the Blomfield box the dewatering would also affect the permanent works (e.g. deformation of the shaft walls and associated structural forces induced in the bored pile walls) and potential movement of adjacent infrastructure. Therefore, the OM was planned as an 'ab initio' application at the project inception to manage the implementation of the dewatering. The performance requirements for the OM, in terms of amber and red trigger levels for groundwater pressure, were specified, as shown in Figure 11.8.

A difficulty which emerged during discussions to use the OM was the risk of wall deformation leading to unacceptable movement of adjacent buildings/infrastructure. The project team considered that this risk could be managed through the amber/red trigger levels for groundwater pressure. However, potential damage assessments for existing buildings/infrastructure were being managed through a separate contract, which created a significant co-ordination challenge. Although far from ideal, the pragmatic way forward was to agree amber/red trigger levels for wall deformation (similar in principle to those agreed for the Moorgate shaft, refer to Chapter 12). This created a dual traffic light system for the OM implementation. In general, this should be avoided due to the increased complexity; for example, potential for confusion if a trigger for one parameter is breached while another remains green, and the potential uncertainties regarding the appropriateness of implementing a particular contingency measure (see Section 11.4.3). In this case, the risk of wall deformation breaching an amber or red trigger, before amber/red triggers were breached for groundwater pressure, was considered by the project team to be extremely remote. Therefore, in practice, only a single traffic light system was likely to operate during the works.

11.4 Implementation of the OM

11.4.1 Dewatering Systems – Potential Failure Scenarios

A dewatering system can fail in several different ways (Roberts and Preene, 1996), as highlighted in Table 11.1. Groundwater flows may be excessively high or low, relative to the dewatering system capacity, which may then lead to failure of the dewatering system. Groundwater flow volumes, and their variation during excavation, depend upon several parameters, including

Table 11.1 Failure of dewatering systems – common scenarios.

Failure mechanism	*Common cause*	*Typical solutions*	*Comments*
Groundwater flow excessively high	Ground permeability unexpectedly high. Disposal of water becomes impractical.	Larger pumps and/or extra wells. Worst case: use cut-off wall and/or permeation grouting.	Careful assessment of plausible range in permeability. Large-scale pump tests.
Well yield too low	Problem across interface zones between well and aquifer, e.g. well screen or filter pack too fine.	Replacement and/or additional wells, different installation or well development methods or well screen system	If low permeability use vacuum assistance. Well design is challenging in uniform fine sands.
Well/aquifer-poor connectivity	Complex soil stratification: well not connected to all high permeability layers, local channels can be problematic.	Highly site specific, depends on aquifer connectivity and depths.	Intensive investigation and good logging. Passive wells can be cost effective.
System deterioration in long term	Bio-fouling or calcium carbonate precipitation.	If hazard identified: specify appropriately.	Monitoring necessary; problematic for longer term (>3 months).
Unacceptable effects in adjacent areas	Settlements induced in soft clays, peats, etc. Groundwater contaminants mobilised.	Cut-off walls or recharge wells.	Risks often overestimated.

ground permeability, hydraulic gradient, connectivity between permeable horizons, their ability to recharge and so on. Permeability test data at the Blomfield site were limited; however, extensive permeability testing (both small-scale lab tests, moderate-scale borehole tests and large-scale pumping tests) had been carried out in the Lambeth Group in adjacent sites. Permeability test data, summarised in Figure 11.9, indicated a permeability range varying across several orders of magnitude. Based on previous experience and careful logging of site-specific boreholes, this variability of permeability was considered to be realistic.

To reduce the risk of failure, it was necessary to design a flexible set of dewatering systems which could cope with the range of conditions likely to be encountered (Figure 11.10). This included locally high-permeability clean sands to lower permeability silts. The conceptual design included both deep wells and ejector wells. It should be noted that the fuzziness in the boundaries between different groundwater control techniques, shown in Figure 11.10, is deliberate and highlights the need for judgement when a boundary is being approached. An outline dewatering design, or base case, was provided as part of the design drawings. This enabled the overall scope of the possible dewatering effort to be assessed for construction planning purposes and to ensure a suitable range of equipment would be mobilised to site. However,

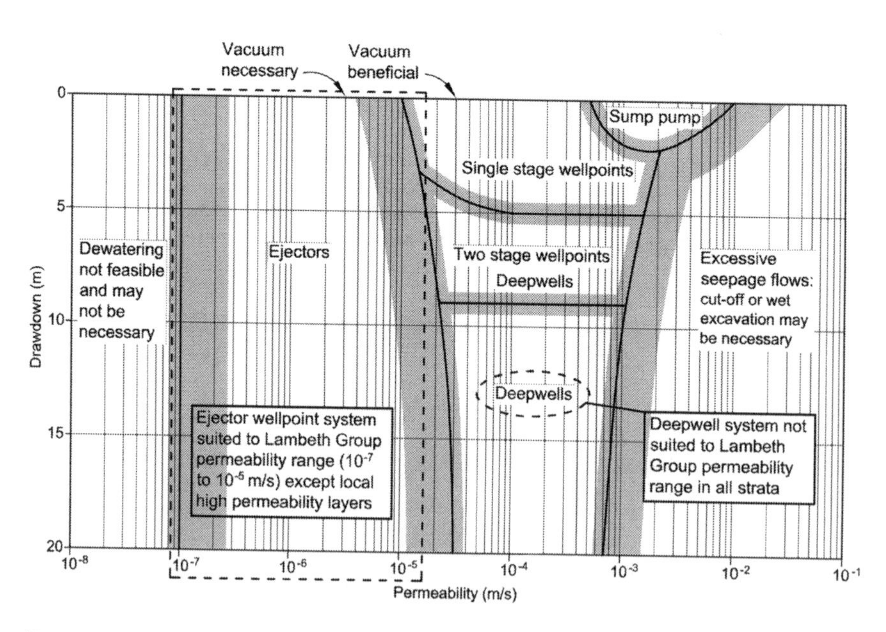

Figure 11.10 Permeability range for various pumped groundwater control techniques (modified after Preene, 2012).

the contractor was permitted to modify the dewatering design provided the OM performance requirements were met. It was considered that the main uncertainty was the degree of connectivity between permeable horizons which depended on the fine detail of soil stratification (as outlined by Terzaghi, see Introduction and Table 11.1). A fault zone was also anticipated across part of the box footprint, although its precise location was unknown. This greatly complicated the assessment of the hydrogeological regime, the likely changes in groundwater pressures during excavation and the effective radius of influence of wells installed in the vicinity of the fault. Dewatering would also be required to construct the SCL tunnels (the tunnels were built under a separate contract). However, it was known that this dewatering would be switched off after the tunnels had been built (and before the box excavation required dewatering to be effective). There was some uncertainty associated with the rise in groundwater pressure which would then take place once the tunnel dewatering had been switched off, and which would need to be managed under the box construction contract.

11.4.2 Instrumentation and Monitoring

The primary monitoring system comprised fourteen fast response vibrating wire piezometers (VWPs) and five standpipe piezometers to monitor the performance of the dewatering system (Figure 11.11). Retaining wall deformation was monitored by thirteen Shaped Accel Array inclinometers. Secondary instrumentation for base stability comprised four magnetic extensometers. The critical observations were groundwater pressure; these had to be supplemented by the observations summarised in Table 11.2 to facilitate interpretation of the groundwater pressure data.

In addition, to the instrumentation it was vital for the soil stratification to be carefully evaluated during the drilling of any boreholes for instrument or well installation, and as the shaft excavation progressed through the Lambeth Group. Therefore, a senior geologist was located on site throughout the OM implementation.

The installation of the dewatering wells and the instrumentation was planned to be carried out across several phases. This enabled the geology and hydrogeology to be better understood. It also enabled the effectiveness of the dewatering system to be assessed in a progressive manner. This phased approach enabled the dewatering design to be modified on the basis of observed performance. It also enabled safety to be assured, while minimising the risk of slowing down excavation.

The installation of the piezometers was carried out in two different phases, to facilitate faster excavation. Two VWPs were installed in the first phase and before box excavation. These piezometers were located in the sandy layers of the Lambeth Group. The holes for piezometer installation

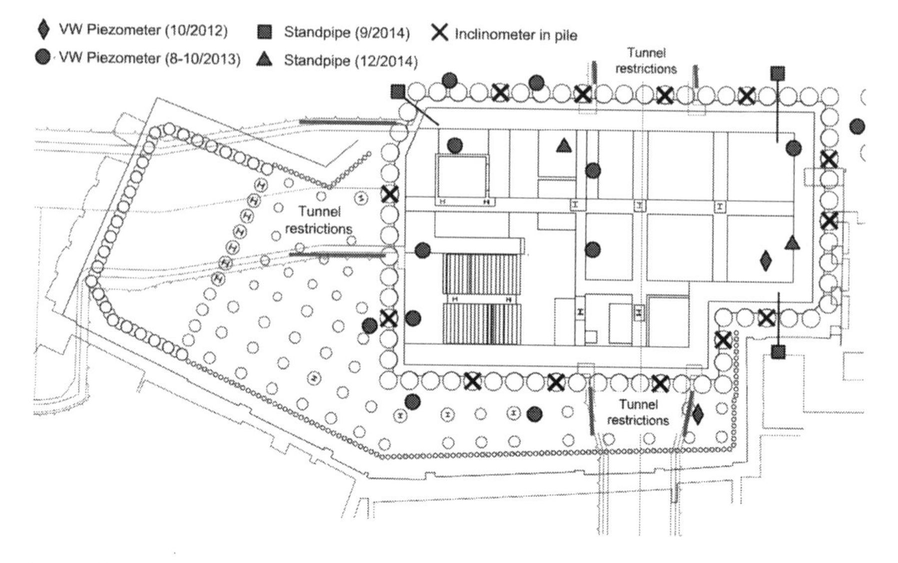

Figure 11.11 Instrumentation locations for monitoring groundwater pressures and wall displacement.

Table 11.2 Observations to support monitoring of groundwater pressure.

Parameter	*Method*	*Comment*
Soil stratification	Logging during installation of all wells and instrument holes	Critical issue due to complex geology of Lambeth Group
Flow from individual wells	Flow metres	Important for selection of remedial options
Water quality and fines	pH/conductivity, check discharge tank	Water treatment necessary, removal of fines
Mechanical performance	Pump pressure, power supply	System maintenance, back-up needed

were drilled using high-quality rotary coring (triple tube wire line with mud flush) and the continuous core was carefully logged by an experienced geologist. This provided important information for the dewatering system design. The piezometer monitoring data provided a set of baseline readings, prior to box excavation influencing groundwater pressures. In the second phase of piezometer installation, 12 VWPs were installed and five standpipes. As shown in Figure 11.11, some of the piezometers were installed behind the wall, by inclined drilling through the gaps between contiguous piles, when the excavation reached 83 m ATD. Limited space meant that monitoring instruments and any wells for dewatering behind the shaft walls had to installed by inclined drilling.

11.4.3 Traffic Light System for the OM

The groundwater control flow chart for implementation of the OM, showing trigger levels and contingency measures, is shown on Figure 11.12. As discussed earlier, the primary focus for the traffic light system was groundwater control. If an amber trigger was breached, then more frequent monitoring was required, and the coverage of piezometers was also re-assessed. If considered to be inadequate, then additional instruments were to be installed. If the red trigger was breached, then excavation would be stopped, and the dewatering system would be enhanced by installing additional dewatering from a tool box of alternative systems (Figure 11.12). Selection of appropriate systems from the tool box mainly depended upon the anticipated local ground conditions and their assessed permeability (Figure 11.10). The ground model (summarising the 3D variations in ground conditions) was progressively developed by the designer's site geologist on the basis of logging all excavation exposures and drill holes for instrument installations, wells and so on. The effectiveness of existing wells could also be enhanced (e.g. converting to vacuum enhanced wells). In the unlikely event of a red trigger for wall deformation being breached, then temporary props would be installed. A key component of the risk management was full time input by the OM design team, who were resident on site. This facilitated all monitoring data and

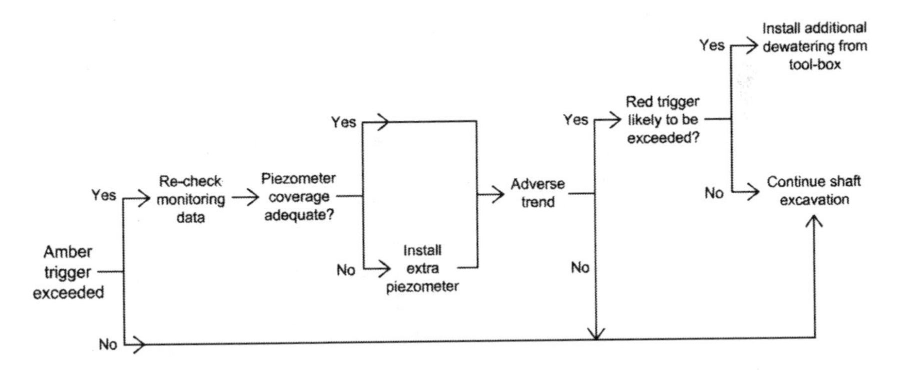

Figure 11.12 Groundwater control flow chart for the OM implementation.

Note: The dewatering toolbox comprised several systems, including deep wells; ejector wells; passive relief wells; trenching and sump pumps. Selection was based on locally assessed conditions. All equipment mobilised to site and available for installation.

site observations to be rapidly integrated and assessed in the context of the design requirements.

The installation of dewatering wells was carried out in two main phases: a first phase, with four deep wells installed; a second phase, with nine inclined and two vertical ejector wells (Figure 11.13).

11.4.4 Dewatering System, Observations Outside of the Blomfield Box

The purpose of the deep wells (installed when the excavation was only about 6 m deep) was to facilitate a better understanding of the permeability and connectivity of the sand/silt layers in the upper units of the Lambeth Group. The ejector wells were installed when the excavation was at 83 m ATD (i.e. when the excavation was 30 m deep and about 7 m above the surface of the Lambeth Group; Figure 11.8). Figure 11.14 shows a plot of groundwater pressures, measured outside the box bored pile walls, plotted against time, for the period between November 2012 and November 2014. During dewatering for SCL tunnel (AP1) construction, into the western end of the box, groundwater pressures had dropped to values of between 90 and 120 kPa by July 2013. During the subsequent construction of the SCL tunnel (RCE), adjacent to the northern wall of the box, groundwater pressures were reduced to between 65 and 90 kPa by November 2013. Following installation of the SCL tunnel

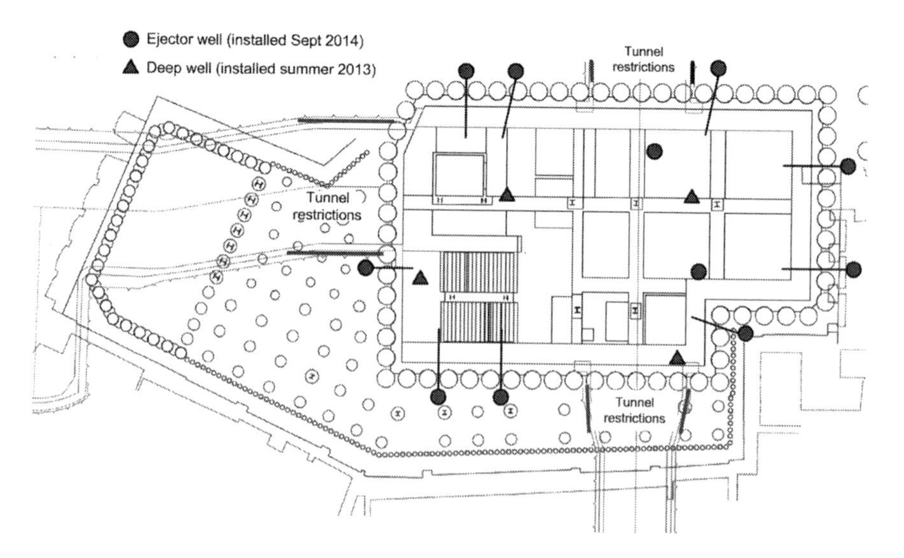

Figure 11.13 Well installations in September 2014, excavation at 83 m ATD.

lining, the tunnel dewatering was switched off and groundwater pressures increased during a six-month period (up to May 2014) to about 110 kPa. The internal deep wells, operating between July and September 2014, had a limited zone of influence, only reducing groundwater pressures outside the box by about 10 kPa (to about 100 kPa). The inclined ejector wells, operating from September 2014, were more effective, leading to a drop of 35 kPa in about 60 days (i.e. to pressures of about 65 kPa, below the amber trigger set for the final excavation stage of 70 kPa). The inclined ejector wells continued to operate effectively throughout the remainder of the works, and the amber trigger for groundwater pressures behind the bored pile walls was not breached. The gaps between the contiguous bored piles were infilled with concrete after each one metre excavation through the Lambeth Group. This provided temporary stability prior to construction of the permanent lining.

11.4.5 Displacement of Contiguous Pile Retaining Wall

The displacements of the box retaining walls are summarised in Figure 11.15. The observed horizontal displacement of the bored pile walls, for the northern wall of the box (at inclinometer IN 101), are plotted for excavation levels of 94.4, 88.6, 76.5 and 71 m ATD. Wall displacements at the

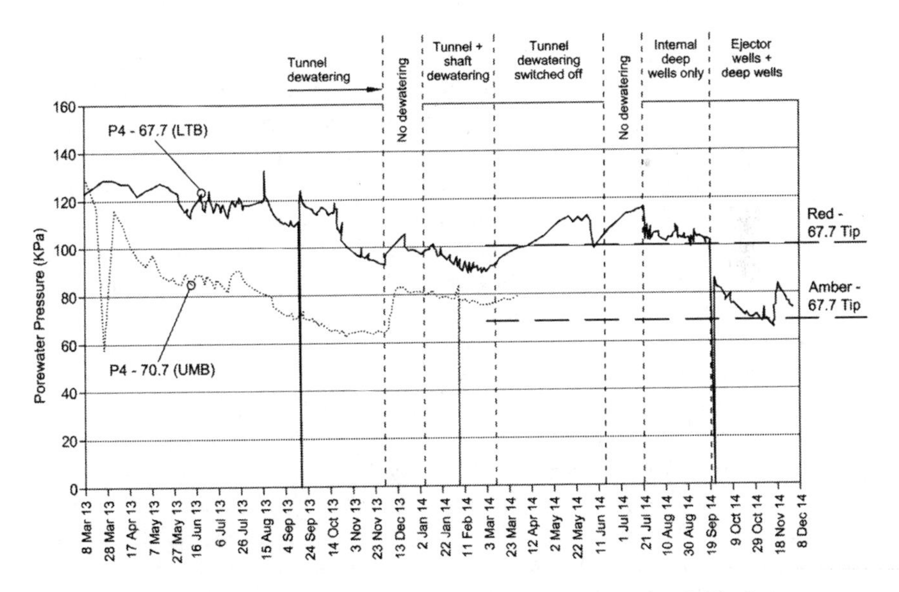

Figure 11.14 Observed groundwater pressures, outside Blomfield Box, up to November 2014.

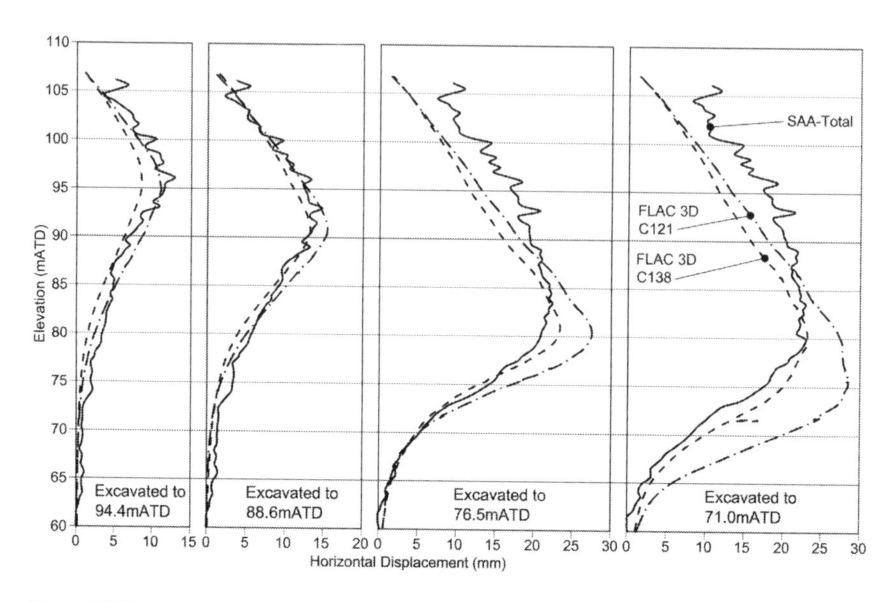

Figure 11.15 Observed wall displacements versus calculated values.
Note: C138, A^*G_o based on G_{hh}; C121 A^*G_o based on average of G_{hh} and G_{vh}.

other north wall inclinometers were rather lower, about 10–20% lower. The wall displacements for the southern wall were typically about 25% lower than those for IN101 in the northern wall, consistent with numerical modelling predictions (probably due to the influence of the adjacent building basements). Observed wall displacements can be compared against calculated values from numerical modelling, shown in Figure 11.15. The numerical modelling used a non-linear small strain stiffness model known as A^*. The A^* model uses field geophysics measurements (G_{vh} and G_{hh}) to define initial ground stiffness and, as discussed by O'Brien and Harris (2013) and O'Brien (2015), has several features which considerably improve the reliability of ground movement calculations compared with alternative advanced soil models (for heavily over-consolidated soils). The observed wall movements were typically about 10–20% lower than the original calculations, labelled as C121 on Figure 11.15 (based on the isotropic A^* model, used successfully for SCL tunnel design, which defined initial ground stiffness as the average of G_{vh} and G_{hh}). A second set of calculations (which used G_{hh} only to define initial stiffness) are also shown, labelled as C138 in Figure 11.15. Observed wall movements during the early stages of box excavation are slightly larger than calculated values derived from G_{hh}, although calculated wall displacement at 71 m ATD was practically equal to observed wall displacement. The amber and red trigger for wall displacement were set at 30 and 45 mm, respectively (the SLS value for the bored piles exceeded 50 mm), these compared with a maximum observed wall displacement of 23 mm and calculated wall displacements of 28.5 and 23 mm (for the C121 and C138 values, respectively). All inclinometers recorded wall displacements substantially below the amber trigger, with a maximum value of about 80% and an average of about 70%.

11.4.6 Dewatering System, Observations Inside the Blomfield Box

The main challenge during the later stages of the OM implementation was the control of groundwater pressures inside the box, below the base of the excavation. When the excavation approached the Lambeth Group, at 76.5 m ATD, the trends for the piezometer readings beneath the eastern side of the box indicated that the red trigger level was likely to be breached if no further actions were taken (Figure 11.16).

Figure 11.16 plots the variation of observed groundwater pressures, inside the box, against time, for the period up to November 2014. Groundwater pressures are shown for sand channels in the upper mottled beds (at 71.7 m ATD) and in the underlying laminated beds (at 68.2 m ATD) of the

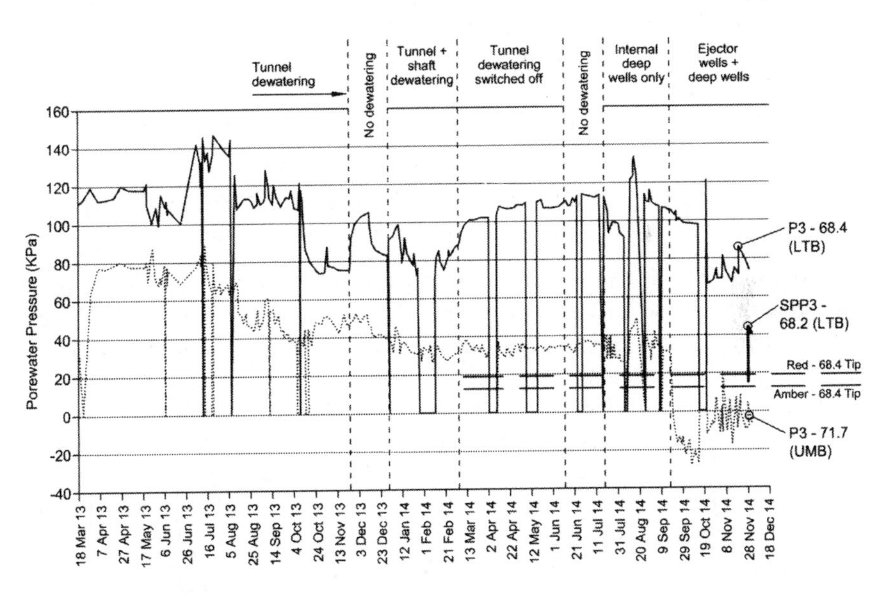

Figure 11.16 Observed groundwater pressures, inside Blomfield Box, up to November 2014.

Note: Passive wells installed in December 2014 were converted to active wells during final phase of works.

Lambeth Group. During 2013, when tunnel dewatering was operating, groundwater pressure variations exhibited a similar pattern to those beyond the box retaining walls. In the upper sand channel, tunnel dewatering reduced the groundwater pressures to slightly below 40 kPa. In the laminated beds, groundwater pressures were reduced to about 80 kPa. However, once tunnel dewatering was switched off, there was a marked contrast between the sand channel and laminated beds, with groundwater pressures increasing within the laminated beds, whereas they remained practically constant within the sand channel. This indicated that the sand channel and laminated beds were part of two independent hydrogeological systems. Those within the laminated beds were being recharged (and were probably part of an extensive 'sheet' of permeable soil with a much larger footprint than the tunnels and box). In contrast, the upper sand channel seemed to be relatively isolated, and once dewatered by the tunnel dewatering it was not vulnerable to groundwater recharge. In the laminated beds, groundwater pressures increased by 30 kPa in 60 days, reaching a value of 115 kPa (when the excavation was approaching 83 m ATD).

The ejector wells located inside the box began to reduce groundwater pressures during September and October 2014. In the upper sand channel, groundwater pressures were reduced below zero (with the vibrating wire piezometers recording small negative pressures between 10 and 20 kPa). However, in the laminated beds groundwater pressures could not be reduced below 75 kPa. Once the excavation had reached 76.5 m ATD, the groundwater pressure in the laminated beds was close to the amber trigger level. It became clear that contingency measures would need to be implemented to avoid breaching the red trigger at formation level (of 25 kPa, i.e. one third of measured values).

11.4.7 Dewatering System, Modifications during Final Phase of Excavation

In addition to the different behaviour between the laminated beds and upper sand channel, there was also different hydrogeological behaviour between the western and eastern halves of the box (Figure 11.17). Groundwater pressures below the eastern area were significantly larger than those beneath the west and it was considered that this asymmetric behaviour was due to the presence of faulting in the Lambeth Group across the south-eastern part of the box. It was considered that another six wells would be needed to reduce pressures in the laminated beds. The practicality of installing six wells was a concern given the confined conditions at that level inside the box.

Installation of all the wells in a single stage was likely to seriously delay the main excavation works, and so the decision was made to split the excavation area into two zones, to allow the first three wells to be installed in the eastern zone while allowing excavation in the western zone. This staged approach allowed well installation and operation to be carried out in parallel

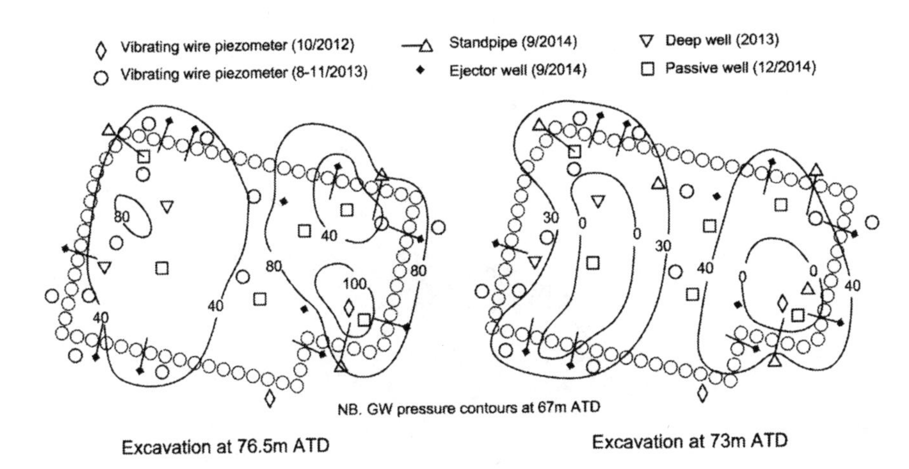

Figure 11.17 Groundwater pressure contours, before and after final modifications to groundwater control systems.

with the excavation. When the excavation approached 73 m ATD, an area of 'wet' very silty sand was identified across the northern half of the box, with stiff clay to the south (Figure 11.18). Narrow slit trenches were installed in the norther half to 70 m ATD with local sump pumps. Once activated these further reduced groundwater pressures to below the amber level within the Lambeth Group and enabled the final formation level of 71 m ATD to be safely reached (Figure 11.17). Once the excavation had reached 71 m ATD, a 300-mm-thick blinding slab was cast (this acted as a temporary base strut) to support the box during the Christmas holiday period. Passive pressure relief wells were extended above the blinding to relieve any groundwater pressure below the blinding slab. The final layout of deep wells, ejector wells, piezometers and standpipes are shown in Figure 11.19.

11.4.8 Summary of Changes Introduced through Implementation of the OM

In total eleven ejector wells and ten deep pumped wells were used (four wells, then six passive wells converted to active wells), compared with the original base case layout of twenty-two ejector wells and twenty deep wells, that is about 50% reduction due to the application of the OM. In part this was achieved because only the laminated beds were subject to groundwater recharge via a remote source, and also due to each installed well having a wider zone of influence than originally assumed. The position of each well could also be optimised relative to observed effectiveness (partly influenced by the fault zone), which minimised the disruption to excavation works. The

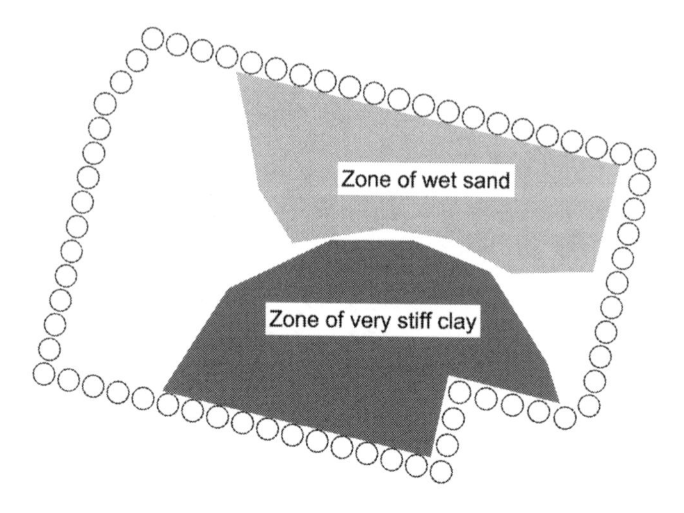

Figure 11.18 Configuration of clay and sand layers at 73 m ATD.

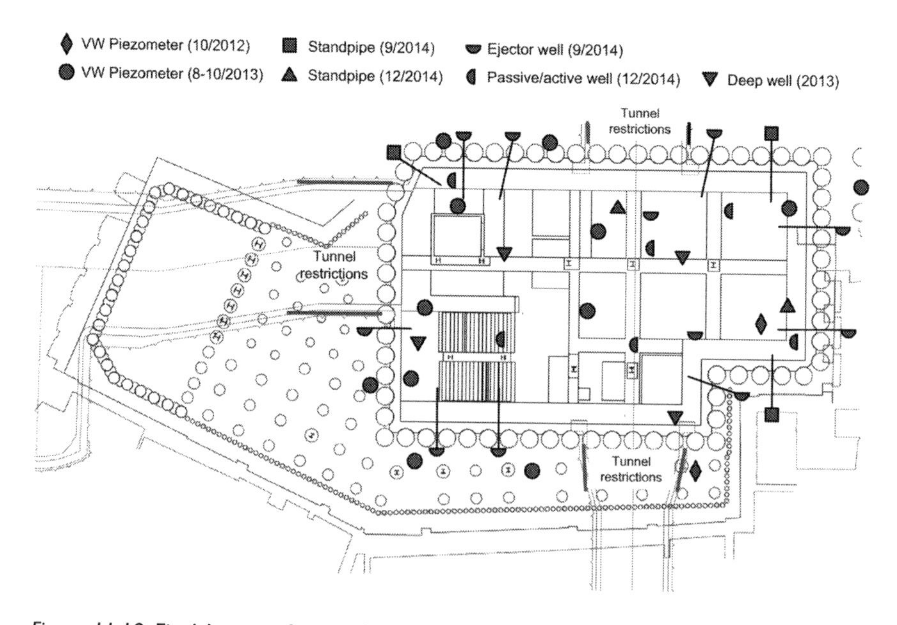

Figure 11.19 Final layout of groundwater monitoring and wells (note slit trenches not shown for clarity).

complexity of the fabric of the Lambeth Group was revealed by careful logging of the excavated material and from logging of all the drilling for instrument and well installations (Figure 11.20). The complex inter-layering of the sand, silt and clay layers, and local faulting could not have been predicted from any ground investigation; as noted in the introduction, these conditions are an example of the 'minor geological details that are unknown' as discussed by Terzaghi. Hence, accurate predictions of dewatering requirements could not practically have been made at the design stage.

11.5 Conclusions

The use of the OM by progressive modification is ideally suited to the implementation of dewatering, especially within the complex geology encountered at the Blomfield Box. The 42 m deep box penetrated 5 m into the Lambeth Group which comprised a complex interbedded sequence of water bearing strata, with groundwater pressures of up to 250 kPa. Within the congested confines of the site, ground investigations were severely constrained, and even if full ground investigations could have been carried out, there would have been considerable residual uncertainty concerning the groundwater conditions and the design of an effective dewatering system. Hence, there was a high risk that an optimistic dewatering design based directly on the

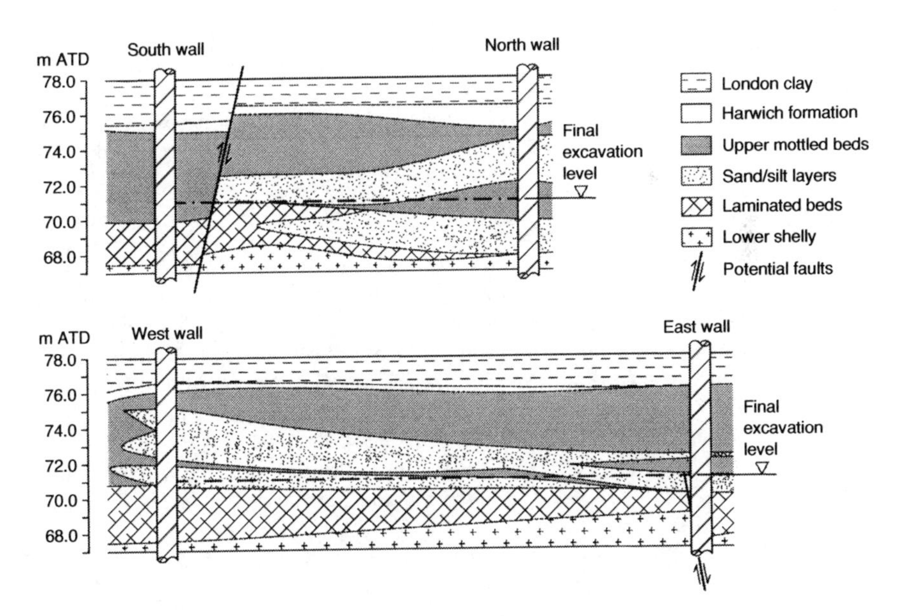

Figure 11.20 Blomfield Box – logged profile through Lambeth Group.

predictions of the most probable conditions would have been unsafe. In practice, observed conditions proved worse than these conditions. An overly conservative design based on the worst credible conditions would have been impractical to install in a single phase while excavating the box in a cost-effective manner. The observed groundwater pressures during the box excavation varied in a highly complex manner, both spatially (due to faulting and the interbedded nature of the sand/silt layers) and with time (e.g. groundwater pressures increased significantly in the laminated beds when dewatering was temporarily switched off), as summarised in Figures 11.14, 11.16 and 11.17. Dewatering for adjacent tunnel construction reduced groundwater pressures by about a half. To maintain the stability of the base of the excavation groundwater pressures had to be reduced to close to zero. Dewatering for the box excavation required the installation of three different techniques, initially deep wells, then ejector wells and finally slit trenches and sump pumping. The timing of well installation and their location was based on observations. Observed wall displacements were relatively small (23 mm or less) and were broadly consistent with calculated values from an advanced 3D numerical model. The careful application of the OM by progressive modification minimised delays to the box construction while maintaining an acceptable level of safety.

References

CIRIA report C583 (2004). Engineering in the Lambeth Group.

O'Brien, A. S. (2015). Geotechnical characterisation for underground structures, Proc of XVI European Conference on Soil Mechanics and Geotechnical Engineering, volume 7, pp. 3729–3734.

O'Brien, A. S. and Harris, D. (2013). Geotechnical characterization, recent developments and applications, 12th International Conference on Underground Construction, Prague. April 2013.

Preene, M. (2012). Groundwater control, Chapter 80, volume 2, pp. 1173–1190. ICE Manual of Geotechnical Engineering.

Roberts, T. O. L. and Preene, M. (1996). The design of groundwater control systems using the observational method, Geotechnique symposium in print – the observational method in geotechnical engineering.

Rowe, P. W. (1968). The influence of geological features of clay deposits on the design and performance of sand drains, Proc of ICE, supplemental volume 1, pp. 1–72.

Rowe, P. W. (1972). The relevance of soil fabric to site investigation practice. 12th Rankine lecture. *Geotechnique*, 22, No. 2, 195–300.

Terzaghi, K. (1929). Influence of minor geologic details on the safety of dams. *American Institute of Mining, Metallurgical, and Petroleum Engineers Technical Publications*, **215**, 31–44.

Terzaghi, K. (1955). Influence of geological factors on the engineering properties of sediments, Economic geology, 50th anniversary volume, pp. 557–618.

Chapter 12

Crossrail Moorgate Shaft (2012–2014)

12.1 Introduction

Crossrail's Liverpool Street station will serve the City of London and provide interchanges with London Underground's Northern, Central, Metropolitan, Circle and Hammersmith & City lines, connections to Stansted airport and National rail services at Liverpool Street and Moorgate stations. The Moorgate shaft is a key structure within Crossrail's Liverpool Street development. The westbound running tunnel passing through the shaft placed the shaft on the critical path for completion of the 8.2 km long Y-drive tunnels being driven by tunnel boring machine (TBM) from Limmo Peninsula in Canning Town to Farringdon. The shaft had to be completed by the end of November 2014 to allow construction of the launch chamber to speed TBM Victoria on her way to Farringdon. The timely passage of the TBM through the Moorgate shaft was fundamental in meeting the Crossrail critical path programme for the opening of the central London section; any delay in the construction of the Moorgate shaft was unacceptable.

The shaft was to be constructed on the footprint of 89–134 Moorgate. The building previously occupying the majority of the shaft site was 91–109 Moorgate (Amro Bank), a 1970's six-storey structure with a single-storey basement. The building spanned over the LU subsurface tracks and was founded on 900 mm diameter bored piles up to 34 m long. When the building had been demolished in advance of the shaft construction works, the piles were left in place. Removal of the pile foundations proved to be highly problematic. The diaphragm wall contractor after prolonged efforts had to develop a bespoke extraction machine to clear the shaft footprint of the redundant piles. These problems led to an 11-month delay in the shaft construction programme. Meeting the original project milestone for handover to the tunnel contractor seemed highly unlikely. At this stage, the designer proposed the use of the observational method (OM) to enable the safe implementation of several potential programme-saving measures.

12.2 Key Aspects of Design and Construction

12.2.1 Overview

The shaft is located within a highly confined urban area, with numerous surface and sub-surface structures located within a few metres of the shaft perimeter (Figure 12.1). The existing Moorgate station ticket hall was enlarged to create an integrated western ticket hall from which a bank of

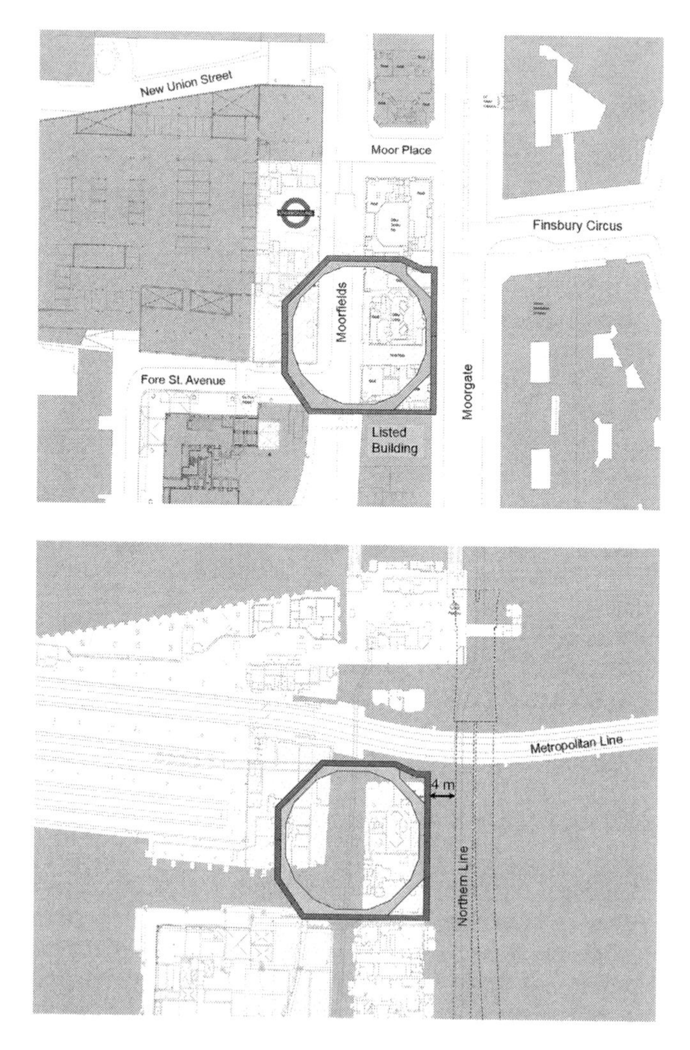

Figure 12.1 Surface and underground infrastructure in vicinity of Moorgate shaft: (a) surface buildings (b) rail infrastructure and underground tunnels.

three escalators descend to an intermediate concourse, whence a further bank of three escalators descends to the platform central concourse 35 m below street level. In addition, the shaft's eight levels of basement also accommodate lifts and emergency escape stairs from platform level, tunnel ventilation, draught relief, smoke extraction and associated plant rooms. This entailed the construction of the deepest shaft on the Crossrail project to 42 m below street level (113 m ATD).

Existing London Underground (LUL) metro tunnels lie to the north and east, a listed building lies to the south and the existing Moorgate station ticket hall to the west of the shaft; the 100-year-old northbound Northern Line tunnel is only 4 m from the shaft wall (Figure 12.1(b)).

The shaft is an irregular polygon in plan, roughly 35 by 35 m, formed by 1.2-m-thick diaphragm walls, 55 m deep. The geometrical complexity of the new Crossrail SCL tunnels which enter the shaft is illustrated in Figure 12.2. The unusual shape was driven by the limited available footprint available for the shaft construction together with the space required for mechanical and electrical equipment which had to be located within the shaft.

The majority of these new tunnels were excavated concurrently with the shaft excavation, together with shallow permeation grouting on the east side of the shaft (to facilitate the construction of a shallow tunnel into the shaft; Figure 12.3).

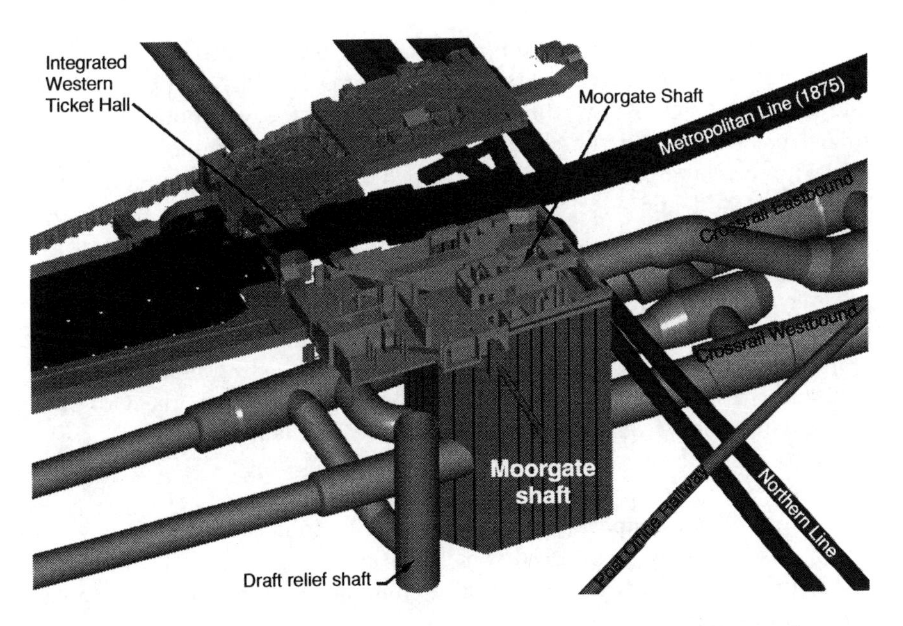

Figure 12.2 Three-dimensional CAD model of Moorgate shaft and adjacent tunnels.

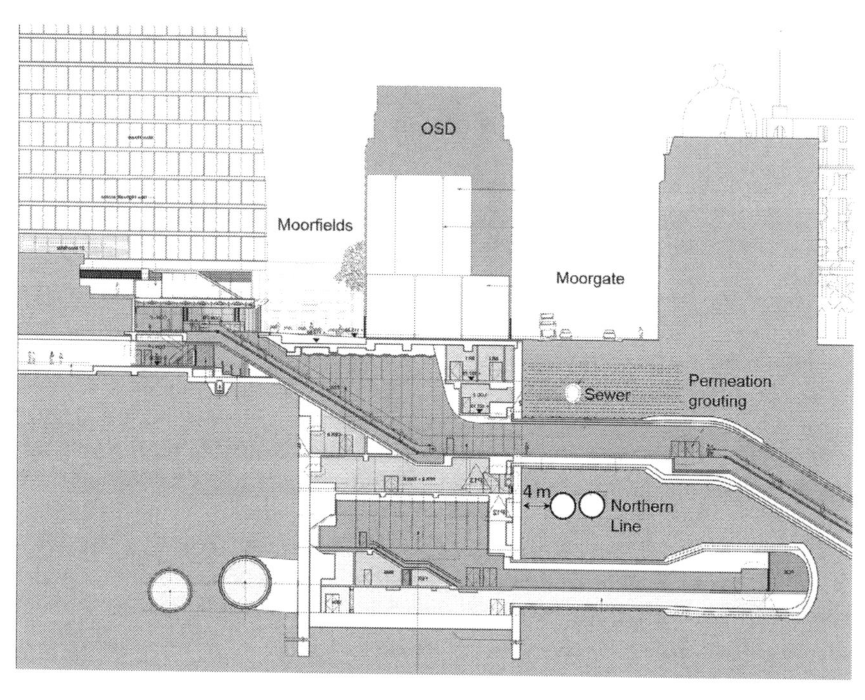

Figure 12.3 Section through shaft showing permeation grouting (east side of shaft).

The shaft was constructed top-down and is permanently supported by seven levels of reinforced concrete ring beams installed between 107.5 and 83.1 m ATD. Below the lowest ring beam the westbound running tunnel passes through the shaft. The design of deep retaining walls and shafts in urban areas is usually governed by the need to limit ground movements and minimise the risks of adverse impacts on existing buildings and infrastructure. These issues were the dominant consideration for the permanent and temporary works design for the shaft. The intent was to limit horizontal deformation of the shaft walls to less than 45 mm, in general, and for the east wall, adjacent to the Northern Line tunnel, the agreed limit was only 30 mm. To minimise the deflection of the shaft walls and associated ground movements and comply with the Crossrail civil engineering design standard (CEDS), the following temporary works measures were originally specified for the design:

1. Two levels of temporary steel props, at 80.5 and 76 m ATD,
2. Two 'slab strips' (reinforced concrete beams 3 m wide by 1.5 m deep), spanning in an east-west direction, to minimise deformation of the adjacent Northern Line tunnel;

3. Below the shaft base level, two cross walls were formed by building two 1.2 m thick unreinforced concrete low cut-off diaphragm walls.

Figure 12.4 shows an outline of the original shaft design, including the permanent and temporary support to the shaft walls. Shaft construction comprised 11 separate construction stages and this requirement, together with the temporary supports outlined earlier, was anticipated to lead to a slow construction process.

12.2.2 Geology

The ground level at the Moorgate shaft is about 113 m ATD. The diaphragm walls were built from the basement level of the building which had previously occupied the site, at a level of about 108 m ATD. The ground conditions comprise made ground, Terrace Deposits, London Clay and the Lambeth Group (Figure 12.5). The Lambeth Group is about 16–18 m thick and overlies the Thanet Sands. No significant sand layers were detected in the Lambeth Group. The large variation in thickness of the Terrace Deposits (between 3 and 13 m) and the corresponding 10 m variation in London Clay thickness is due to the presence of a drift filled hollow. This was understood to have been formed by the former River Walbrook, and this geological feature raised concerns about local softening of the London Clay. Figure 12.6 provides a plot of SPT '*N*' against elevation through the London Clay, and Figure 12.7 provides a plot of in situ shear modulus at very small strain, G_o (derived from field geophysics testing) versus depth, for the London Clay and Lambeth Group.

12.3 Achieving Agreement to Use the OM

12.3.1 Design Objectives and Constraints

To mitigate the delay, the designer proposed the use of the OM. The main objectives were to shorten the time required for shaft construction, while maintaining acceptable deflections of the shaft walls. Acceptable shaft wall deflection was driven by the need to protect adjacent infrastructure from unacceptable movement rather than structural capacity of the shaft walls. It was anticipated that these objectives could be achieved by elimination of some of the temporary propping and by simplifying the construction sequence by combining several construction stages. The Crossrail project management team were highly concerned about the perceived risks associated with elimination of the temporary propping and particularly the potential for damage to the nearby Northern Line tunnel. They were also deeply concerned about any changes which may increase the risk of damage to existing buildings and other adjacent infrastructure. Potential

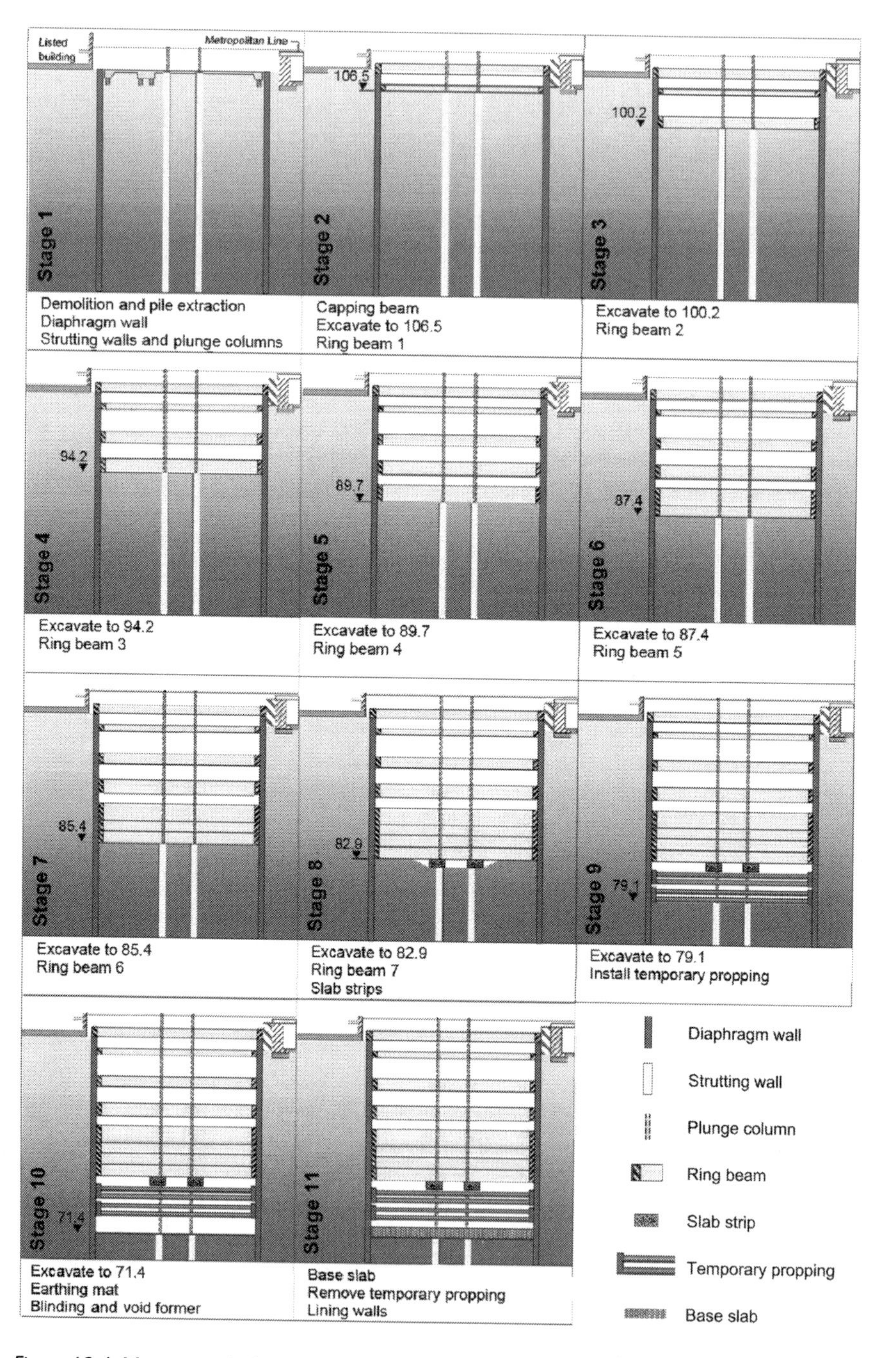

Figure 12.4 Moorgate shaft construction sequence for the original design.

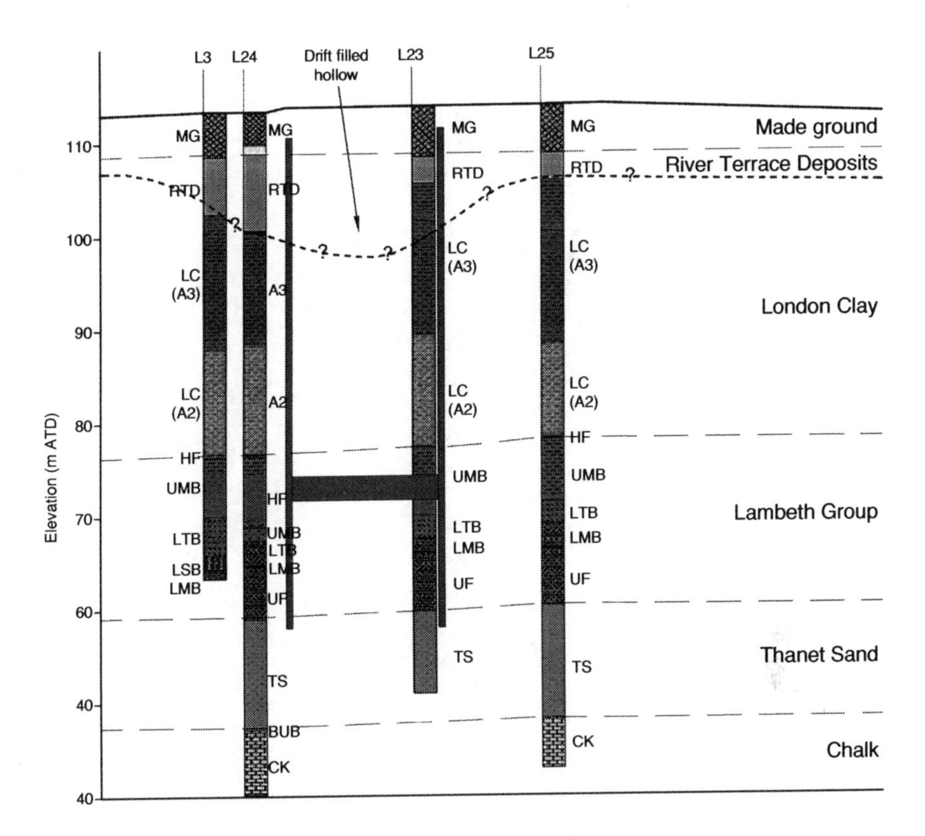

Figure 12.5 Simplified geological profile at Moorgate shaft.

damage assessments for all buildings and infrastructure near to the shaft were the responsibility of another consultant. This inevitably increased the challenge in achieving agreement to use the OM (and the associated requirement to potentially make changes during construction to the conservative base case design). An additional challenge was that there was a strict requirement for the shaft design to be checked and formally signed off by an independent checking consultant, the Cat 3 checker. Hence, it was not feasible, within the available timescales to develop an alternative design (e.g. with reduced temporary works requirements) and for it to be checked prior to starting shaft construction.

12.3.2 Creating Opportunity

Achieving agreement to use the OM involved several steps, where the first involved challenging the project design standard CEDS. This standard

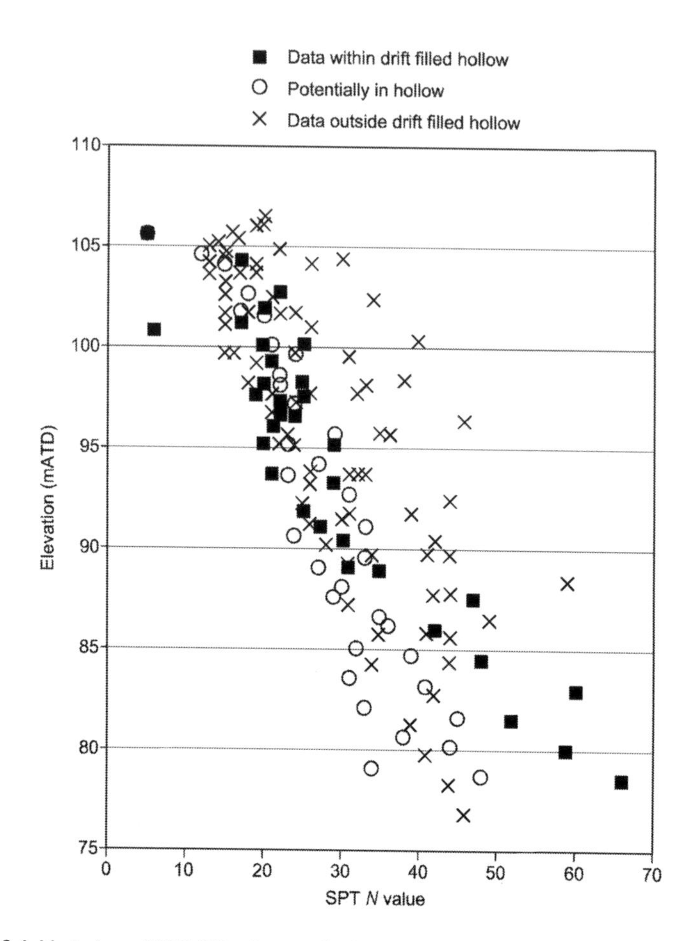

Figure 12.6 Variation of SPT '*N*' values with elevation.

required the loads, during construction, on shafts and retaining walls to be based on active earth pressures calculated on long-term effective stress conditions together with full groundwater pressures. In practice, because of the thickness and low permeability of the London Clay, there was little likelihood of full effective stress earth and groundwater pressures being applied on the shaft walls during the construction period. Within the uppermost few metres of London Clay (adjacent to the water bearing Terrace Deposits) and the lower few metres of London Clay (adjacent to sandier horizons within the Lambeth Group), the earth pressures may (over a relatively short timescale, perhaps several weeks) locally transition towards effective stress conditions. However, within the anticipated timescales for excavation and propping (typically less than two months per stage), it was eventually agreed

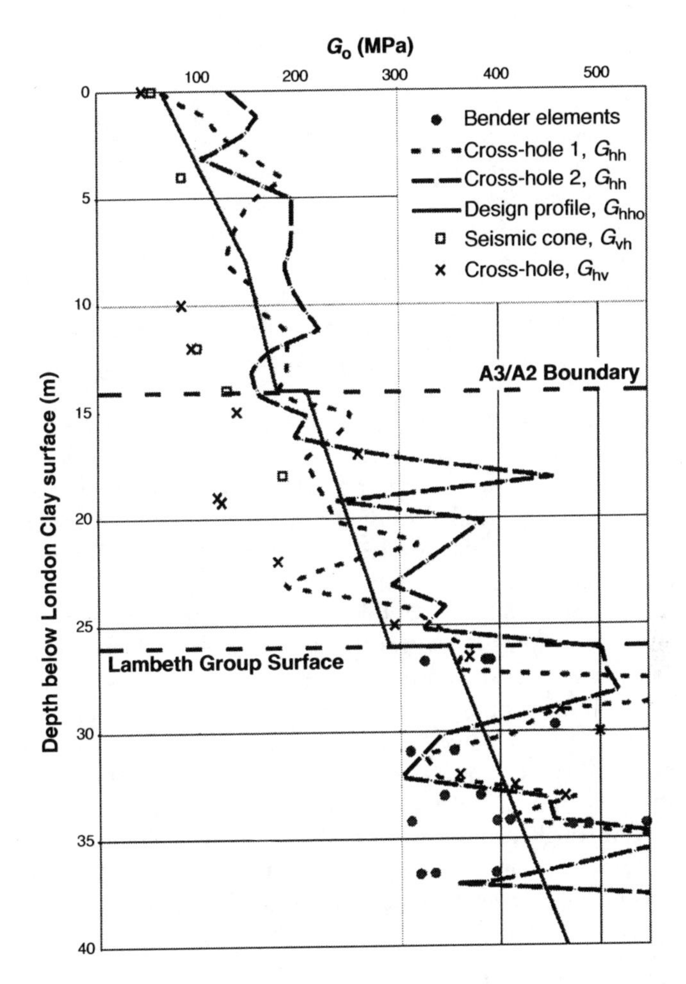

Figure 12.7 Variation of shear modulus, G_o, at very small strain

Note: G_{hh}-shear modulus in horizontal direction; G_{vh}/G_{hv}-shear moduli in vertical direction.

that the bulk of the London Clay was likely to remain in an undrained state, provided that the overall duration of the excavation between 100 and 71.4 m ATD was less than one year. Relatively simple analysis (using 2D PLAXIS with a linear elastic Mohr–Coulomb model) indicated that assuming undrained conditions within the London Clay (on the active and passive sides of the wall) would lead to reduced wall displacements and lower structural forces on the shaft walls. This change was not sufficient, on its own, to enable a radical change to the shaft design.

The application of an advanced non-linear soil model (A^*) within a 3D numerical analysis (O'Brien and Liew, 2018) did indicate that there was a good chance of substantially reducing the temporary works requirements. Following completion of the non-linear 3D modelling (and the agreed departure from CEDS), it was confirmed, with the Cat 3 checker, that the temporary prop at 76 m ATD could be omitted (and this became the base case design). It was recognised that it may also be possible to simplify and speed up construction by combining several construction stages and also omit the temporary prop at 80.5 m ATD, so that the diaphragm wall spanned 12 m between the lowest ring beam and the final excavation level at 71.4 m ATD. However, because of the sensitive nature of the adjacent infrastructure, modifying the temporary works solely on the basis of advanced analysis was not acceptable and would not provide adequate assurance to the multiple stakeholders involved in the project. Nevertheless, given the severe impact of any shaft construction delays on overall completion of the central tunnelled section of Crossrail, it was also agreed that some design and construction modifications would be necessary to have any chance of reducing project delays. The highest priority was the omission of the temporary propping at 80.5 m ATD. Prop installation at this level would take several weeks and while in place would severely constrain access for the lowest level of excavation and subsequent construction of the base slab. The TBM could not transit through the shaft with this propping in place. Hence, it was necessary to develop a methodology that evaluated value engineering opportunities on the basis that the probability of omitting the 80.5 m ATD propping could be maximised. The designer therefore recommended that the OM should be implemented through progressive modification, since this meant that construction could start on the original fully assured design basis, the base case design. This was eventually agreed by all stakeholders.

12.3.3 Verification Process

Design changes would only be made if recorded wall displacements were sufficiently low and analysis (calibrated by back analysis of previous stages) indicated a low risk of breaching agreed trigger levels for wall displacements during the next stage of excavation. The trigger levels were agreed by all stakeholders, prior to commencement of the works, and could only be changed if all stakeholders agreed. To facilitate this process, three key stages were identified as verification points (Figure 12.8). At each key stage, or verification point, a detailed analysis of all the available monitoring data and construction activities (including those being undertaken by adjacent tunnelling contractors) would be carried out. If this indicated a low risk of unacceptable wall movement, then planned temporary works could be changed (following discussion and agreement with stakeholders). Because of

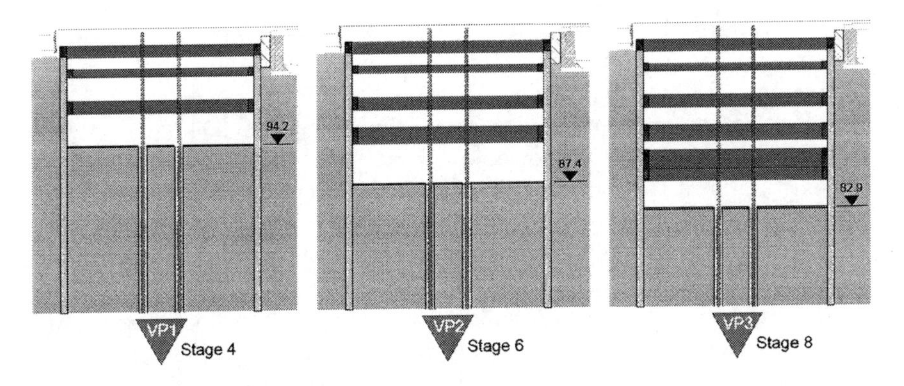

Figure 12.8 Verification process – three VP levels.

the use of these verification points, this OM application was usually referred to as the verification process (VP) within the project team; the use of this term was more comfortable for some stakeholders who were not familiar with the OM.

12.4 Implementation of the OM

12.4.1 Key Steps and Activities

For the implementation of the OM at Moorgate, all the agreed contingency measures, by basically implementing the base case design, were included at the outset. The contingency measures could only be omitted if observations, during previous construction stages, indicated that unacceptable wall displacements would not occur during subsequent construction phases. For the project management team, the benefit of this approach is that it provided a bounded programme. Any agreed changes to temporary works would then provide a benefit by shortening the shaft construction programme. The construction sequence drawings showed the base case design, but with the project manager having the authority to instruct omission of propping and modifying the construction sequence should the outcome of the VP be favourable. The key requirements for implementation of the process were:

1. real-time monitoring of wall displacement (as the primary control system);
2. close collaboration between contractor, designer and client;
3. the use of an advanced non-linear numerical model, carefully calibrated on the basis of monitoring of previous construction stages;
4. full-time presence on site of designer's representatives and their active input into the site OM team;

5. careful analysis of cause/effect between the various construction activities;
6. following each stage, producing fully assured construction drawings, so favourable outcomes (from OM implementation) can be implemented with minimal delay;
7. effective communication with the various project stakeholders (Cat 3 checker, consultant responsible for potential damage assessment of existing infrastructure, building and infrastructure owners, etc.).

The primary instrumentation system included thirteen in-place inclinometers (using Shaped Accel Arrays, SAA) to monitor wall displacement. Empty inclinometer tubes were also installed adjacent to each inclinometer to provide redundancy in the monitoring scheme. To verify the inclinometer readings, mini-prism survey points were installed on the face of the ring beams and monitored on a daily basis. As the excavation progressed below the lower ring beam and into the Lambeth Group, laser extensometer surveys were carried out to monitor wall convergence. Thermistor temperature sensors were installed to monitor the variations in the ring beams and the ambient temperature. The temperature measurements and associated survey points on the ring beams proved to be invaluable, as outlined later.

A detailed flow chart was prepared and presented to the Crossrail project management team, the contractor and external stakeholders in order to explain the implementation of the OM. A simplified version of the flow chart is shown on Figure 12.9. As noted earlier, it was a project requirement for the Cat 3 checker to be involved in the OM implementation. In general, the inclusion of a Cat 3 checker is a significant constraint on the OM process. Cat 3 checkers often have little incentive to assist in successful implementation of the OM. This is exacerbated if, as is often the case, the Cat 3 team is inexperienced. Hence, a Cat 3 checker's views and experience of OM should be assessed before they are involved in the OM implementation. In this project the Cat 3 checks were completed in an efficient manner. Good personal relationships were established at an early stage between the designer and Cat 3 checker and this facilitated a supportive checking process.

It should be noted that the three formal verification points effectively replaced the amber trigger level (Figure 12.9). The objective was to assess the risk of exceeding the red trigger. If this was judged to be a sufficiently low risk, then programme beneficial changes to the temporary works would be made. The red trigger levels (for maximum lateral wall displacement during shaft excavation to formation level at 71.4 m ATD) were agreed in consultation with Crossrail's asset protection engineer and LUL. The wall displacement values were constrained by movement limits set for adjacent buildings and infrastructure, which were defined by the consultant responsible for the potential damage assessments. The implementation of the OM for shaft construction was only managed through a single parameter, the shaft wall displacement. This was an important simplification,

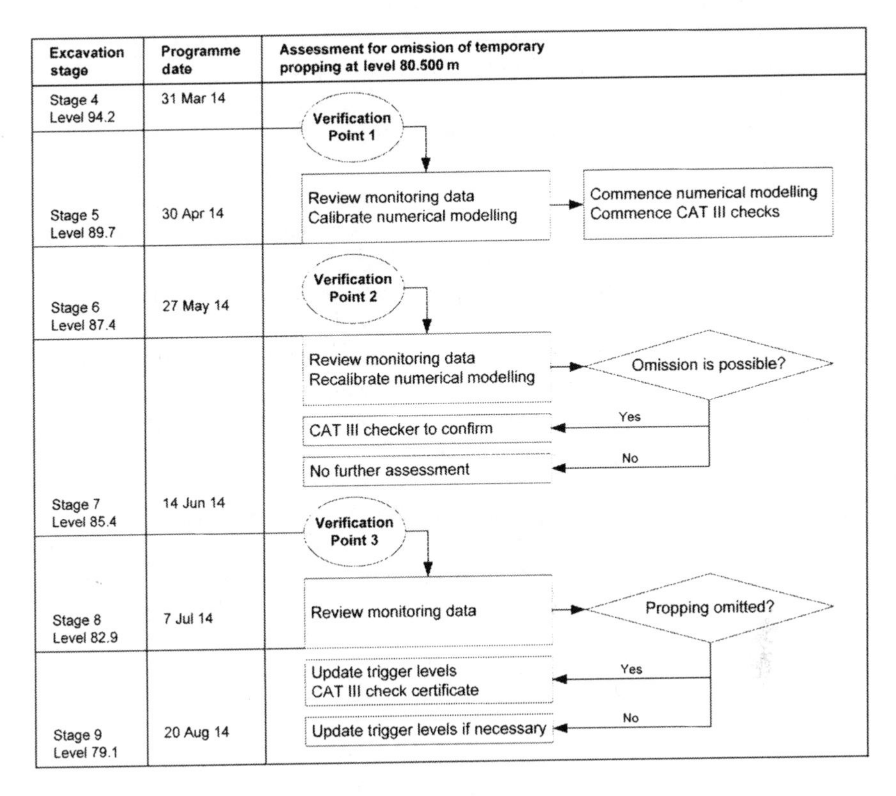

Figure 12.9 Flow chart for the OM implementation.

because other works in the vicinity were being managed under separate contracts. The other works would also cause some movements to adjacent buildings/infrastructure, and the cumulative effects of these movements were assessed by the consultant responsible for potential damage assessments. From a technical perspective, this arrangement was not ideal. A better option would have been for trigger levels to be directly specified, for the structures being protected, but this was not practical given the complex contractual arrangements. For the north, west and south walls, the red trigger was set at 40 mm, whereas for the east wall the red trigger was set at 30 mm (due to the proximity of the 100-year-old LUL Northern Line metro tunnels which were adjacent to the east wall). At each VP stage the following activities were planned to be undertaken to assess the risk of exceeding the red trigger level during subsequent excavation stages, in particular for the final excavation stage between the lowest ring beam and the shaft formation level at 71.4 m ATD:

(a) Wall displacement data was processed. Daily monitoring data were assessed, and anomalies were identified. Stage-by-stage monitoring data were plotted and compared with previous wall displacement predictions.
(b) The advanced numerical model was progressively updated to include: actual as built sequence for adjacent SCL tunnelling, and prediction of future SCL tunnelling effects; geometry of drift filled hollow (as revealed during the early excavation works); mobilised stiffness of ring beams and cross-walls (the latter was particularly important); changes to ground stiffness model to allow for stiffness anisotropy effects.
(c) Produce revised wall displacement calculations from the updated numerical model plotted against monitoring results for each completed stage, produce revised calculations for subsequent construction stages.
(d) Category 3 checks were carried out in parallel, with the monitoring data being provided to the checker.

Advanced numerical modelling played an important role in this OM application. The shaft and SCL tunnels were represented in the FLAC 3D model, by approximately 55,000 shell elements (the approximately 270,000 brick elements representing the soil layers are not shown; Figure 12.10).

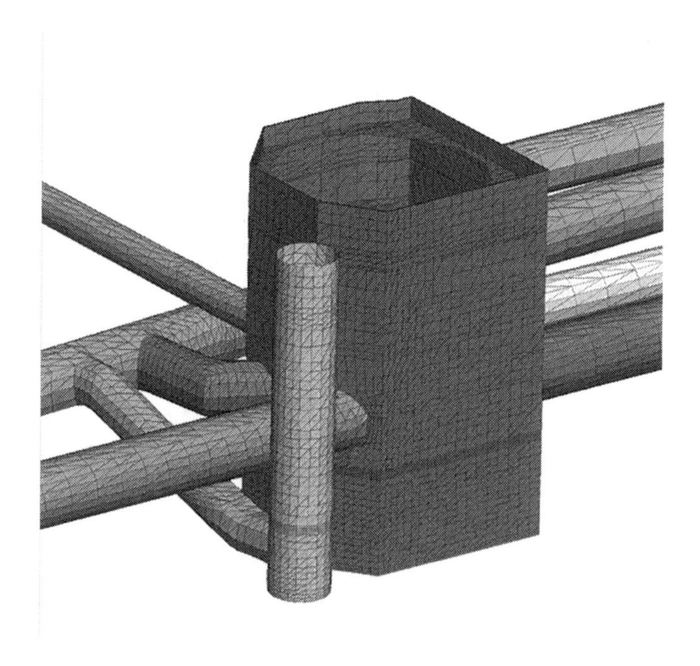

Figure 12.10 FLAC 3D model for shaft and adjacent tunnels.

It is now well recognised that the stiffness behaviour of natural soils is highly non-linear, and to make reasonably reliable ground movement predictions of over-consolidated soils, such as the London Clay and Lambeth Group, it is necessary to simulate small strain stiffness non-linearity accurately (e.g. Jardine *et al.*, 1986). The designer chose to use an in-house bespoke small strain stiffness model known as A^* (described by Eadington and O'Brien, 2011; O'Brien and Liew, 2018). The advantage of the A^* model, compared with other small strain stiffness models, is that the key input parameters can be directly measured, and both field and laboratory data can be used within a holistic framework which allows, for example, the important effects of sample disturbance on laboratory derived measurements to be rationally taken into account. Several back analyses of well-documented case histories of underground construction had demonstrated the A^* model reliability, and it was being used for the adjacent Crossrail SCL tunnel design. Monitored SCL tunnel lining movements were broadly consistent with calculated values from the A^* model.

12.4.2 Verification Point 1

Verification Point 1 (VP1) was started on 10 April 2014 when the excavation level had reached 94.2 m ATD. The observed wall displacements for the north, south and west walls were lower than the calculated values from the 3D FLAC model. In contrast, the east wall displacements were significantly higher than expected. This discrepancy raised several questions:

1. Were the east wall inclinometers faulty?
2. Was the permeation grouting, which was being carried out adjacent to the east wall, causing higher earth pressures leading to larger wall displacements?
3. Were the two unreinforced concrete D-walls (below excavation level) ineffective?

Because of the aforementioned concerns, more detailed investigations were initiated to assess the cause of the unexpectedly large east wall displacements. It was decided to combine construction stages 5 and 6, since the observations indicated that this change was unlikely to affect wall displacements below the 82.9 m ATD level (below the lowest ring beam) and this change would speed up construction. However, the rest of the base case design was maintained.

12.4.3 Verification Point 2

Verification Point 2 (VP2) was started when excavation had reached 87.3 m ATD on 18 May 2014. For the north, west and south walls, the trend from VP1, for observed movements to be less than calculated values, was maintained. Based on this trend, the initial soil stiffness of the A^* model was increased from the isotropic value (based on the average of field geophysics shear moduli, G_{vh} and G_{hh}) to the horizontal shear moduli (G_{hh} only; Figure 12.12). Figure 12.11 shows the assumed non-linear stiffness curves for the Lambeth Group at 74 m ATD, compared with field and lab measurements; further detail is given in O'Brien and Liew (2018). This modification increased the initial shear modulus by about 40%, although stiffness at large strain (0.3–0.4%) was unchanged (this is because the other A^* model inputs, which dictate the rate of stiffness degradation, were unchanged). This revised stiffness model led to calculated wall displacements, for the north, south and west walls being improved compared with observed values. Figure 12.12 shows a comparison for the south wall between observed and calculated wall displacements (derived from three different analyses – 3D FLAC analyses using A^*G_o iso and A^*G_{ohh} and a PLAXIS analysis using a linear elastic Mohr–Coulomb model). The A^*G_{ohh} model provides a reasonably close match of both the deflected shape and maximum wall displacement. The

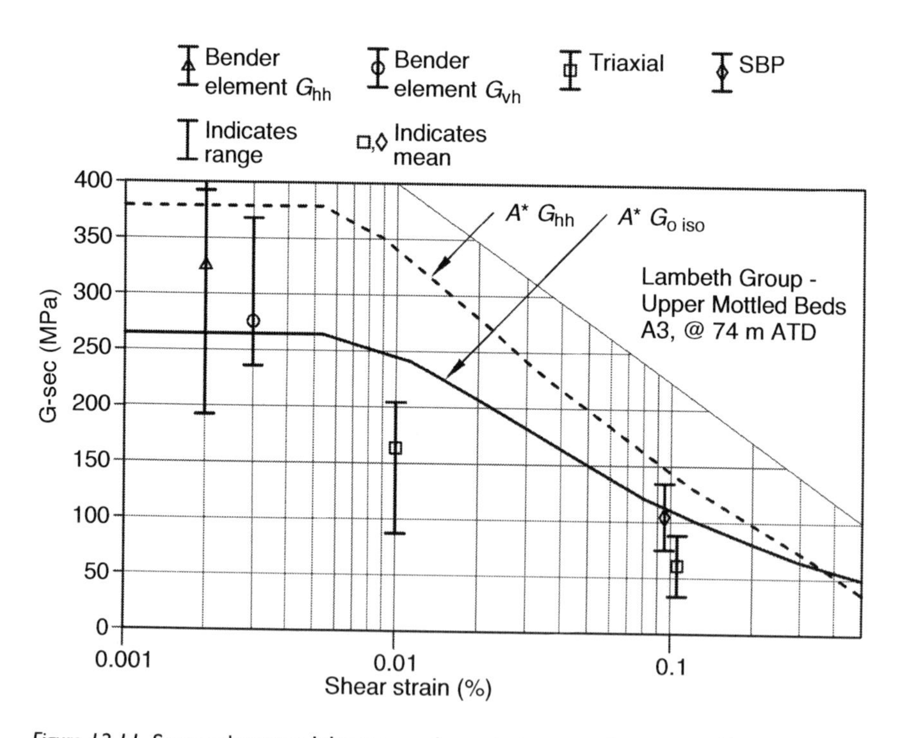

Figure 12.11 Secant shear modulus versus shear strain – test data versus A^* model.

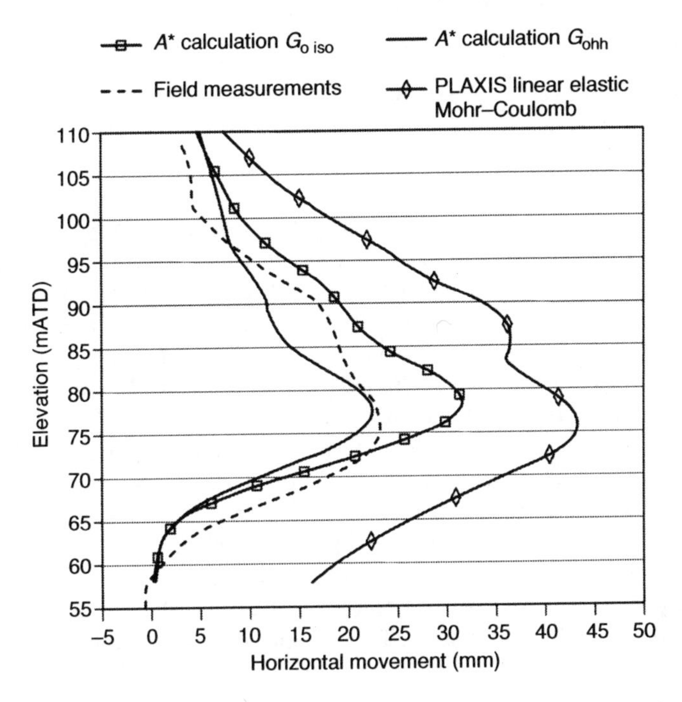

Figure 12.12 Calculated versus measured wall movement, south wall, excavation depth 42 m.

maximum wall displacement derived from the A^*G_o iso model is about 35% too high. The calculated maximum wall displacement given by the conventional linear elastic Mohr–Coulomb model is about double the observed value and the deflected shape is incorrect (with excessively high toe displacement). Hence, based on the observed performance of the north, south and west walls, there was a high level of confidence that the revised soil stiffness model (A^*G_{ohh}) was closely simulating actual ground stiffness conditions. In contrast, for the east wall, the revised soil model significantly under-predicted wall displacement. The ground conditions were expected to be practically the same for all the shaft walls and therefore it seemed likely that some aspect of structural behaviour or construction activity was not being simulated appropriately within the analysis. Hence, actual conditions were not being replicated and simply modifying the ground stiffness was not sufficient on its own.

Detailed checks indicated that the inclinometers in the east wall were functioning correctly. Because of the critical importance of better understanding the displacement of the east wall (and the likely effects of shaft excavation on the busy Northern Line tunnel, only 4 m from the eastern wall) the scope of VP2 was increased to include assessments of:

(a) potential effects of locally reduced London Clay strength (it was conjectured that there may be some effects due the formation of the drift filled hollow);
(b) potential reduction in cross-wall stiffness;
(c) potential effects of permeation grouting, inducing locally high pressures at the back of the east wall.

Sensitivity studies indicated that (a) would not explain the observed increased eastern wall displacement (and it was also considered unlikely that this type of feature would only affect the east wall and not the other walls). The cross-wall panels that abut the shaft D-wall panels were constructed individually; there was a concern that there could be clay inclusions at the junction between the ends of the cross-walls and the face of the shaft D-wall panels. Field inspections of the cross-wall/D-wall junctions were carried out when the excavation reached 87.3 m ATD, and a continuous layer of clay (circa 200–300 mm thick) was observed between the D-wall and cross-wall. The clay had not 'failed', but obviously the effective stiffness of the cross-wall panels was significantly reduced compared with its full theoretical value. To assess this, the FLAC models were run with varying cross-wall stiffnesses (between 0 and 100%; Figure 12.13). A reduced cross-wall stiffness (estimated to be about 10% of its

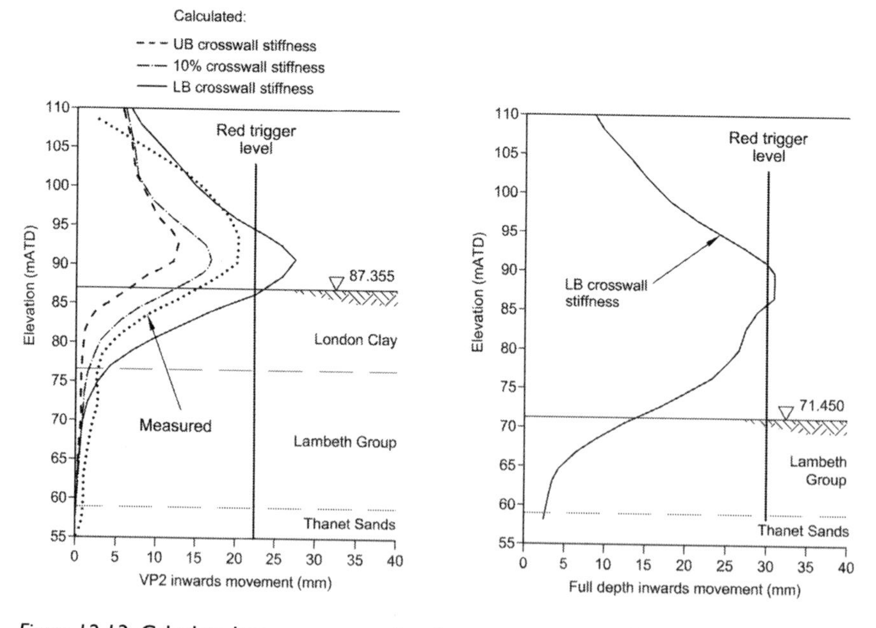

Figure 12.13 Calculated versus measured wall movement, east wall at VP2.

theoretical stiffness) would only explain the increased east wall displacement below the excavation level of 87 m ATD (refer to Figure 12.13). The large wall displacement at about 105 m ATD (about 7 mm higher than calculated) was believed to be due to some other effect, most likely the permeation grouting. When shaft excavation was temporarily halted (between VP1 and VP2), the effects of grouting on the east wall were observed. It was also noticed from a detailed evaluation of all the monitoring data that there appeared to be a temperature-related effect on the shaft walls.

Based on the VP2 stage evaluations, it was decided to combine construction stages 7 and 8, since the observations and updated analyses indicated that this change would not jeopardise the omission of the 80.5 m ATD temporary propping. By combining stages 7 and 8, the excavation could be done in one continuous dig and the ring beams cast resulting in a significant reduction in construction time for these stages. The VP2 assessment did identify a worst-case scenario (due to the reduced cross-wall stiffness) that the red trigger for the east wall could be exceeded.

12.4.4 Verification Point 3

Verification Point 3 (VP3) started when excavation reached 82.9 m ATD on 2 July 2014. The scope of VP3 was increased to include the following additional activities:

(a) Grouting pressures acting on the east wall were simulated by application of a local lateral pressure of 200 kPa between elevations of 105 and 100 m ATD on the east wall.
(b) Influence of temperature variations on wall movements was assessed – this included modifying the ring beam stiffness (to simulate the temperature induced movement).
(c) The stiffness of the ring beams was modified to account for the restraining effect of a temporary construction jetty/crane on the north wall of the shaft.
(d) Sensitivity studies to assess the influence of varying the properties of the Lambeth Group (based on local variations identified during instrumentation installation and pore pressure changes in the clay/silt layers).
(e) Discussions with LUL and Crossrail's asset protection engineer to evaluate the observed movements of the Northern Line tunnels due to SCL tunnelling and shaft construction and the sensitivity of the tunnel lining to further movements.

The back analysis of the VP3 observations, in particular, an assessment of the effects of (a) and (b), demonstrated that the high movements observed at the east wall (compared with those observed at the north, south and west walls) during the initial excavation phases were primarily due to locally high grouting induced pressures acting on the east wall. Temperature-induced movements of the upper reinforced concrete ring beam are summarised in Figure 12.14. Ring beam movements at the east wall correlated with measured temperature variations (using temperature sensors mounted on the upper ring beam at an elevation of 106.5 m ATD) during May and July (during a period of minimal construction activity). Inward and outward wall movements (up to about 3 mm) reflected temperature decreases and increases, respectively. These movements were largely reversible; however, temperature-induced movements had to be taken into account when interpreting the observed wall displacements. The practical importance of this analysis was that it demonstrated that the influence of grouting was a localised effect, which only influenced the movement of the eastern wall above a level of circa 95 m ATD, and importantly had negligible effect on the critical excavation phases below a level of 82.9 m ATD. The assessments carried out for (c) and (d) also led to minor improvements in the calibration of the 3D FLAC model. The discussions with LUL and Crossrail's asset protection engineer led to an agreement on the relaxation of the red trigger level for the east wall from 30 to 40 mm. The monitoring data showed that the potential effects on the Northern Line tunnels due to shaft excavation induced movements were relatively modest compared with the SCL tunnelling

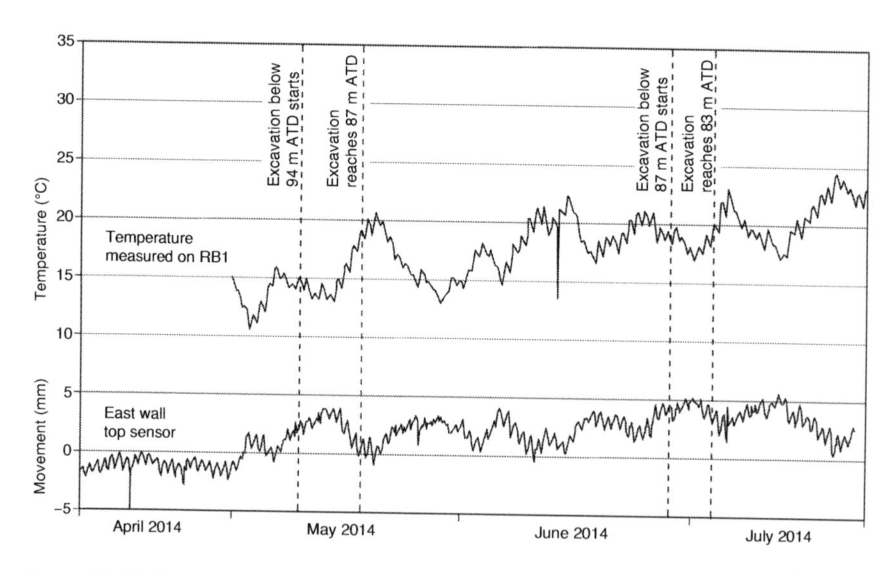

Figure 12.14 Temperature variations and upper east wall movement.

(above and below the Northern Line tunnels). The distortion induced by the SCL tunnelling generally increased the tunnel diameter between crown and invert (due to settlement/heave), whereas the shaft excavation induced minimal settlement/heave of the Northern Line tunnel. The shaft excavation caused small lateral movements of the LUL tunnel (less than 10 mm), which, to an extent, counteracted the SCL-induced tunnel distortion.

The calibrated 3D FLAC model, based on the modifications introduced during VP2 and VP3, provided calculated wall displacements which were close to the observed wall displacements. Figures 12.15 and 12.16 compare the calculated and observed wall displacements for the east and north walls, respectively. The introduced primary modifications were: the increased ground stiffness; influence of temperature fluctuations; reduced cross-wall stiffness and grouting pressures, where the latter two modifications had the most effect on the east wall. As a result of using the OM and the increased confidence this generated across the whole project team (client, designer, contractor and project stakeholders, including the asset protection consultant, checker and infrastructure owners such as LUL), it was possible to gain agreement, by the end of VP3, to:

1. remove the temporary props at the 80.5 m ATD level;
2. remove the slab strips (reinforced concrete beams).

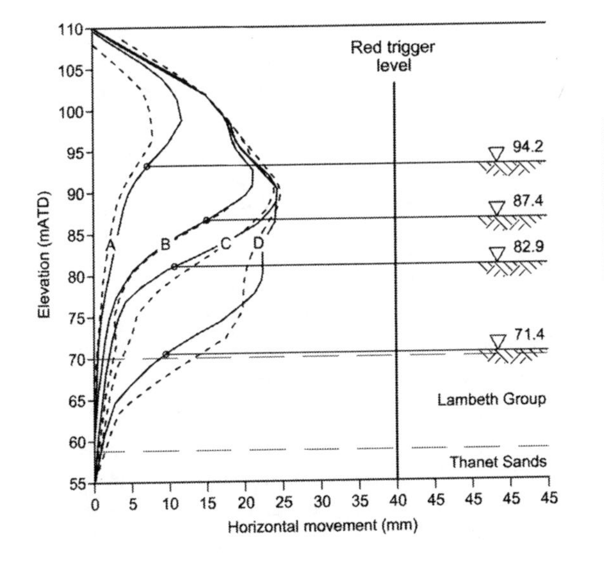

Figure 12.15 Calculated versus measured wall movement, east wall at VP3.

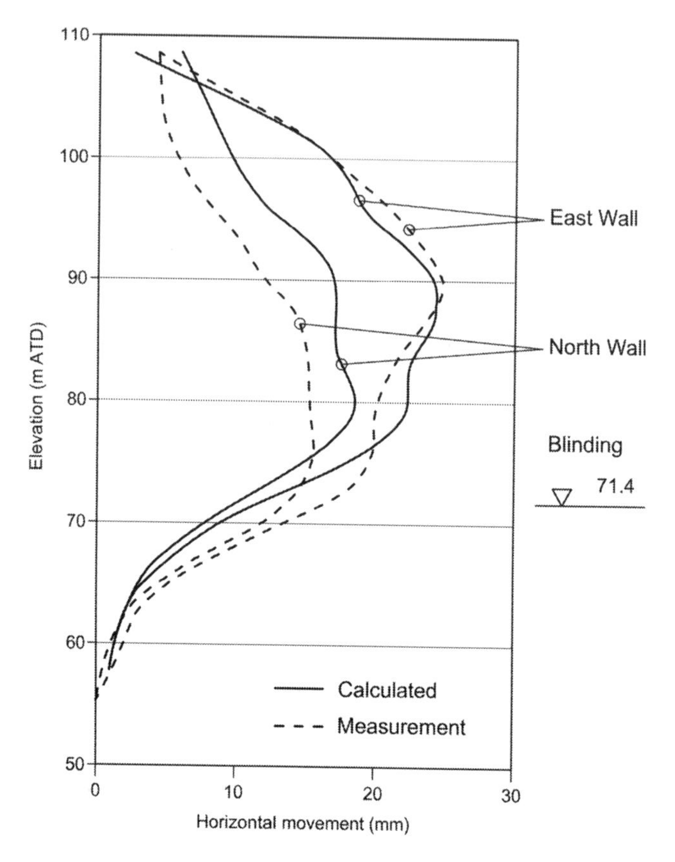

Figure 12.16 Calculated versus measured wall movements, north and east walls, excavation depth 42 m.

12.4.5 Summary of Changes Introduced through Implementation of the OM

The changes in the temporary works, which were introduced as a result of using the OM, are summarised in Figure 12.17. All the temporary props between 83.5 m ATD and formation level at 71.4 m ATD (a vertical span of about 12 m) were omitted and the number of construction stages for the final four levels of ring beams was halved. When the excavation reached formation level at 71 m ATD, a 300 mm thick, unreinforced concrete blinding strut (with a mix design to achieve early strength gain) was cast. This provided temporary structural support (through diaphragm action), while the 2 m thick reinforced concrete base slab was constructed. These changes enabled the construction programme to be accelerated substantially and

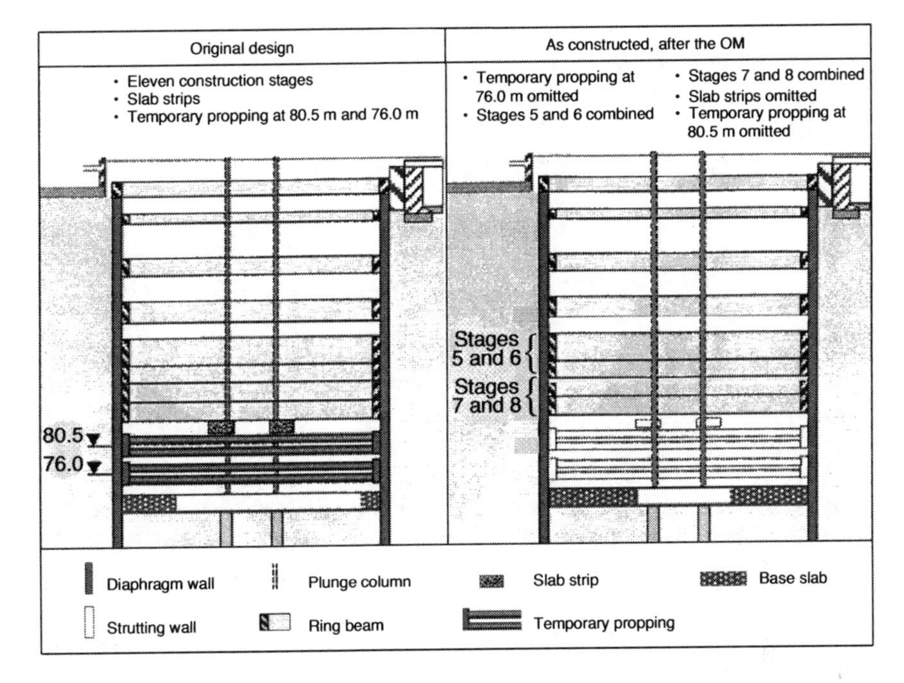

Figure 12.17 Changes to construction sequence and temporary wall support due to implementation of the OM through progressive modification.

reduced the programme by 14 weeks and avoided delay to the Crossrail TBM which had to traverse through the shaft. The final observed displacement of the east wall was similar to that calculated from the calibrated advanced numerical model (Figure 12.16). The movement of the adjacent Northern Line tube tunnel was acceptably small and also consistent with calculated values from the calibrated model. The shaft formation level was reached on 1st October and the 1,800 m^3 concrete pour for the base slab was completed on 1 November 2014, two weeks ahead of the original planned completion date.

12.4.6 Reflections on the Role of Analysis and Bayesian Updating

The iterative application and recalibration of numerical modelling based on measurements during construction, as described earlier, is a practical application of Bayesian updating, for example, refer to Wu (2011). The application of Bayes theorem in the context of implementation of the OM has been discussed by several authors (e.g. Lacasse and DiBiagio, 2019). Terzaghi (1961) when describing the OM stated:

> *"Base the design on whatever information can be secured. Make a detailed inventory of all the possible differences between reality and assumptions. Then, compute on the basis of original assumptions, various quantities that can be measured in the field. On the basis of the results of such measurements, gradually close the gaps in knowledge and, if necessary, modify the design during construction".*

Bayesian updating can be considered to be a mathematical tool for quantifying the uncertainty described by Terzaghi. The substantial computing power that is now available enables sophisticated analyses to be carried out that would have been unthinkable in previous decades; nevertheless engineering judgement is still of primary importance. Hence, it is the authors' view that Bayesian updating should only be used as a support tool for applications of the OM where the analytical effort is justifiable. Most applications of the OM do not require this level of analysis. As discussed in this case history, even leading-edge analysis would have been of limited reliability without calibration against construction observations. Judgement played a key role when assessing the interactive effects between various construction activities and their influence on soil–structure interaction and then implementing appropriate changes to the numerical model. The analysis inputs at the start of the OM reflected an assessment of moderately conservative conditions (although these were less conservative than the conventional design basis required under the project design standard), then at each verification point, the inputs were updated towards most probable conditions on the basis of the field measurements. As discussed earlier, this assessment of the differences between moderately conservative and most probable conditions was far broader than just an assessment of ground behaviour. The primary value of the calibrated numerical model was to reassure various stakeholders that the works could be progressed safely without the planned contingency measures.

12.5 Conclusions

Liverpool Street station's 42 m deep Moorgate shaft was on the critical path for completion of Crossrail's central tunnelled section. The start of shaft construction was delayed by the prolonged period of time required to extract the foundation piles of the building that had previously occupied the site. This delay would have led to knock on delays for shaft completion and, critically, the transit of the running tunnel TBM through the shaft and jeopardised overall completion of tunnelling. To mitigate these delays, the OM was proposed to facilitate changes to the original temporary works and construction sequence. The project procurement generated a complex set of interfaces between several different construction contracts which were implemented simultaneously. Also, there were a large number of influential stakeholders who were seriously concerned about the potential adverse effects of underground construction on their buildings and infrastructure. The application of the OM through progressive modification facilitated agreement to commence the implementation of the OM. Construction commenced with Crossrail's fully assured base case design. It was possible to demonstrate through observational feedback that there was a very low probability of critical wall displacements being exceeded, when originally planned temporary props were omitted and the number of construction stages was reduced. Implementation of the OM through progressive modification provided assurance to stakeholders and ensured approvals from Crossrail's independent checker and asset protection engineer were obtained. Gaining approvals was challenging; nevertheless, it was possible to gain agreement without delays to the works on site. Close collaboration between designer, contractor and client is always crucial for any application of the OM, and in this case, it enabled many challenges to be overcome within acceptable timescales. The on-site presence of the OM designer enabled monitoring data to be assessed and compared with numerical modelling calculations in a timely and efficient manner.

The use of the OM at the Moorgate shaft enabled the perceived risks associated with several complex soil–structure interaction issues to be managed in a demonstrably robust and safe way which reassured all the affected project stakeholders. The following key issues were safely managed:

1. The effect of shallow permeation grouting in river terrace deposits which caused atypical local displacement of the east wall of the shaft.
2. Due to construction installation difficulties, clay inclusions were observed at the interface between cross-walls (installed by diaphragm wall methods) and the shaft D-walls. These defects significantly reduced the effective stiffness of the cross-wall struts (specifically installed to limit the east wall movement).
3. Concerns about the potential for excessive movement of the 100-year-old Northern Line metro tunnel located only 4 m from the east wall of the shaft.
4. The omission of deep temporary props. Although temporary props stiffen a retaining wall structure, the installation of props towards the base of a deep shaft is a relatively slow and hazardous process, which can also be counter-productive in terms of limiting retaining wall displacement due to time-related ground movements (Limehouse Link, Chapter 4, provides another example of the potentially counter-productive nature of deep temporary props within a constrained excavation).
5. Reduction in the number of construction stages compared with the base case design.

The implementation of the OM at Moorgate was highly successful and enabled the required objective of an accelerated construction programme that allowed the Crossrail Project critical milestone for shaft handover to the TBM tunnel contractor to be met. Crossrail Director (Central Section) said:

> *"This achievement is a culmination of an extraordinary story of collaboration between contractor, designer and CRL (the Crossrail project delivery team). Looking back six months ago there were many doubts about the ability to excavate, construct and bottom-out the shaft on programme and the teams did it!"*

TBM Victoria transited through the shaft in April 2015 and was launched on her way to Farringdon on 13 April 2015.

References

Eadington, J. and O'Brien, A. S. (2011). Stiffness parameters for a deep tunnel – developing a robust parameter selection framework, Proceedings of the 15th ECSMGE, Athens.

Jardine, R. J., Potts, D., Fourie, A. B. and Burland, J. B. (1986). Studies of the influence of non-linear stress-strain characteristics in soil-structure interaction. *Geotechnique*, **36**, No. 3, 377–396.

Lacasse, S. and DiBiagio, E. (2019). Two observational method applications: an ideal solution for geotechnical projects with uncertainty, Proceedings of ASCE Geo-Congress 2019, GSP 314, pp. 117–128.

O'Brien, A. S. and Liew, H. L. (2018). Delivering added value via advanced ground investigations for deep shaft design in urban areas, D.F.I conference 2018, Rome.

Terzaghi, K. (1961). Past and future of applied soil mechanics. *Journal of Boston Society of Civil Engineering*, **68**, 110–139.

Wu, T. H. (2011). 2008 Peck lecture, the observational method, case history and models. *ASCE Journal of Geotechnical and Geoenvironmental Engineering*, **137**, No. 10, 862–873.

Chapter 13

Reflections on the Advantages and Limitations of the Observational Method

13.1 The Legacy of Terzaghi and Peck

The fundamental and far-reaching contributions of Terzaghi and Peck to the development of applied soil mechanics are legendary and profound. Right from the start, the observational method played a key role. Ralph Peck's seminal paper, presented as the 9th Rankine Lecture in 1969, was entitled 'The advantages and limitations of the observational method in applied soil mechanics'. This was the first formal use of the term 'observational method' and the paper set out the key criteria for its application. Peck notes in the introduction, the OM is an inherently natural way for engineers to address uncertainty. This was very evident to both Terzaghi and Peck and there was a corresponding reluctance to formalise it (Terzaghi and Peck, 1967; Peck, 2001). In a later introduction to his Rankine Lecture paper (Dunnicliff and Deere, 1984), Peck admitted that he was not altogether happy with the results and still felt, some 15 years later, that his efforts to formalise the OM were too contrived and rigid. He returned to this theme in one of his last published papers on the OM (Peck, 2001). He observed:

> *"It did not occur to me at that time that what I came to call the observational method had been at the heart of Terzaghi's approach on the Chicago Subway project, and that the notably beneficial results had come about without any formal framework or any application of theory".*

This paper stressed the importance of simplicity with the message directly emphasised by the title: 'The Observational Method can be Simple'. Apart from the Chicago Subway, it featured two other key case histories from the early 1940s – the Cleveland Ore Yard and the Newport News Shipway. These projects were forerunners in the application of the OM where quite simple field observations led to substantial improvements. Peck concluded:

> *"... the observational method paid-off handsomely, without more than the most elemental theory and with only qualitative predictions.*

There were few refinements, and no elaborate computer modelling to be 'validated' by exotic remote-reading sensors. These refinements are not the essence of the observational method. They have their place but they should not deflect attention and resources from the essence of the method".

13.2 Advantages and Limitations of the OM

13.2.1 Lessons Learned from the Case Histories

The case histories featured in this volume demonstrate both the power and the flexibility of the OM in ensuring safety and adding value through substantial savings in cost and time. Generating key data from the actual performance of real structures, it also stimulates innovation (Powderham, 2002). These additional benefits derive through:

i) stronger connection of design to construction;
ii) increased safety during construction;
iii) improved understanding of soil/structure interaction;
iv) improvements in the use and performance of instrumentation;
v) higher-quality case history data;
vi) greater motivation and teamwork.

These are compelling benefits and yet it has often been noted that the OM is significantly underused (Peck and Powderham, 1999). This neglect is typically imposed by contractual constraints or adverse perceptions of undue risk. However, despite the many advantages, the OM, like any other method, is not a universal panacea and does have important limitations.

The key to success in applying the OM is in identifying the critical observations and having the means to obtain and act upon them in a safe and timely manner. This latter aspect relates to a cardinal limitation of the OM: If it is not possible to change the design during construction, then this renders the OM to a non-starter. The OM is most effective where the critical observations can be reliably obtained through simple measurements. The deflection of a retaining wall is a common example of a critical observation. It will result from many complex and interacting factors, such as construction methods and sequences, pore pressure, permeability, anisotropy and geometry. However, if critical observations are correctly judged and focussed on, then the resulting simplicity enhances the effectiveness of the whole process. In other fields, such as mathematics, such critical observations are described as emergent phenomena or emergent behaviour (Stewart, 1998). Of course, it is important to appreciate the nature and range of the complexity and obtain, in parallel, appropriate supplementary information. An example of this would be ongoing condition surveys in the protection of a building affected by adjacent excavation. The critical observations may be derived from settlement data, but there would be unknown factors influencing the building response. They may be beneficial, such as free-body movement, or adverse, such as a lack of structural continuity or local sensitivity. Such potential factors also need to be anticipated and accommodated. The case history for the Mansion House (Chapter 3) illustrates such aspects. Another example where simple measurements were used to address complex soil/structure interaction is described in Chapter 10. In this application of the OM, pile cap displacements were monitored to assure the safety of raising the arch at the new Wembley Stadium.

13.2.2 The Importance of Simplicity

It is essential to apply the key principles and to understand the method's limitations. One common limitation is self-imposed or at least generated by the contractual or political environment of a project. This may involve the proliferation of non-critical instrumentation and the imposition of supplementary management systems – often with overly prescriptive requirements. Such complications can substantially compromise the effectiveness of the OM. Simplicity is central to the OM – a simplicity made comprehensive by good judgement. The essence of this is eloquently captured by Peck in the concluding paragraph of his 2001 paper quoted in full in Section 13.1. The temptation to over-complicate the method should be resisted. In its essence, the OM provides a simple way to address and resolve complexity. Its success risks being significantly and ironically compromised if it becomes unnecessarily elaborate. All of the featured case histories involved the resolution of very complex conditions. The Boston Central Artery Tunnel Jacking (Chapter 7) provides a cogent example with many layers of interactive complexity including both the ground conditions and the protected infrastructure. The railway involved a sensitive network of seven interconnecting tracks and was subject to seven distinct sources of ground movement. These involved, apart from the tunnel jacking, global ground freezing and extensive jet grouting. And yet, all of this complexity was resolved on a simple basis through one set of OM control traffic lights focussed on the allowable limits on the movement of the tracks. In each of their projects, the authors have found that each application of the OM, even those addressing seemingly similar situations, require a bespoke approach. Thus, as noted in Chapter 1, apart from over-complication, an unduly prescriptive approach should also be avoided.

13.2.3 Key Limitations

The following are not conducive to the safe and effective application of the OM. If it is not possible to eliminate or resolve these factors, the method should not be applied:

- Inability to reliably obtain critical observations
- Inability to implement timely contingency plans
- Lack of stakeholder support
- Progressive failure or collapse
- Sudden or brittle failure

While superficially these factors may appear to express separate issues, they all boil down to a combination of the first two – the issue of reliably obtaining the critical observations in a timely way. In the OM, a fundamental criterion for what constitutes a critical observation is the associated time factor. It must be possible to detect adverse trends in advance of an unacceptable event and be able to implement a planned contingency to safely deal with it. That, essentially, is at the heart of the OM. Considering the last three factors in turn:

1. Lack of stakeholder support: This is vital because application of the OM demands clear communication and close ongoing teamwork, particularly in the environment of large and complex projects. Lack of support or commitment from any member of the project team could easily undermine this essential communication and hence critically impair the management of risk. Any lack of trust or ownership of specific tasks or roles would unacceptably compromise the ability to obtain reliable and timely measurements.
2. Progressive failure or collapse: Progressive failure, potentially leading to complete collapse, results from a lack of redundancy and alternative load paths in a structure or soil mass. An example to prevent progressive failure is the provision of adequate redundancy in a temporary propping system for a retained excavation. A typical criterion of such redundancy is that any individual prop can be removed without any of the remaining props or the protected structure becoming unsafely overloaded or overstressed. Under conditions leading to collapse, a stiff or rigid structure is made vulnerable to becoming a mechanism by a local failure. Once initiated, progressive failure towards collapse cannot generally be halted and certainly not in a controlled timely manner. Its onset may be subtle and not necessarily revealed by measurements until it is too late. Thus, it would not be possible to obtain the critical information in a timely way to enable the safe implementation of effective contingent measures. Even in unusual cases where severe overstress, which would lead to collapse, can be contained, it marks a limitation of the OM. A dramatic example of this was the application of the OM in the Limehouse Link Basin (Chapter 4). Here, the critical observations of a sheet-piled wall clearly showed very adverse trends in its deflected profile. These movement trends were such that there would have been insufficient time for construction of the base slab. The demands of monitoring wall flexure combined with the risks

associated with the control of water levels were considered too onerous. Here, the point had thus been reached where the main benefit sought through the OM (i.e. the elimination of the mid-height propping) was outweighed by the demands of applying it. The original OM design objective was therefore abandoned. However, on a positive note, despite these limitations, the OM, in this example, was still able to deliver some substantial advantages. The measured performance from the OM facilitated a change in the propping sequence applied during the second phase. In this, the excavation to the base slab formation level proceeded in advance of the lower props which instead of the original sequence were installed immediately after the construction of a blinding strut. This simple change made excavation easier, quicker and less restrictive. It was consequently safer and also brought significant environmental benefits.

3. Sudden or brittle failure: This may involve a sudden catastrophic collapse such as the buckling failure of a compression member. The blinding struts used for the Heathrow Airside Road tunnel (Chapter 9) provide a useful example of the associated limitations. The initial innovation was the development of laser controlled excavation sequences to facilitate the installation of an intermediate blinding strut as the contingency measure. When this innovation was adapted to solve the interface with the Piccadilly Line tunnel, it required the actual use of blinding struts rather than their potential use as a contingency. With these compression members as an integral part of the construction, the OM could not be applied since failure would be through the buckling of a blinding strut. Such failure is sudden and occurs with little advance warning. It would therefore be impossible to monitor effectively and implement a timely contingency. Consequently the design of the blinding struts had to be demonstrably robust in advance. It should be noted that a brittle response within a protected structure may be acceptable under some circumstances. For example, the generation of shear or diagonal tension cracks in a masonry building would locally constitute a brittle response. However, this is likely to be inevitable when such a structure is subjected to imposed movements, yet it is widely accepted that the OM is applicable to the protection of masonry structures. It is generally possible to ensure that the resulting crack widths should lie within agreed levels of acceptable damage, which in turn relate to ease of repair. Moreover, such damage typically develops over a significant time scale, thus allowing timely intervention to protect the building from the damage reaching unacceptable levels. It is when brittle failure suddenly and directly causes unacceptable damage or leads to progressive failure that it would prevent the effective application of the OM.

13.2.4 Heathrow Terminal 5 Tunnels – An Illustrative Case History

A key example of the limitation caused by the inability to reliably obtain critical observations arose during the design development for Terminal 5 at London's Heathrow Airport. It was the largest building project in Europe during the first decade of this century, with a construction cost of nearly £4.3 billion (Shanghavi *et al.*, 2008). The extensive underground works amounted to nearly a fifth of the overall budget. The excavation for the main T5 basement was reported to be the largest ever in the United Kingdom, with a plan area of 396 m by 176 m (equivalent to 10 football pitches) and, at 22 m, one of the deepest. This huge excavation took place above the recently constructed tunnel extensions to the Heathrow Express (Hex-Ex; Figure 13.1).

Due to the concerns regarding the unacceptable effects that the basement construction could have on these tunnels, the excavation was planned in stages to minimise the amount of total unloading at any one time. This included strategically placed temporary spoil heaps. Despite these mitigating measures, the sheer magnitude of the works was still considered to present potentially critical risks. Consequently, to avoid very onerous and expensive physical preventative works, it was proposed to

Figure 13.1 Scale of excavation for Terminal T5.

use the OM. The objective was to closely monitor the effects on the tunnels and demonstrate that expensive additional works were not required.

The tunnels were constructed in London Clay with precast expanded linings which were therefore unbolted. Three key distinct critical risks were identified (Figure 13.2). Two of these were the distortion in shape and the free-body movement of the tunnel. It was agreed that practical and reliable observations could be obtained for these. The third factor was the de-stressing of the individual rings and segments of the tunnel lining from the reduction in overburden. However, it was not considered reliable to try to establish a critical observation to manage the associated risks to the integrity of the tunnel lining. First, given the complexity of the erection process for the linings, it was considered impractical to determine the initial hoop compression in each ring, and then even more difficult to reliably measure the reductions in these forces as the excavation above the tunnels progressed. The associated hazard was that the lining would become unacceptably destressed with the risk of a key segment becoming loose or even fully dislodged and falling out of its ring. The presence of this risk would have prevented the application of the OM in general and led instead to the need for major global changes in the design and to construction sequences.

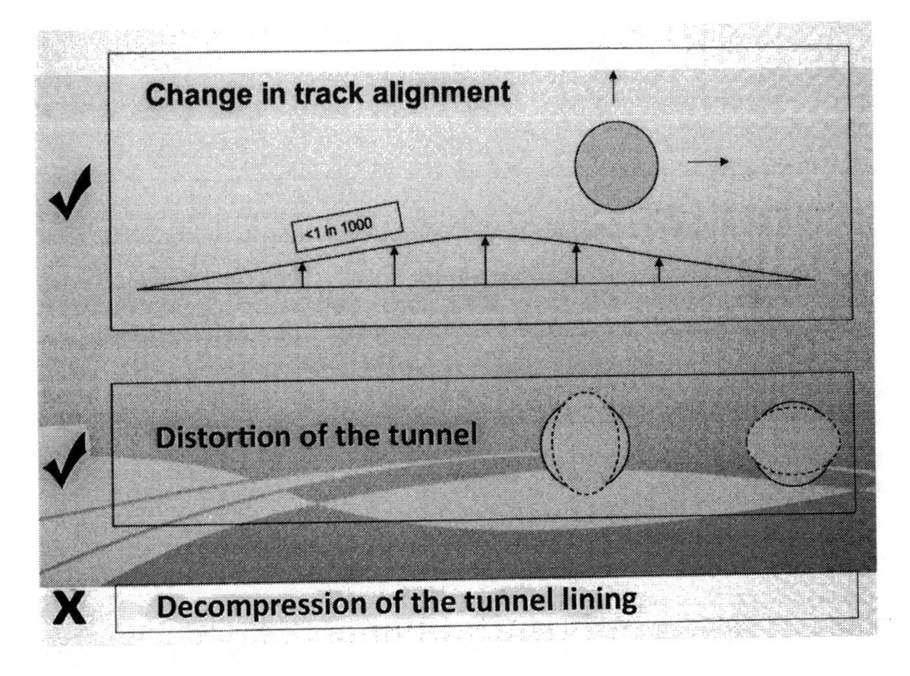

Figure 13.2 Critical observations.

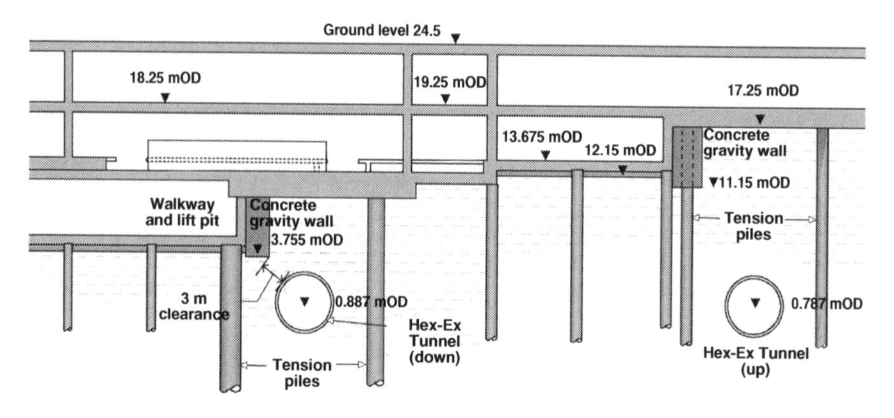

Figure 13.3 Basement cross section showing Hex-Ex tunnels.

However, if this hazard could be eliminated, there would then still be a viable case to implement the method. It was consequently decided to incorporate some limited preventative works that would ensure in advance that the compression in the tunnel lining would be maintained above an acceptable threshold throughout the period of excavation.

This was achieved by constructing holding down slabs and gravity walls directly above the affected tunnels (Figure 13.3). It still left the bulk of the original excavation to be undertaken in an efficient way and the OM was successfully applied ensuring that the two remaining critical observations, the tunnel distortion and free-body movements, were maintained within acceptable limits.

This case history thus provides an example of a critical limitation that would have prohibited the implementation of the OM as a whole. Its successful elimination was achieved by relatively minor design changes enabling the agreement to implement the OM for the bulk of the work. Since these critical factors all derived from the excavation induced movement, they were measured and evaluated individually but managed together using one set of OM traffic lights. These were placed prominently in clear view above the site offices enabling the status of the OM implementation to be constantly displayed (Figure 13.4). As can be seen, the green light is illuminated. It never reached amber during the whole period of construction.

13.3 Progressive Modification – Solving the Issue of the 'Most Probable'

One of the perennial debates regarding the OM is that relating to the most probable conditions – what they constitute and how they are used in

Figure 13.4 Traffic lights – green light on.

design. Peck (1969) originally introduced this term as one of the key ingredients in an application of the OM. The third step in his sequence of eight procedures was *"Establishment of the design based on a working hypothesis of behaviour anticipated under the most probable conditions"*. He further qualified this statement by noting: *"The degree to which all of these steps can be followed depends on the nature and complexity of the work"*. This immediately implies that any implementation of the OM should be developed on an appropriately flexible basis and requires the exercise of engineering judgement. This is an eminently reasonable approach. However, there has been a tendency to treat Peck's third step too simplistically with the assumption that it essentially means 'establish the design directly on the most probable conditions'. It followed from such an interpretation that there could be an associated risk of compromising safety. This over simplification ignores the inherent flexibility of the actual statement in which the design is not directly based on the most probable conditions but rather, as Peck carefully states, on a working hypothesis of anticipated behaviour. Peck was also careful not to prescriptively recommend what factors of safety should be applied. An approach to addressing these issues was proposed by Powderham and Nicholson (1996) (see discussion in Section 14.1). Nevertheless, the concern has persisted – for example, as noted in the CIRIA Report 185 (Nicholson *et al.*, 1999). It states:

> *"In light of the difficulties in directly implementing designs based on Peck's most probable conditions many designers have started construction with 'predefined' design parameters. Subsequently, they have reviewed performance and upgraded parameters by back-analysing the performance and rerunning predictions. Powderham (1994) defined the term 'progressive modification' for this approach".*

The report concluded with a clear preference for adopting progressive modification rather than Peck's approach. However, the authors have found that, in practice, given the inherent and indeed essential flexibility in Peck's proposals, they find no real conflict with it in their applications using progressive modification. Moreover, along with the tendency to be overly prescriptive, there is a lack of clarity as to what constitutes the most probable conditions. The criteria tend to be unduly focussed on the soil parameters. This bias is influenced by the fact that soil parameters readily fit into such categories as most probable, moderately conservative and worst credible. It is less easy to establish the comparable criteria for most probable conditions in a structure – for example, a masonry building. However, in terms of the OM, it is not just the ground conditions but those that comprehensively relate to the entity being protected and these are centred on soil/structure interaction. This applies whether the structure is manmade, like a tunnel or building, or to naturally occurring materials as in slope stability. A key issue is to achieve a balanced assessment of anticipated behaviour as expressed through soil/structure interaction. It is a classic example of the challenges encountered at the interface between soil mechanics and structural engineering (Burland, 2006). Thus, to summarise the issues relating to the term 'most probable':

1. It needs to embrace the combined effects of soil/structure interaction and not just evaluations of certain parameters of either the soil or the structure.
2. It is not a theoretical concept and in practice will typically depend on many complex factors both known and unknown.

3. Consequently to establish a working hypothesis of behaviour anticipated under the most probable conditions requires experience and engineering judgement.
4. There will therefore be a potentially wide range of views of what actually constitutes most probable conditions both within specialisms and beyond. The views of a geotechnical specialist, a building owner, or a contractor will be influenced by more than just technical factors. Various stakeholders will have different perspectives of risks.

The aforementioned issues consequently place difficult constraints to achieving consensus on what are likely to be the most probable conditions that will govern actual performance. A design based on most probable conditions still tends to be associated with low levels of safety. Muir-Wood (2000) observed that commencing the OM based on estimates of the most probable conditions, even if possible to identify, would usually lead to a more expensive outcome than commencing the OM on a more conservative basis – especially in tunnelling applications. Muir Wood (2004) and Szavits-Nossan (2006) reinforced this important point. As discussed in Sections 1.2 and 14.1, this drawback can be exacerbated if the design is based too prescriptively on the most probable conditions. A bespoke and flexible approach to each application of the OM is required and demands engineering judgement. It is therefore the authors' recommendation that applications of the OM should always be implemented through progressive modification wherever possible. They do not consider this to be a radical departure from Peck's approach but rather an interpretation and development of it applied through their own experience. Peck was well aware of the benefits of such an approach and his Rankine Lecture featured a classic example of an application of the OM using progressive modification for the Cape Kennedy Causeway. Both Muir Wood (2004) and Szavits-Nossan (2006) recommend the flexibility of progressive modification.

As noted in Section 1.2, one of the major advantages of progressive modification is that it is not necessary to commence the OM on site with design having potentially low factors of safety. This advantage alone makes achieving agreement to implement the OM more attainable. There is generally even no need to start with a design developed on some prescriptive basis to be moderately conservative. It is an inherent benefit of progressive modification that it typically offers the flexibility to commence construction from a very low level of risk. To consider a few of the examples featured in this book: Right from the start for the Channel Tunnel cut and cover tunnels (Chapter 2), the application of the OM was developed with safety set as a cardinal tenet of the progressive modification approach, yet also offering the potential to maximise savings. For the Mansion House, the progressive construction stages of the OM started with an estimated level of negligible risk to the building. Moreover this assessment was based on conservative

Greenfield assumptions without assuming any mitigating benefits from soil/ structure interaction. The OM design at Limehouse Link (Chapter 4) effectively commenced on site with the original base case design which, in context, was considered very conservative since it was the basis to justify the introduction of the OM. This major advantage of the progressive modification approach is also evident in the other case histories featured in this volume.

References

Burland, J. B., (2006), Interaction between structural and geotechnical engineers, the structural engineer, 18, April, Presented at IStructE/ICE Annual Joint Meeting 26, April 2006, London, UK.

Dunnicliff, J. and Deere, D. V. (editors) (1984). *Judgment in Geotechnical Engineering – The Professional Legacy of Ralph B. Peck*, p. 122, Wiley, New York.

Muir-Wood, A. M. (2000). *Tunnelling: Management by Design*, pp. 63–66, E & FN Spon, London.

Muir Wood, A. M. (2004). *Civil Engineering in Context*, p. 99, Thomas Telford, London.

Nicholson, D., Tse, C.-M. and Penny, C. (1999). The observational method in ground engineering: principles and applications, CIRIA Report 185, London, p. 214.

Peck, R. B. (1969). Advantages and limitations of the observational method in applied soil mechanics. *Géotechnique*, **19**, No. 2, 171–187.

Peck, R. B. (2001). The observational method can be simple. *Proceedings of Institution of Civil Engineers, Geotechnical Engineering*, **149**, No. 2, 71–74, Thomas Telford, London.

Peck, R. B. and Powderham, A. J. (1999). *Talking Point, Ground Engineering*, British Geotechnical Society, February edition, London, UK.

Powderham, A. J. (1994). An overview of the observational method: development in cut and cover and bored tunnelling projects. *Géotechnique*, **44**, No. 4, 619–636.

Powderham, A. J. (2002). The observational method – learning from projects. *Proceedings of the Institution of Civil Engineers, Geotechnical Engineering*, **155**, No. 1, 59–69, UK.

Powderham, A. J. and Nicholson, D. P. (1996). *The Way Forward: The Observational Method in Geotechnical Engineering, Institution of Civil Engineers*, pp. 195–204, Thomas Telford, London.

Shanghavi, H., Beveridge, J., Darby, A. and Powderham, A. J. (2008). Engineering the space below Terminal 5. *Royal Academy of Engineering, Ingenia*, No. 34, March, London, UK, 12–20.

Stewart, I. (1998). *Life's Other Secret – The New Mathematics of a Living World*, Allen Lane, The Penguin Press, London, UK.

Szavits-Nossan, A. (2006). Observations on the observational method, XIII Danube-European Conference on Geotechnical Engineering, Slovenian Geotechnical Society, Ljubljana, 2006. str. pp. 171–178.

Terzaghi, K. and Peck, R. B. (1967). *Soil Mechanics in Engineering Practice*, 2nd Edition, pp. 294–295, John Wiley & Sons, New York.

Chapter 14

Some Observations on the Way Forward

14.1 The 1996 Overview

As a precursor to reviewing the current status of the observational method (OM), the overview given in 'The way forward' (Powderham and Nicholson, 1996) provides a useful perspective. This paper followed a review of the *Symposium in Print* published in Géotechnique in December, 1994 and thus some 25 years after Peck's Rankine Lecture.

To quote from the 1996 overview:

> *"Despite strong endorsement and enthusiasm for wider application of the OM, a range of concerns have also been expressed together with divergent views on the status and definition of the method. Some consider that the method is basic to engineering (particularly geotechnical) and therefore common practice. Others consider a more specific definition is necessary. The latter is compatible with Peck (1969) and indeed is the authors' view. The observational method has a special meaning and very specific procedural requirements. On this basis, proper use of the observational method is not common practice. Clarification is a key issue. In determining the way forward, the following should be addressed:*
>
> *(a) Establish a clear definition of method including objectives, procedures and terms, with a clear emphasis on safety.*
> *(b) Increase awareness of the method's potential and benefits, particularly to clients, contractors and regulatory bodies.*
> *(c) Remove contractual constraints.*
> *(d) Identify potential for wider use.*
> *(e) Initiate focused research projects.*
> *(f) Improve performance and interpretation of instrumentation systems.*
> *(g) Establish extensive database of case histories".*

Now, a quarter of a century later, while progress has been made, several of these key issues still need addressing. The three most important items in the list (b), (c) and (d) remain to be adequately resolved and yet present the most common constraints to using the OM. These contractual issues are considered in detail in Section 14.2.

Items (a) and (e) were addressed through the CIRIA research project 'The Observational Method in ground engineering – principles and applications' (Nicholson *et al.*, 1999). With regard to (f), there have been substantial improvements in the range and reliability of instrumentation. However, its ready and more economic availability has also encouraged its proliferation. This has created the risk of data overload at the expense of clarity and the need to identify and focus on the critical observations. This book's prime aim has been to illustrate and advocate the application of the OM through detailed accounts of a wide range of case histories and there is a growing literature on the subject. However, an extensive and accessible database has yet to be established.

The issue of how to address the most probable conditions in applications of the OM remains controversial. On this subject, Powderham and Nicholson (1996) noted:

> *"Terms used must be clear and defined. Consideration of Peck's step (c), 'Establishment of the design based on a working hypothesis of behaviour anticipated under the most probable conditions', will serve to highlight areas that need to be addressed. Issues here are:*
>
> 1. *what is the basis of design?*
> 2. *what are the criteria to establish the 'most probable' conditions?*
> 3. *what is the factor of safety assumed?*
> 4. *how does the design differ from conventional practice?*
> 5. *what are the requirements for site investigation?*
> 6. *what is an acceptable level of risk?*
>
> *Such issues are not easy to resolve definitively. They are at the centre of engineering practice, judgement and understanding".*

It is fundamental that such questions need to be raised and resolved in any application of the OM. In this context, Peck's proposed set of eight ingredients for a complete application of the OM provides a very useful checklist. The full set is provided for reference in the Introduction to this book. The featured case histories provide examples of how such issues have been considered and addressed. However, there has been a growing tendency to reduce Peck's step (c) down to 'Establishment of the design based on the most probable conditions'. Expressed in this shortened form, and as noted in Section 13.3, it is viewed as being too optimistic and therefore likely to bring the need for contingency measures. When taken in this way without qualification, the concern is understandable and is raised, for example, by Nicholson *et al.* (1999). But, as emphasised in the extract from the 1996 overview earlier, there are at least six issues to consider in addressing Peck's step (c). Moreover, as we note in Section 1.2, if Peck's steps are considered flexibly and as a guiding checklist, the concern of over optimism is quite resolvable. Peck (1969) is also careful to observe that *"The degree to which all of these steps can be followed depends on the nature and complexity of the work"*. Such a caveat in turn implies the need for judgement. This vital issue is comprehensively addressed by Parkin (2000) in his book *Engineering Judgement and Risk*. He opens with *"Engineering practice is a patchwork of codes, rules of thumb, applied science, management processes, and policy. In all of this judgement is king"*. Parkin proceeds to note that engineering judgement is under attack. This issue is also reflected in some of the negative attitudes towards the OM seen, for example, in the trend to replace judgement through more prescription and the imposition of additional detailed procedures.

14.2 Commercial and Contractual Environment

14.2.1 Overview

Although the OM has been implemented under a wide range of different contracts, contractual conditions are often the key constraint to using the OM. Table 14.1 provides an overview of different contract forms used in the United Kingdom and their compatibility with the implementation of the OM. The optimum environment for the OM comprises:

1. Close co-ordination between the designer and contractor enabling opportunities to be identified and developed.
2. A non-adversarial environment, which facilitates teamwork.
3. Inclusion of a value engineering (VE) clause, so the benefits derived from OM can be shared equitably between parties.
4. Trust between the client, designer and contractor and the fostering of longer-term relationships between them.

Issues 2 and 4 above are a particular challenge given the nature of the industry. In the United Kingdom, there have been several industry reviews (e.g. Latham, 1994; Egan, 1998; Wolstenholme, 2009) which have highlighted the counter-productive nature of adversarial contract relationships.

Technically, the concept of the OM to deliver benefits through careful observation during construction is inherently natural and practical. Unfortunately, from a commercial perspective such synergy between design and construction can be notably vulnerable. Under a conventional contract, a contractor bids on a project based on a fixed design specified in the contract documents and on the premise that it will be built as designed. The

Table 14.1 Overview of contract types and potential synergy with the OM.

Contract type	*Key features*	*Collaboration between designer and contractor*	*Opportunity for OM*
Traditional (design/bid/build)	Client appoints designer, design completed, successful contractor builds design	Very limited. Typically, designer separated from contractor	Very limited, unless a VE clause is introduced – for example at Limehouse Link
Design and build	Client's designer prepares a reference design. D&B team prepare tender design and if successful completes final design and builds project	Potentially good, but intense time pressure during tender may limit opportunity to build rapport and trust	More opportunity than traditional, but client approvals and independent checkers may hinder. May not be in commercial interests of contractor or designer to pursue, unless incentivised through a VE clause
Early contractor involvement (ECI) + the NEC	Reference design developed by appointed contractors during stage 1 of ECI process. Detailed design developed during stage 2	Typically, better than conventional D&B, due to reduced time pressures and opportunity to seek and develop innovation	Very good potential, since additional time during stage 2 of ECI provides greater opportunity to build trust between parties. For example, the use of partnering and the 'single team' approach developed under the NEC for the Heathrow cofferdam recovery solution
Alliancing	Delivery framework which focusses on a co-operative process to promote trust, risk and responsibility sharing. Alignment of commercial interests are encouraged	Very good. Innovation promoted through more collaborative environment Positive interaction between all parties encouraged	Potentially excellent. The alignment of commercial interests and risk sharing is conducive to OM. Development of longer-term relationships, should further enhance OM opportunities

introduction of the OM within such a contract immediately presents commercial risks from the need to allow design changes during construction. Such risks tend to fall predominantly upon the contractor who can consequently be exposed to the double disadvantage of less return but more ownership of the design. Risk allocation is reasonably well defined in a conventional contract where most of the design risk is taken by the client and most of the construction risk is carried by the contractor. Without a VE clause, or other equitable process for sharing benefits, there is little incentive to promote the OM if a stakeholder is disadvantaged. The issues related to design changes, buildability, construction sequencing and clarity of contract documents are discussed in detail by Tran and Molenaar (2012).

14.2.2 Design and Build

These forms of contract offer greater potential to adopt the OM where design and construction are inherently more closely inter-related and the contractor has significant ownership of the design. However, intense time pressures (especially during tender phases) and fragmentation of design effort within an adversarial environment may often inhibit the adoption of the OM. Stakeholder approval, especially of the client, may be difficult to achieve. Implementation of the OM requires greater effort by the designer and the contractor, and it may not be in the commercial interests of either party to pursue the OM unless there is an appropriate financial incentive.

14.2.3 Developing Greater Collaboration

Alternative contract forms which allow time to create a more collaborative environment and greater alignment of commercial interests, such as the New Engineering Contract NEC, Early Contractor Involvement (ECI) and Alliancing, offer more potential to use the OM. With NEC4 (2017), ECI clauses can be incorporated into the contract.

The first clause in all NEC contracts requires the parties to: 'act as stated in this contract and in a spirit of mutual cooperation.' Much has been written about the practicable ability to strictly enforce the last part of that sentence, but the clause sets an important aspiration. The latest version of NEC, (NEC4 published in 2017), includes several important changes compared with NEC3 which could facilitate greater use of the OM. These are discussed in detail by Gerrard (2017). In NEC3 only the target contracts commercially incentivised the contractor to propose changes to the employer's works information (known as 'client's scope' in NEC4). In NEC4 a VE clause has now been introduced into the priced contract options (termed A and B for lump sum and bill of quantities). If a VE proposal is accepted by the client, the cost savings are shared between the client and contractor through a compensation event. This enables both to share the benefits, and this, for example, would be achieved by introducing the OM into a project. Greater collaboration between the parties can also be encouraged by use of the following secondary options (available in NEC3 and NEC4):

- Option X12 allows the linking of the NEC contracts of two or more partners to facilitate multi-party partnering or alliancing for a single project or programme of projects. This option also provides for a joint project steering group of partners, through a single document stating how the partners will collaborate for the good of the project.
- Option X20 enables key performance indicators (KPIs) to be brought within the contract and financial benefits on the attainment of KPIs can be built in.

Hence, the above provisions within NEC4 can create a more collaborative environment and greater alignment of commercial interests, all of which can encourage greater use of the OM. However, as noted by Patterson (2010), the NEC is significantly different to conventional contracts, and a culture shift, training and effective procedures are needed to enable the NEC to operate properly.

The Policy Paper Construction 2025: Industrial Strategy: government and industry in partnership (2013) highlighted the following:

(a) Lack of integration – both between design and construction and construction management and its execution. This leads to lost opportunities for innovation.
(b) Collaboration, knowledge sharing and learning (such as in successful applications of the OM) is lost once a project team disbands at the end of a project.
(c) The benefits of a strong supply chain and development of longer-term relationships.

The lack of the aforementioned issues, which inhibit greater use of the OM, can potentially be overcome through alliancing forms of contract and which, as noted above, can be introduced through NEC option X12.

14.2.4 Value Engineering Clauses

As demonstrated at Limehouse Link (Chapter 4), one of the most effective stimuli to adopt the OM is to include a VE clause in the contract, the key features of which are as follows:

1. The parties (client, designer and contractor) receive a fair share of the savings, based upon the level of risk taken by each party.
2. Each party is encouraged to propose advantageous changes. Each idea should demonstrate a clear benefit, such as improvement in safety, cost or time savings.

3. The client can reject a proposal, but the reasons must be given, and a resubmission should be permitted provided it addresses the reasons for a previous rejection.
4. Peer review (as discussed in Section 14.5) is preferable to independent checking but either should have fixed approval periods, following submission of design modifications.

The main parties involved in VE will be the client and the main contractor. The contributions by the designer and independent reviewer or checker must also be recognised (especially since their specialist senior staff will be involved in any applications of the OM) and they should receive an equitable increase in their fees. Similarly, sub-contractors who play an important role should also receive a fair share of the cost/time savings, if they are taking a share of the risks.

14.3 Instrumentation and Monitoring

14.3.1 General Considerations

By definition, observations are essential in the application of the OM. In this, the identification, timely collection, interpretation and communication of the critical observations are vital. Every instrument deployed should answer a specific question. Instrument selection should be based on an understanding of the advantages and limitations of different instrument types for the specific function required. Some redundancy should be included to ensure continuity of readings. However, it is important to avoid installation of excessive amounts of instrumentation which seriously risks data over-load and disrupting the focus on the critical observations. It is also important to establish primary and secondary systems, to avoid an over-reliance on one type of system. The over-riding requirement is to design an instrumentation and monitoring system which is as simple as possible while also being reliable, robust and adequately comprehensive. Table 14.2 provides a simple checklist of key issues which should be considered when developing a monitoring plan. Design of instrumentation and monitoring is often not given sufficient priority. To quote Peck from his foreword to Dunnicliff (1988):

> *"Every instrument installed on a project should be selected and placed to assist in answering a specific question. Following this rule is the key to successful field instrumentation. Unfortunately it is easier to install instruments, collect readings, and then wonder if there are any questions to which the results might provide an answer".*

In recent years, there have been impressive improvements in the quality of instrumentation. For example, in the United Kingdom the CSIC at Cambridge

Table 14.2 Instrumentation and monitoring requirements.

Requirement	*Comments*
Identify critical observations	The primary criteria for controlling the works have to be identified. The reliability, accuracy and precision of the measurements must be compatible with the assigned traffic light zones and timescales. Trends need to be monitored and related to construction activity.
Establish the instrumentation system	What instruments will be installed, when and where. Should be simple, reliable and use proven technology. Need to be installed early enough to establish background readings. Critical readings to be checked with an independent system. Robust protection and adequate redundancy to allow for construction damage or instrument malfunction.
Identify additional observations to enable monitoring records to be interpreted.	Essential for comprehensive records of construction activities to be produced on a regular basis and at sufficiently short intervals during critical stages. These should be visual and present information unambiguously (e.g. construction photos and dimensioned sketches – both in plans, and long and cross sections).
Define roles and responsibilities within project team.	Project team will include representatives from the designer and contractor. Frequency of monitoring review meetings must be defined (e.g. daily during critical phases). Specific roles and responsibilities should be clearly established. Actions should be defined if trends demonstrate opportunities for beneficial design changes or, if adverse trends develop, the implementation of contingency plans.
Method statement	Should give outline of how daily/weekly reports should be prepared. The monitoring plan should be clear and unambiguous. Construction control and simple flow charts should be produced to ensure key measurements and observations are recorded on a regular, planned and consistent basis. The OM team should lead preparation of these documents and ensure clear ownership of process. Examples of construction control charts and flow charts are shown in Figures 3.9 and 9.12.

has promoted the innovative use of emerging technologies in real-time sensors and data management (Mair, 2016). These new instrumentation and data management systems can provide extensive real-time monitoring and promise to provide important new insights into ground and soil/structure interaction behaviour. Soga *et al.* (2015) describe the results of fibre-optic monitoring which provided considerable detail in this regard from a variety of construction induced effects.

As noted in Table 14.2, measurements from instruments are not sufficient on their own. It is important to establish a comprehensive record of contemporary construction activities and condition surveys of adjacent structures that may be adversely affected by the construction. When planning the instrumentation and monitoring requirements for an OM application, it is important to remember the advice by Peck (1972): *"The emphasis should be on observation rather than on instrumentation. Observation is a much broader term and, in my view, includes instrumentation. Even with the most sophisticated Instrumentation, other types of observation are essential"*.

14.3.2 Performance Limits and Traffic Lights

In the traffic light system for the OM, performance limits for critical observations are typically set by the three zones of green, amber and red. The amber trigger level is the line of demarcation between green and amber zones and for the red trigger level is that between amber and red. The red trigger level consequently marks the amber limit. Exceeding these trigger levels initiates a contingency action – as shown in the various construction control and flow charts in the case histories (e.g. in Figures 3.9 and 9.12). The red limit is set at or slightly below the maximum acceptable limit for the relevant critical observation – for example, the maximum allowable deflection of a structural member.

Operating the traffic light system is essentially a simple process and the associated performance limits are a vital part of managing risk through the OM. Establishing these limits provides a cogent example of where the OM is not served well by a prescriptive approach. To determine effective performance limits that can be managed through the OM in a timely and safe way requires experience and judgement. It necessitates an appreciation of the particular risks involved and how the selected instrumentation will fulfil the function required. To be adequately comprehensive, trigger levels are not easily defined by one set of specific parameters – for example, the geotechnical properties of the soils. Consider the deflection of a retaining wall. It is a common example of a critical observation in applying the OM. It was the key metric in several of the featured case histories (e.g. Chapters 2, 4, 5, 9 and 12). And, as discussed in Section 13.2, the deflection of a retaining wall is, in itself, an example of the resolution of complexity – being an emergent phenomenon of a complex system. This quality is central to its role as a critical observation. But how, in this example, should the trigger levels for a retaining wall be established? One simple way could be to determine the maximum allowable deflection from its safe structural capacity and then, preferably with some margin, set the red limit for this. The resulting range of deflection can then be divided proportionately to assign the green, amber and red zones. A starting point could be for the three zones to be of equal divisions. For example, if the maximum allowable deflection of the wall was say 65 mm, then the red limit could be set, allowing some margin, say at 60 mm, the red trigger level at 40 mm, and the amber trigger level at 20 mm. However, as emphasised, the final assignment of these limits requires careful judgement. The case histories have been chosen to illustrate a range of different situations.

The typical considerations in determining trigger levels include:

1. Although the underlying philosophy is the same, the range of factors that need to be assessed for the critical observations will vary with each specific application of the OM.
2. Continuing with retaining wall example, it is important to appreciate that its allowable maximum deflection may be governed by the criteria to protect an adjacent structure from the associated ground movement, rather than by the wall's structural capacity.
3. The nature of potential failures and their consequences need to be carefully assessed. Establishing trigger levels may be relatively straightforward if the critical factor is solely associated with ground failure. For example, in the Blomfield Shaft case history in Chapter 11, the red limit for base instability was established to ensure an acceptable factor of safety was maintained specifically for this risk. However, it should be noted that failure mechanisms involving soil/structure interaction are much more common and generally more complex involving a wider range of factors.
4. The need to appreciate the accuracy, precision and repeatability of the systems of instrumentation selected.
5. Is the observation relative or absolute? For example, for a retaining wall is the actual deflection being directly measured or is it the convergence between opposite walls? The latter situation, while providing a critical observation, will also need additional observations to assess the actual deflections of each wall. How this is resolved will be influenced by the particular instrumentation being used.
6. Based on an assessment of the most probable conditions what is the likely trend and is it predicted to stay within the green zone?
7. Will any trends developing towards the amber or red trigger levels allow enough time to implement the planned contingencies?

8. Is there enough margin in the green and amber zones to avoid the likelihood of false or frequent breaches of the trigger levels arising from the limitations of instrumentation performance combined with background noise? Such a lack of bandwidth is very undesirable and risks creating 'alarm fatigue'.
9. An example, of the need for flexibility in addressing this issue of practical margins within zones, occurred with the ramp sections of the ART in Chapter 9. As the allowable wall deflection progressively decreased in line with the reducing retained height, dividing the monitoring into the three traffic light zones of green, amber and red became increasing problematic. The point was reached where the three zone widths became so narrow that the fluctuations in the instrument readings meant that the amber trigger risked being falsely breached continuously. The OM had proved very successful for the cantilevered walls of the ramp in the preceding and more critical deeper sections. It was therefore clearly evident, since the shallowest sections presented the lowest risk, that the construction of the ramp could be completed without the temporary propping of the base case. It would have been ironic if a requirement to have the three zones based on some prescriptive criterion had led to the elimination of the OM in the sections of the ramp with lowest risk. In the event, the simple and effective solution was to reduce the zones down to two thus having just the green and the red. The construction was successfully completed with neither temporary props nor any false alarms.
10. Another issue, particularly when tracking trends, is the need to have increasing zone widths with time. Figure 9.14 shows an example of this for the TBM chamber of the west portal of the Heathrow ART.
11. There are also cases where the value of the red limit is imposed rather than being decided by the OM team. The Boston tunnel jacking case history (Chapter 7) provides an example of this. Here, and most appropriately, the limits of performance were set by the railway authorities and the OM team had to work within them. This case history also demonstrated how a flexible approach, adopted by all stakeholders, created the most effective environment to achieve maximum benefit from the OM. Through partnering and close collaboration, the red limit, on the basis of measured performance, was incrementally increased throughout the construction period eventually reaching up to three times the value originally imposed.

14.4 Progressive Modification – A Comprehensive Approach

As discussed in Section 1.2, implementing the OM through progressive modification addresses many stakeholder concerns in an effective, flexible and safe manner. It also provides a robust way to address concerns relating to the issue of the most probable conditions by commencing construction stages with a demonstrable level of safety acceptable to all stakeholders. It enables and enhances close tracking of trends and thus facilitates the timely implementation of design changes during construction. These may be either corrective contingencies or beneficial changes depending on whether the trend is adverse or advantageous. Progressive modification was adopted in all of the case histories featured in this volume and is also the approach recommended by Nicholson *et al.* (1999) and Szavits-Nossan (2006). (See also discussion in Section 13.3.)

While applications of the OM generally involve temporary works, they also frequently relate to the temporary conditions of permanent works during construction. This may involve existing adjacent structures as well as the new works under construction. The increasing need for the development of underground space in urban areas, including metros and other transportation systems, often presents the risk of damage to whole ranges of important and sensitive infrastructure within the zone of influence of the new construction. The use of the OM to enable the safe replacement or stabilisation of old structures is also likely to become increasingly important, given the extensive and ageing infrastructure in many countries worldwide. The Irlam railway embankment (Chapter 8) involved an OM application for the complete replacement of an ageing bridge. Lacasse and DiBiagio (2019) provide an example where the OM was used to extend the operational life on an offshore gravity platform in the North Sea. Jamiolkowski (2014) describes how the OM was used to reduce the risk of instability at an operational tailings dam in Poland. The stabilisation of the leaning tower of Pisa (Burland *et al.*, 2013) is a compelling example of how an application of the OM through progressive modification helped to successfully address the stabilisation of an ancient monument of international importance.

14.5 Peer Review

High-level peer reviews of designs have been used extensively in the United States for many years. They are typically undertaken by a small panel of suitably qualified and experienced engineers and focus on the overall concepts and key risks. They can be fulfilling for both the designers and the reviewers and have proved particularly effective for applications of the OM. Peck (1969) includes the case history of the application of the OM for the Cape Kennedy Causeway. He was a member of the review team and described the assignment as "*one of the most satisfying of my professional experience*". (Dunnicliff and Deere, 1984). That is quite an accolade considering the diverse range of amazing projects that feature in Peck's long career.

A similar process for project review known as Peer Assist (PA) has been developed in the United Kingdom. Its key features are:

- Presentations to the PA reviewers by the project team.
- Focus on risks through highlighting and sharing concerns.
- Following the review the project team produces the draft review report for evaluation and approval by the PA reviewers.

PA is applicable to any project and any scope of review and is particularly effective for complex or challenging projects. It is very flexible and can focus reviews on specific aspects in a timely way through the design development, be it safety, technical or commercial. This, of course, does not remove the need for an overall project review, which can also be undertaken through PA.

Adopting PA has created extra benefits including:

- Stimulating greater awareness and interaction within the project team and other stakeholders.
- The interaction encouraged by PA has stimulated re-evaluation of initial assumptions and helped misconceptions to be addressed early and effectively. Examples have included clarifying the brief and deliverables, duty to warn and a wide range of potential innovation.
- Mentoring: Because PA is so direct and focussed on key issues, a strong communication between the reviewer(s) and project team tends to be rapidly established. It has proved particularly helpful to younger members of a team by creating access to experienced staff and tapping into wider knowledge and experience.
- Stimulating innovation through enhancing the balance between risk management and creativity.

The focus that PA brings to risk management and safety has led to its endorsement in the United Kingdom by the Institution of Civil Engineers, the Institution of Structural Engineers and the Health and Safety Executive through their promotion of the Standing Committee on Structural Safety (SCOSS). A Guidance Note and Model Form of Agreement for Independent Review through PA are available on the SCOSS website: www.scoss.org.uk.

14.6 Recommendations for OM Practitioners

1. Visit the site early during the design development and establish an ongoing site presence for key members of the OM team. This was a cardinal rule of both Terzaghi and Peck and should be so for every practitioner. It is an essential part of observation and assessing conditions.
2. Relate design to construction on an informed basis by early involvement with contractors and by consulting relevant specialists. Gain an appreciation of the construction plant, methods and sequences.

3. Ensure a balanced approach to soil/structure interaction by identifying the key observations both for the geotechnical and structural aspects.
4. Treat each application of the OM as bespoke. They should never be considered as routine. 'Beware the oddball' (Peck, 1998).
5. Recognise that achieving agreement to use the OM typically presents a substantial challenge.
6. Promote teamwork and engender trust. Seek awareness and understanding of stakeholders' priorities and concerns. They will vary between the parties.
7. Develop concise construction control and flow charts to demonstrate the specific process for the OM in each application to all stakeholders. These charts enhance focus on key requirements and aid communication. Examples are provided in the featured case histories.
8. Produce method statements to clearly establish ownership of roles and responsibilities.
9. Keep it simple – but not simplistic. Exercise engineering judgement and avoid undue complication. Ensure focus on the critical observations, carefully monitor their trends but also maintain surveillance of secondary issues and effects. Do not let systems dull your senses (Peck, NCE News, 2000).
10. Wherever feasible consider implementing the OM through progressive modification. The featured case histories cover a wide range of site conditions and construction and were all implemented through progressive modification.
11. Welcome peer reviews (or PA). These are intentionally interactive and should focus on the overall concept and key risks and are thus significantly distinct from an independent design check.
12. Study case histories and appreciate lessons learned. Seek to make your own case histories detailed and informative.
13. Promote the OM.
14. To adapt a phrase from the Renaissance: 'The OM offers the gift of opportunity.'

References

Burland, J. B., Jamiolkowski, M., Squeglia, N. and Viggiani, C. (2013). *The Leaning Tower of Pisa, Geotechnics and Heritage*, Taylor and Francis, London.

Construction 2025: Strategy. (2013). Policy paper on Joint strategy from government and industry for the future of the UK construction industry.

Dunnicliff, J. (1988) *Geotechnical Instrumentation for Monitoring Field Performance*, John Wiley and Sons, New York.

Dunnicliff, J. and Deere, D. V. (editors) (1984). *Judgment in Geotechnical Engineering – The Professional Legacy of Ralph B. Peck*, p. 122, Wiley, New York.

Egan, J. (1998). *Chairman, the Construction Task Force, Rethinking Construction*, p. 40, DETR, London.
Jamiolkowski, M. (2014). Soil mechanics and the observational method, *Géotechnique*, **64**, No. 8, The Institution of Civil Engineers, London, 590–619.
Lacasse, S. and DiBiagio, E. (2019). Two observational method applications: an ideal solution for geotechnical projects with uncertainty. *Proceeding of ASCE Geo-Congress, GSP*, **314**, 117–128.
Mair, R. J. (2016). Advanced sensing technologies for structural health monitoring, *Forensic Engineering*, **169**, No. FE2, Proc The Institution of Civil Engineers, London, UK, 46–49.
Latham, M. (1994). *Constructing the Team*, HSMO, London, July, pp. 130. Final report of the government/industry review of procurement and contractual arrangements in the UK construction industry, HMSO, London.
New Engineering Contract (NEC4). (2017). *The Institution of Civil Engineers*, Institution of Civil Engineers, London.
Nicholson, D., Tse, C.-M. and Penny, C. (1999). The observational method in ground engineering: principles and applications, CIRIA Report 185, London, p. 214.
Parkin, J. (2000). *Engineering Judgement and Risk*, p. 224, Thomas Telford, London.
Patterson, R. (2010). NEC Contracts as an enabler to partnering. The partner, May 2010.
Peck, R. B. (1969). Advantages and limitations of the observational method in applied soil mechanics. *Géotechnique*, **19**, No. 2, 171–187.
Peck, R. B. (1972). Observation and instrumentation, some elementary considerations. *FHWA, 131*, **4**, No. 2, 1–5, New York.
Peck, R. B. (1998). Beware the oddball, Urban geotechnology and rehabilitation, Seminar, ASCE, Publication No. 230, Metropolitan Section, Geotech Group, New York.
Peck, R. B. (2000). Don't let systems dull your senses. *NCE, ICE News*, **11**, May 2000.
Powderham, A. J. and Nicholson, D. P. (1996). *The Way Forward: The Observational Method in Geotechnical Engineering, Institution of Civil Engineers*, pp. 195–204, Thomas Telford, London.
Soga, K., Kwan, V., Pelecanos, L., Rui, Y., Schwamb, T., Seo, H. and Wilcock, M. (2015). The role of distributed sensing in understanding the engineering performance of geotechnical structures. geo eng for infrastructure and development. *Proceeding of 16th Euro Conf of SMGE*, **1**, 13–48.
Szavits-Nossan, A. (2006). Observations on the observational method, XIII Danube-European Conference on Geotechnical Engineering Ljubljana: Slovenian Geotechnical Society, 2006. str. pp. 171–178.
Tran, D. and Molenaar, K. (2012). Critical risk factors in project delivery method selection for highway projects, Construction Research Congress 2012: construction challenges in a flat world, pp. 331–340.
Wolstenholme, A. (2009). Never waste a good crisis: a review of progress since Rethinking Construction and thoughts for our future, Constructing Excellence: www.constructingexcellence.org.uk/.

Index